SIR WILFRID LAURIER SECONDARY SCHOOL

Department _____ Book Number _____

STUDENT'S NAME	HOUSE UNIT	SCHOOL YEAR	SEMESTER	SUBJECT TEACHER

SIR WILFRID LAURIER SECONDARY SCHOOL

STUDENT'S NAME	HOUSE UNIT	SCHOOL YEAR	SEMESTER	SUBJECT TEACHER

HUMAN SOCIETY
CHALLENGE & CHANGE

Frederick Jarman
Central High School of Commerce, Toronto

Helmut Manzl
Oakville-Trafalgar High School, Oakville

John Wiley & Sons

Toronto   New York   Chichester   Brisbane   Singapore

Copyright © 1988 by John Wiley & Sons Canada Limited
All rights reserved.

No part of this publication may be reproduced by any means, stored in a retrieval system, or transmitted in any form or by any means, electronic, mechanical, photocopying, recording or otherwise, without the prior written permission of the publisher.

Care has been taken to trace ownership of copyright material contained in this text. The publisher will gladly receive any information that will enable them to rectify any reference or credit line in subsequent editions.

Canadian Cataloguing in Publication Data

Jarman, Frederick E., 1945-
   Human society

For use in secondary schools.
Includes index.
ISBN 0-471-79694-8

1. Sociology. 2. Anthropology. 3. Psychology.
I. Manzl, Helmut, 1947- . II. Title.

HM66.J37 1988   301   C88-094174-X

Dedication

For David and Ona

DESIGN: Brant Cowie/Artplus Limited
TYPE OUTPUT: Tony Gordon Limited
PHOTO RESEARCH: Francine Geraci

Printed and bound in Canada by John Deyell Company
10 9 8 7 6 5 4 3 2 1

ACKNOWLEDGMENTS

This project has been on-going for the past several years. We wrote this book because we wanted to develop materials that we considered to be relevant and interesting to our students. Our thanks are first to our students who field tested the manuscript and made valuable suggestions.

Many different people have been involved in this project. The sensitive nature of some of the topics made us very careful about the selection and coverage of materials.

Several editors worked on this project but special thanks to Susan Howlett who worked and agonized with us and made valuable changes and suggestions. Her insight and long hours of editing this book have resulted in a very fine text.

Lou Pamenter was in charge of co-ordinating the entire package and displayed her usual competence by bringing all of the components together in a very short period of time.

Allan Hux provided important suggestions to help improve the reading and skill activities.

Janet Thomas was very patient at the word processor and without her help, this book would not have been possible.

Librarians Sylvia Jessops and Sandy Lessner co-operated in their usual friendly manner to provide research materials that were needed yesterday.

Media teacher Sharon Foster made important observations on the media chapter.

Dorothy McKim, Head of Guidance, also provided much appreciated materials on the subject of psychology.

To all of these people and the many others who have not been mentioned, we extend our sincere thanks and appreciation.

The authors assume the sole responsibility for any errors that have escaped capture.

Fred Jarman
Helmut Manzl

NOTE TO STUDENTS AND TEACHERS

This book has been designed to stimulate students to be more aware of themselves and their relationships with others. Hopefully this awareness will lead them to a greater understanding and tolerance of people's differences – both individually and in groups.

It is suggested that teachers select materials, activities, and assignments that best meet the needs and interests of their students. It is not expected that all the activities at the end of each chapter will be used – only those which are most relevant. All materials used have been successfully tested in classroom situations.

Throughout each chapter, progress check questions have been included to test the comprehension of the main points covered. Career profiles and case studies have been included to expand on topics and understandings.

At the end of each chapter is an overview that summarizes and expands upon the main ideas covered in the chapter. Vocabulary and language skills are emphasized. There are lists of key words and personalities that have been used in the chapter. An important feature of this text is the emphasis on skill development. These skill development activities encourage students to think, interact, research, report, and communicate. They challenge students to ask questions on important topics and issues.

Students are asked to organize information and key points that are presented in the text and in suggested sources. This will enable them to identify important points, to make connections, and to expand upon their comprehension of thinking, attitudes, and behaviours. Students are also asked to assess and give their own opinions on important issues.

You will find that, rather than exploring social issues of the future in a separate chapter at the textbook's end, we have included activities at the end of various chapters in which future issues are relevant. As well, the topics of aggression and violence are best discussed in context. We have provided students with the means to question those behaviours at points in the text where they are most applicable.

We hope that through reading, research, analysis, and discussion this book will give them a greater understanding of human society.

Table of Contents

UNIT 1   THE NATURE OF THE HUMAN SPECIES

CHAPTER 1   Sociology, Psychology, and Anthropology   2

    1.1   The Individual in Society   2
Looking for Answers • The Individual in a Global Society • Questions and Answers from the Past

    1.2   Progress Through Challenge   6

    1.3   The Behavioural Sciences: Sociology, Psychology, and Anthropology   8
The Scientific Method • Can We Trust Our Perceptions? • Studying Behaviour

    1.4   The Role of the Sociologist   11
The Development of Sociology • Sociology Today

    1.5   Psychologists and Behaviour   14
Methods in Psychology • Case Study: Cause and Effect • Educational Psychology • Clinical Psychology

    1.6   Anthropology   20
Cultural Anthropology • Physical Anthropology • Charles Darwin • Fossil Discoveries

    Overview   26

CHAPTER 2   On Being Human   32

    2.1   The Animal Kingdom   32
Classification of Animals • Human Groupings

    2.2   Heredity and Environment   37
Heredity • Instinctive Behaviour • Inherited Diseases • Genetic Adaptations to the Environment • Environment • Case Study: Obesity – Nature or Nurture?

    2.3   Human Needs   44
Basic Needs • Higher Needs

    2.4   Learning to be Human   47
Culture • Socialization • Feral Children • Isolates • Case Study: The Wild Boy

2.5 Our Human Ancestors *52*
  Early Human Cultures • From Hunters to Farmers
  Overview *57*

UNIT 2 SOCIAL BEHAVIOUR

CHAPTER 3 Socialization *64*

3.1 How are People Socialized? *64*
  The Need for Support • Case Study: Harlow's Monkey Studies

3.2 Socialization and Norms *68*
  Why Do We Conform?

3.3 Learning and Conditioning *70*
  Classical Conditioning • Operant Conditioning • Learning to Fit In • Case Study: Piaget's Theory

3.4 The Importance of Roles *77*
  Role Models • Ascribed Roles, Achieved Roles, and Life Chances • Role Sets • Fronts and Settings • Role Conflict, Role Strain, and Role Ambiguity

3.5 Gender Roles *82*
  Gender Expectations • Career Choices • Socialization and Stereotypes

3.6 Parenting *86*
  Overview *91*

CHAPTER 4 Personality *98*

4.1 Who are You? Describing Personality *98*
  Describing Personality in Terms of Traits • Describing Personality by Type

4.2 Assessing Personality *103*
  Personality Rating • Personality Inventory • Interviews • Projective Tests • Behavioural Assessment • Responses and Interpretation

4.3 The Family and Personality *107*
  Birth Order • Influence of Gender • Family Size

4.4 Some Theories of Personality Development *111*
  Freud's Psychosexual Theory of Personality Development • Carl Jung's Theory of Personality • Erik Erikson's Theory of Personality Development

4.5 The Sense of Self *123*
  Career Profile: Erik Erikson • The Concept of Self in Two Societies
  Overview *127*

CHAPTER 5   **Women and Equality**   *133*

    5.1   Prejudice, Stereotypes, and Discrimination   *133*
        Case Study: Attitudes Toward Minorities

    5.2   Attitudes Toward Women   *138*

    5.3   The Early Women's Rights Movement in Canada   *140*

    5.4   Canadian Pioneers in the Struggle for Equality   *144*

    5.5   Discrimination in the Work Force   *146*
        Pay Equity • Equity in Hiring and Promotions • Discrimination and the Law • Case Study: Human Rights Commission • Equity and Mental Health • Case Study: The Use of Sexist Language

    5.6   The Socialization of Women   *154*
        Gender Roles – Biology or Socialization • Early Socialization

    5.7   The Canadian Working Woman   *160*
        Career Profile: Rosalie Abella

    Overview   *166*

CHAPTER 6   **Aboriginal and Ethnic Groups**   *175*

    6.1   Racism and Prejudice   *176*
        The Myth of Race • The Reality of Racism • Case Study: Apartheid in South Africa • Causes of Prejudice

    6.2   The Aboriginal Peoples of Canada   *186*
        The Collision of Several Cultures • Prejudice and Discrimination • The Native People Today

    6.3   Canada, A Multicultural Society   *193*
        Canadian Immigration Policies • Multiculturalism • Prejudice in a Multicultural Society • Some Ethnic Minorities in Canada • The Quest for a Just Society

    Overview   *207*

UNIT 3   COMMUNICATION

CHAPTER 7   **Human Communication**   *214*

    7.1   What is Language?   *214*
        Language and Society • Language and Values • Language and Social Identification • Language and Social Class

    7.2   Types of Symbols   *219*
        Case Study: Koko the Gorilla

7.3 How Do We Express Ourselves? 225
Listening • Illiteracy

7.4 Becoming Literate 229

7.5 The Development of the English Language 231
Language is Shaped by Change • Career Profile: Helen Keller • Language and Culture • International Languages

Overview 237

CHAPTER 8 The Influence of the Mass Media 243

8.1 Radio 244

8.2 Films 246

8.3 Television 246
Rock Videos

8.4 Printed Materials 248
Newspapers • Magazines • Books

8.5 Media Literacy 249
Career Profile: Marshall McLuhan

8.6 Advertising in the Media 253
Children and Advertising • Discrimination in Advertising

8.7 Violence in the Media 255
Young Children • Teenagers • Case Study: Terry Fox

8.8 Related Questions 262
Censorship • Cultural Sovereignty

Overview 264

UNIT 4 CULTURE

CHAPTER 9 What is Culture? 274

9.1 Defining Culture 274
A Text for Living • A Program for Survival • A Way of Life

9.2 The Development of Culture 278
The Human Body • The Human Brain • Language

9.3 How Anthropologists Study Culture 282
Ethnography • Ethnology • Classifying Culture

9.4 Analyzing Culture 285
Social Structure • Social Interaction • The Environment of a Culture • The Economic Structure of a Culture • Career Profile: Biruté Galdikas • Social Control • The Religious Customs of a Culture

Overview 297

CHAPTER 10 How Anthropologists View Cultures 303

10.1 The Tasaday Tribe 304
Subsistence Patterns • Social Structure • Religious Customs and Beliefs • The Future of the Tasaday

10.2 The Masai 308
Subsistence Patterns • Masai Customs And Social Structure • Religion • The Masai and the Modern World • Career Profile: Franz Boas

Overview 315

UNIT 5 SOCIAL INSTITUTIONS

CHAPTER 11 The Family 322

11.1 Defining a Family 322
Types of Families • The Elements of a Family

11.2 Functions of the Family 324
The Sexual Function • The Reproductive Function • The Economic Function • The Educational/Socialization Function

11.3 The Changing Nature of the Family in Canada 327
Women in the Workforce • The Rise of Single-Parent Families and Reconstituted Families

11.4 The Post-Modern Family 331
Socialization and the Family • Changing Roles of Husbands And Wives • Non-Family Child Rearing

Overview 336

CHAPTER 12 Love, Marriage, and Divorce 341

12.1 Defining Marriage 342

12.2 Arranged Marriages 343

12.3 Developing Marriage Customs in Canada 344

12.4 Love and Marriage *345*

The Privatization of Courtship • Courtship Today • Marriage: A Weakening Institution? • The Controversy over Romantic Love • Case Study: The Importance of Love in a Relationship

12.5 Divorce in the Twentieth Century *353*

Seven Hypotheses to Explain the Increase in Divorce • Case Study: A Marriage Break-up • The Effects of Divorce on Children • Teenagers and Divorce • The Effects of Divorce on Women • The Effects of Divorce on Men

Overview *363*

CHAPTER 13 Schools: Looking Back and Looking Ahead *369*

13.1 The Purposes of Education *369*

Education and Jobs • Education and Status • Education and the Social Order • Education and the Transfer of Culture • Education and the Need for Personal Fulfillment

13.2 The History of Schools in Canada *372*

Compulsory Education • Career Profile: Wallace Lambert • The Expansion of the System • Enrolment In Schools

13.3 The Challenge to the Schools *375*

Case Study: Summerhill • Alternatives to the System • Government Reports on Education

13.4 Educational Reform in the 1970's *377*

A Lack of Skills • Inequality in Schools

13.5 A New Call for Reform in the 1980's *379*

Re-evaluating the Schools • Case Study: Report Card on Our Schools • The Schools and the Future • Trends for the 1990's

Overview *385*

UNIT 6 YOU AND YOUR SOCIETY

CHAPTER 14 Behaviour Patterns *394*

14.1 Defining Normal and Abnormal Behaviour *395*

Definitions of Abnormality

14.2 Defining Mental Health *396*

14.3 Defence Mechanisms *397*

Types of Defence Mechanisms • Role of Defence Mechanisms

14.4 Mental Disorders *403*

Emotional Disorders • Thought Disorders • Why are There Mental Disorders? • Treatment of Mental Disorders

Overview *408*

CHAPTER 15 Being A Part Of Society *413*

15.1 Social Influence *414*

15.2 Conformity *414*
Why Do We Conform? • How Do We Conform?

15.3 Classic Conformity Experiments *416*
Sherif and the Auto-Kinetic Phenomenon • Social-Comparison Theory • The Latané Experiments

15.4 Non-Conformity *420*
Case Study: The Hutterites

15.5 Why Do Some People Feel Alienated? *424*
The Alienated

15.6 Suicide *426*
The Social and Psychiatric Causes of Suicide • Who is at Risk? • Preventing Suicide • How to Help

Overview *432*

CHAPTER 16 Current Social Issues *437*

16.1 Aggression *437*
Violence in the Home

16.2 Aging *441*
Career Profile: Gordon Sinclair

16.3 Aids *445*

16.4 Drugs *447*
Recreational Drug Use in Canada • Career Profile: Steve Collins • Society and Drug Abuse • Case Study: The Addiction Research Foundation

Overview *455*

INDEX *462*

Photo Credits

Hafid Alaoui, 216; Associated Press, 178; *Brantford Expositor*, 279; California Institute of Technology Archives, 46; Canada Mortgage and Housing Corporation, 47; Canadian National Institute for the Blind, 234; Canapress Photo Service, 5, 19, 21, 63, 74, 115, 134, 140, 163 right (Claude Willoughby), 164, 179, 180, 182, 223, 224, 233, 247, 250, 252, 260, 276, 305 (Helmut R. Schulze), 309 (Raymond Lewis and Denys Dawnay), 311 (Peter Kain), 423, 435, 439, 444, 450; Creative and Editorial Services, Tom Shields, 1, 65; Francine Geraci, 205 top, 291; Government of Canada/Ministry of Communications, 264; Cindy Green, 446; Jeremy Jones Photography, 347; Michael and Sandy McLean Martineau, 353; Metropolitan Toronto Library Board, 22, 25, 42; Miller Services Limited, 3 (H. Armstrong Roberts), 7, 34, 38, 45, 69, 82, 88, 101 (H. A. Roberts), 116, 119 (H. A. Roberts), 121, 122 (R. Vroom), 126 (H. A. Roberts), 155 (H. A. Roberts), 157 (Harold M. Lambert), 160 (H. M. Lambert), 163 left (H. M. Lambert), 188 (Frederik Stevenson), 192 left, 192 right (Richard Harrington), 206 (Vivienne della Grotta), 213, 226, 228, 273 (H. A. Roberts), 277 (Camerique), 287 (H. A. Roberts), 327 (H. A. Roberts), 333 (H. A. Roberts), 335 (H. A. Roberts), 358 (H. A. Roberts), 386, 393 (John F. Phillips), 396 (H. A. Roberts), 401 (Ellefsen Photographe Ltée), 406 (H. A. Roberts), 415 (H. M. Lambert), 421 (H. M. Lambert), 429 (H. A. Roberts), 443 (H. A. Roberts); N. C. Press Limited, 215 bottom; Public Archives of Canada, 198 (PA 10255), 202, 244 (PAC 139063), 286 (PAC 11616); Royal Ontario Museum, Department of Ethnology, 314 (HN 1367, HN 1173); SSC Photocentre-Centre de Photo ASC, 40 (73-1907), 218, 236, 321 (D. Bancroft), 324 (Terry Pearce), 370 (Crombie McNeill), 378 (Ted Grant); Toronto Board of Education, Archaeological Resource Centre, 372; *The Better Half* (KFS) Copyright 1988: Cowles Inc. Reprinted with permission – The Toronto Star Syndicate, 171 and *Shoe* (TMS) Copyright 1988: TMS. Reprinted with permission – The Toronto Star Syndicate, 408; Vancouver Public Library, 203; Daniel Wood, 292.

UNIT

1

THE NATURE OF THE HUMAN SPECIES

CHAPTER 1

Sociology, Psychology, and Anthropology

What are the behavioural sciences, and how do they deal with the subject of human behaviour?

INTRODUCTION Has a day ever gone by when you have not questioned why someone said or did something? What was the meaning of that person's words or actions? You have likely questioned the reasons for your own behaviour at times. Why did you speak or act in a particular way?

There is an endless variety of ways in which we can and do behave and interact. Several scientific **disciplines**, or branches of learning, are concerned with human behaviour. In this chapter we shall take a look at how three of these disciplines — sociology, psychology, and anthropology — began and how they approach and deal with the subject of human behaviour.

1.1 The Individual in Society

LOOKING FOR ANSWERS

Who am I? What am I? Why am I here? Where am I going? Never before have such questions been asked as often as they are today. We want to understand more about our existence and the reasons behind our thoughts and behaviour. In some ways we are fortunate because we live at a time when many of our questions can be answered.

Some people are seeking the answers to the question of how a safer, healthier world can be achieved, for themselves and for others. They question the accepted ideas and opinions, and ask for proof or evidence. If problems exist, they want to know what they can do to bring about a solution.

Some people are curious about the reasons behind attitudes, opinions, behaviour, and interests, and how people are affected by them. Many people wish to have more control over their lives. They are asking questions and investigating possible answers in the hope that greater understanding will lead to more control over their own lives. Perhaps one reason why many people are challenging traditional thinking by asking questions is that we are living in a time of rapid change.

THE INDIVIDUAL IN A GLOBAL SOCIETY

Our modern technology can provide a machine or a method for tackling almost every task or problem. Technology has accelerated the pace of life and made life easier and more convenient. However, technology has also created new problems and greater potential for destruction and death.

Crowd playing "Earthball"

For almost every new technological advance there is a new environmental problem to address. Wars have never been fought on such a grand scale before. With modern technology, disputes between countries can easily flare into global confrontations.

Many people believe that countries and societies are a reflection of their individual citizens and members. If this is so, then gaining more understanding of why individuals think and act the way they do may increase our tolerance of each other. Greater tolerance and understanding may, in turn, lead people to treat each other better. Such improvements at the individual and the community level, if reflected at a global level, would make the world a better and safer place to live.

QUESTIONS AND ANSWERS FROM THE PAST

In the distant past, long before written language was developed, life was difficult. Much effort was devoted to simply surviving. Humans were constantly having to deal with the harshness of the elements: floods, earthquakes, volcanoes, drought, cold; with the struggle to find food and shelter; and with life-threatening injuries and diseases. Their **cultures**, or ways of life, were shaped by their **environment**, or the physical surroundings where they lived and the society with which they lived.

It is easy to imagine that as these early people gathered around a campfire in the evening, there would be some who would wonder about how life had been created and what happened to a person who died. Many early cultures relied heavily on their religious leaders and medicine-men for answers, since natural events such as earthquakes, droughts, or tidal waves were thought to be the work of gods and spirits. The people believed that their medicine-men could communicate with and influence the powerful forces that controlled life on earth. Ceremonies were performed to attract the attention of the spirits. Festivals and celebrations were held and sometimes animals or people were sacrificed to please the gods.

Different cultures developed different **theories**, or explanations based on thought and observation, to answer their questions. Some of these theories were incorporated into stories to be retold many times and passed down from generation to generation. Eventually some of the stories

This stone sculpture from the Indus Valley probably represents a deity from an early but well-developed civilization, circa 2500-1700 B.C.

would be accepted as true, becoming **legends**. Later, **symbols** were devised for recording the legends and other information. A symbol is something that is used to represent or suggest something else. The symbols were painted or chiseled into rock. This was a more reliable method of passing on knowledge and experience. The Algonquin Indians have a legend that explains why people must die. The legend describes how The Great Hare gave men and women a package which they were forbidden to open. It contained eternal, or unending life. Someone, out of curiosity, opened the package and immediately eternal life flew out and escaped.

Inuit in Greenland believed that in the beginning there was not only no death but also no light. It was said to be impossible to have one without the other. However, one person felt that without the sun, life was worthless. So the light of day was provided in exchange for **immortality**, or eternal life.

Jews, Christians, and Moslems believe in one God whose word was written down long ago in the Old Testament. The Book of Genesis, one of the scriptures in the Old Testament, describes their theory of creation. It tells of creating the world and then making the first humans, Adam and Eve, to live in that world. Because they ate an apple from the Tree of Knowledge Adam and Eve were cast out of the Garden of

Eden where life was perfect, and into the world of work and suffering.

Other cultures have given rise to different religions such as Hinduism and Buddhism. Each of these religions has its own answers to questions regarding life and death.

PROGRESS CHECK

1. Why do you think humans ask so many questions about themselves?
2. What are some of the effects of technology?
3. Why is it considered important to understand why people think and act?
4. What were some of the questions asked by our ancestors?
5. What is the Judeo-Christian-Moslem theory of creation?

1.2 *Progress Through Challenge*

With the passage of time, some people began to have doubts about the traditional explanations and theories offered by science and religion. They observed the world around them and found evidence that did not support the older and more established ideas. Early explorers, like Christopher Columbus (1446–1506), discovered that the oceans were not full of terrible monsters that would swallow them. Their ships did not sail off the edge of the Earth. In the sixteenth century (1500's) a Polish thinker and scientist, called Nicolas Copernicus (1473–1543), challenged the firmly established belief that the sun moved around the Earth. He had reached the opposite conclusion through careful observations of the movements of the Earth and other planets around the sun.

Throughout the seventeenth century (1600's) scientific investigations continued to call into question the traditional explanations. The developing sciences of biology, astronomy, chemistry, and physics were stimulating new discoveries. Sir Isaac Newton (1642–1727), an English scientist, formulated the Law of Gravity from simple observations, such as watching an apple fall from a tree. He thought that gravitation might also explain why the moon was held in place as it moved around the Earth.

A small but growing number of people began to believe that it was possible to discover scientific truths by observing the world around them. Their observations and the information they collected formed the basis for their conclusions and theories. These theories could then be applied in various areas to explain other phenomena. People were changing the way they looked at themselves and at the world in which they lived.

The eighteenth century (1700's), called the **Age of Enlightenment**, was a time of growing human awareness. Some thinkers began to deal with the problems of society, and there was hope that the world could be made a better place to live. Voltaire (1694-1778), a French philosopher, criticized the evils he saw in French society. He believed that the king and his nobles and the Church were corrupt and were standing in the way of changes that would benefit the ordinary people. Voltaire and other thinkers of the Age of Enlightenment wanted society to be ruled by reason, not by faith, privilege, and wealth. Concepts such as "liberty" and "equality" became popular.

The **Industrial Revolution** was also occurring in Britain and Europe in the late eighteenth and early nineteenth centuries (1700's and 1800's). New methods and machines were being developed that made it possible to produce greater quantities of goods at cheaper prices. James Watt (1736–1819), helped to invent the steam engine (1764), and this new cheap source of power improved transportation and manufacturing processes. Other inventors used Watt's discovery to invent the steamboat and the locomotive. As machinery was converted to steam power, factories became increasingly efficient, capable of producing more goods at cheaper prices.

Opening of the South Eastern Townships and Kennebec Railroad at Sherbrooke, Quebec, July 2, 1874.

Cheaper prices enabled people to buy more goods. The increased demand for goods made it possible for factories to expand and more people moved to the cities to work in the factories.

On the farms, steam-driven machinery helped to plant and harvest crops more efficiently. More food could be grown by fewer people to feed the expanding urban populations. Technology was changing the world.

PROGRESS CHECK

1. Why did some people begin to doubt earlier explanations and beliefs?
2. With what problems did thinkers of the Age of Enlightenment attempt to deal?
3. What developments of the Industrial Revolution changed society?

1.3 The Behavioural Sciences: Sociology, Psychology, and Anthropology

Science is the branch of knowledge based on systematic observations of the physical universe. Science has two major divisions. One is the **natural sciences**, which deal with the biological, chemical, and physical environment. The other division is the **social sciences** which deal with history, economics, politics, sociology, psychology, and anthropology.

THE SCIENTIFIC METHOD

Each of the scientific disciplines mentioned above has its own particular field of inquiry, but there is a unifying element among all of them. They all use the **scientific method** in their investigations. This means that they use a method of collecting data and arriving at conclusions that is logical, objective, and **empirical**, or based on experiment and observation.

The basic way that scientists approach a problem that they wish to solve is by first thinking about the problem and the information they already have about it. Then they propose a hypothesis. The **hypothesis** is a statement of a cause and effect relationship that the scientist wishes to test. If the hypothesis is true, then the effects (events) the scientist is predicting will occur and will be observed. When the events occur as predicted, the hypothesis is proven. Once proven, a hypothesis is called a theory. If the hypothesis is disproven, the scientist will propose a new one based on new observations.

Scientific knowledge is continually being expanded. New information frequently casts doubt on old theories. Often theories must be changed or even discarded as more research uncovers new facts.

At all times scientists must try to be objective, that is, they must try not to allow their own prejudices and biases to influence their investigations. Other scientists should be able to repeat the same investigations and arrive at the same results.

CAN WE TRUST OUR PERCEPTIONS?

Sometimes our **perceptions**, the way we view things, are inaccurate and very often they are different from other people's. We therefore have to be sure of our observations before we base conclusions on them.

Look at Figure 1.1 and write down what you see. Compare your answer with the rest of the class. Did some see an old woman and some see a young woman? Examine the dots in Figure 1.2 and determine which inner dot is larger. Measure the two inner dots to prove your perception.

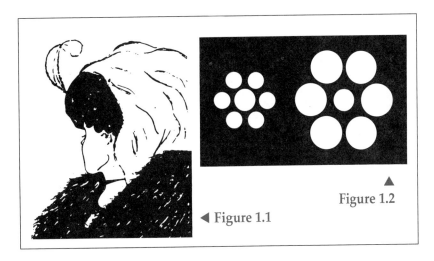

◀ Figure 1.1

▲ Figure 1.2

To make sure that their hunches, feelings, and perceptions are correct, scientists test them using the scientific method. In the social sciences the main methods used to test hypotheses are controlled experiments, observation, sample surveys or questionnaires, and case studies.

Studying Behaviour

In this book we deal with the group of social sciences known as the **behavioural sciences**. This group includes sociology, psychology, and anthropology.

As nineteenth century experts studied the world around them, some became fascinated by human beings and their cultures. They tried to learn more about people's thoughts, actions, and behaviours. The behavioural group of the social sciences arose out of these studies.

Sociology is the study of human groups or societies and the way in which individuals relate to and are influenced by these groups. Experts or professionals in the field of sociology are called sociologists. They study the cultures of different groups by examining such **social institutions** as the family, school, church, media, and government. These social institutions are established to meet the needs of the society and to exercise control over its members. Sociologists try to determine how these social institutions influence the thinking and behaviour of members of the group.

Psychology is the study of behaviour. Those who are professionals in the field of psychology are called psychologists. Psychologists study the forces that determine behaviour, for example inner forces such as the emotions and the intellect, and external influences such as parents, friends, neighbourhood, and society.

An area of study connected to psychology is **psychiatry**. This is the study of the prevention, diagnosis, and treatment of mental and emotional disorders. Professionals in the field of psychiatry are called psychiatrists. They are physicians who have specialized in psychiatry. Unlike psychologists, psychiatrists are medical doctors, and they can therefore prescribe drugs. Psychiatrists are trained to help people who find it difficult or impossible to relate to other people in what is considered to be a normal way.

Anthropology means literally "the study of human beings". Anthropology deals with the physical and cultural origins and development of humankind. As this science was developing, it tended to concentrate on societies and cultures that were unable to read or write, or **nonliterate**. More recently, however, anthropology has been concerned with the so-called civilized societies as well as what are called primitive societies. Professionals in the field of anthropology are called anthropologists. Anthropologists study human populations of the past and present to understand their culture.

Sociologists, anthropologists, and psychologists are called social scientists because they are interested in the same subject: the social relationships of human beings. They all ask questions about human nature and behaviour. These three social sciences, also called the behavioural sciences, are relatively new. Sociology, for example, began to be studied as a separate discipline only in the nineteenth century. Since then, much has been learned and discovered about human origins, behaviour, and societies by scientists working in all three disciplines. The new knowledge about human thinking and behaviour continues to expand through their work.

PROGRESS CHECK

1. What steps are involved in the scientific method?
2. Do we all perceive things in the same way? Why or why not?
3. What do psychologists and psychiatrists study and how do they differ?

1.4 The Role of the Sociologist

Sociologists are interested in all the elements that contribute to making a society. They look for patterns and try to understand the behaviour of individuals interacting in society. They examine the values and attitudes of particular groups or institutions such as family, church, and school. Another aspect that interests them is why people become members of different groups. They also want to know how individuals are affected by the group and by their social institutions.

Sociologists believe that the various institutions in society influence the behaviour of their members, and that the institutions hold society together. As groups change, society changes, and sociologists examine the effects of the changes on individuals and on society.

Sociologists are not social workers, people who go into homes or institutions to work with distressed individuals. Instead, sociologists are concerned with human group behaviour. They want to know what causes certain groups to form and what are the consequences of membership in these groups.

THE DEVELOPMENT OF SOCIOLOGY

Sociology began to emerge as a science in the nineteenth century, with the development of new scientific methods of research. August Comte (1789–1857), known today as the father of sociology, was a French philosopher who believed that a scientific approach could be taken to all fields of knowledge. He felt that society was more powerful than the individual. Therefore, in order to understand the individual it was necessary to study the society in which the person lived. To Comte, the family was the most important institution in any society.

Comte invented the word "sociology" to identify the study of human behaviour in society. He believed that since societies are created by people, they can be changed by people. This notion that people can be in control of their societies was a new one. People were used to being told what they must do by the king, by the church, and by their superiors.

Comte's thinking was greatly influenced by the French Revolution (1789) and changes to society that occurred because of the Revolution. He felt that much damage had been done because the changes took place too quickly. He believed that changes could be positive if they were accomplished carefully and cautiously.

Another Frenchman who had a large influence on the newly developing discipline of sociology was Emile Durkheim (1859–1917). He expanded on the ideas of Comte and wrote that society was needed by people. The rules, schedules, and goals of society were necessary to keep people under control. Durkheim thought that people were naturally bad, and would take advantage of others if they were not controlled by society.

To Durkheim, one of the most important duties of society is the preservation of order. The society holds its members together through similar beliefs, customs, and traditions, and trains them to believe and act in an approved manner. Religious beliefs in particular help to hold people together in a society and make the society work. Durkheim also believed that if order breaks down, people are more likely to commit suicide.

> *"If we should withdraw from men their language, sciences, arts, and moral beliefs, they would drop to the level of animals. The characteristic attributes of human nature come from society. But on the other hand, society exists and lives only in and through individuals. If the idea of society were extinguished in the individual mind, and the beliefs, traditions, and aspirations of the group were no longer felt and shared by individuals, society would die."*
>
> Emile Durkheim

Karl Marx (1818–1883) also had a great influence on the development of sociology. He believed that the economic system determines the nature of society. For example, he felt that those who own the wealth also control the social institutions, such as education, law, religion, and media. Societies, from his perspective, make laws to benefit the wealthy. Only when the economy changes does society change.

Marx noted that the types of jobs people perform affect the way they think and act. Employers think differently from employees because their jobs are different. They have more status and more power. Marx believed that employers would take as much advantage of their workers as they could, and eventually the working classes would become desperate. Finally their desperation would lead to a revolution, and the wealth of the employers would be taken away by force. The workers would then control the wealth with the result that the social institutions would change to benefit the workers.

Religion, Marx felt, was evil because it convinced poor people that they should accept their situation. Traditional religions taught that suffering in this life would be rewarded by heaven in the next life. Marx argued that religion prevented or slowed down a workers' revolution.

Although a revolution, such as the one that occurred in Russia, did not occur in western industrial societies, Marx's ideas have influenced social scientists throughout the world. In particular, the idea that the economy determines the nature of a society's institutions has been studied by sociologists.

SOCIOLOGY TODAY

The study and understanding of human behaviour is important in a rapidly changing world. Rapid economic and technological changes affect the traditional institutions that make up society. The family, school, church, media, and business are performing different functions as a result of these changes. Consequently people's attitudes, habits, and lifestyles are being altered by economic and technological developments.

Fifty years ago our society was very different. Today, computers, satellites, videos, and lasers are only some in the wide range of changes that have had an impact on Canadian society. Sociologists help us to understand the impact of such changes on human behaviour and relationships. They are now employed in many areas of society such as law, education, medicine, business, and government. Professionals in these areas depend on the sociologists to analyze and explain what effects their decisions will have on people. Sociologists can also help to find solutions to special problems in society such as crime, delinquency, and alienation.

PROGRESS CHECK

1. What interests sociologists?
2. How did the study of sociology develop?
3. According to Durkheim, what forces keep people together in a society? Do you agree with his opinion that people are "naturally bad"?
4. Who might be offended by Marx's ideas? Why?
5. What types of work do sociologists do today?

1.5 *Psychologists and Behaviour*

If you met a psychologist you might jump to the conclusion that this person had the ability to detect your deepest secrets merely by being in your presence. This is not the case. Most psychologists are not interested in analyzing personalities. What does interest them is the development of a science that will tell us such things as how humans think and learn, what motivates them, and what determines and influences human behaviour and skills.

Motivation, what regulates our behaviour in the pursuit of goals and satisfaction of needs, is of great interest to psychologists. Our personal attitudes, interests, and aspirations are involved with motivation.

Psychologists are also interested in how humans learn and develop. They study the effects of television and other products of technology on human behaviour. They also examine the ways in which we are influenced by the group and by society.

Some psychologists, called environmental psychologists, study how people relate to their environment. They might investigate how we are influenced by environmental factors, for example, overcrowding; how we respond to sensory deprivation (lack of stimulation of the senses); or how we react to the lack of gravity and confinement experienced in space travel.

Through the investigations of psychologists we have learned a great deal about human mental and emotional problems. Our understanding of problems such as alcohol and drug addiction and crime has been greatly increased. Some conditions that were untreatable in the past can now be prevented and treated.

METHODS IN PSYCHOLOGY

Psychologists use many different methods to study the behaviour of organisms. They are the same general methods used in other fields of science: observation under natural conditions, directed observation under controlled conditions, case-studies, interviews, questionnaires, tests and measuring techniques, and experimentation under controlled conditions.

An example of observation under natural conditions might be a study of aggression in children carried out in a playground. The psychologist could use a video camera to film the children's interactions and record their verbal communication without the children knowing that they were being observed. The same study, using directed observation, might involve providing the children with particular toys to see how they would use them in their play. Some psychologists conduct controlled experiments with animals such as rats or monkeys, and then relate the behaviour of these animals to human behaviour.

One of the first laboratories for psychological experimentation was opened by a psychologist named Wilhelm Wundt (1832–1920), at Leipzig, Germany in 1879. An important study performed by Wundt looked at how environment influences emotions. Wundt listened to a pattern of clicks made by a metronome, an instrument that beats time with a pendulum or beating rod. He discovered that the rate at which the metronome clicked could produce feelings of pleasure or discomfort. When he used a slow rate of clicks, he noticed he felt slightly tense as he waited for the next click in the series. Immediately after the awaited click he felt a slight relief. A fast rate of clicks produced a feeling of excitement and edginess. As the rate was reduced he felt more relaxed. After testing other people in this manner he concluded that people could pass through at least three different emotions — agreeableness to disagreeableness, relaxation to strain, and calmness to excitability. He suggested that all conscious feelings could be measured experimentally using cause and effect relationships. The investigation of cause and effect relationships is still one of the main methods of study in psychology.

CASE STUDY

Cause and Effect in the Classroom

If psychologists were interested in the effects of the class change bell on student behaviour, they might set up two experimental groups: one group would hear the bell every 45 minutes, while another group would not. After a period of time the psychologists would give a question sheet to both groups, asking about each student's feelings at those times of the day which corresponded with the bell ringing. Eventually the psychologists might discover that the bell had a negative effect on the students. It might have interfered with their concentration, or even made them more aggressive or hostile. The group that did not hear the bell might be more relaxed and ready to work in the next class. The students' moods might depend on whether the bell rang or did not ring.

The results of this experiment could also have the opposite effect. Bells might benefit the students, and give them a better sense of direction, enabling them to order their time and energies more efficiently.

Regardless of the results, the psychologists could help us to understand the effects of bells on students. This could lead to the creation of a better learning environment.

Discuss the effects of bells in schools. Do you think they have positive or negative effects?

EDUCATIONAL PSYCHOLOGY

Educational psychologists are hired by boards of education to develop better learning materials and methods of teaching. They are also hired as counsellors to help students choose careers and to help them deal with any problems they are experiencing.

A famous American psychologist, B.F. Skinner, greatly influenced how we view the learning process in schools. Skinner performed experiments with animals and applied the results to human beings. One experiment involved teaching pigeons how to bowl. The bowling ball was a marble and the aim was to teach the pigeon to peck at the marble in such a way that it would move in a particular direction and knock down a set of miniature bowling pins. Every time the pigeon knocked the marble towards the pins it was rewarded with food. Soon the pigeon learned to associate knocking the marble towards the pins with food. Through this system of rewards the pigeon learned to bowl. This type of learning is called **operant conditioning**. The pigeon was conditioned to respond in a particular way by rewarding it whenever it responded in a particular way — in this case, pecking the marble toward the pins.

From this and other experiments, Skinner believed that behaviour in humans can be shaped and controlled by the environment through a system of rewards. When a child first imitates a word, the parents will show their approval by smiling or paying more attention to the child. To receive more rewards of praise and attention, the child will repeat the word and try to imitate more words. In the same manner, students seek approval from teachers, friends, and family by doing well in school. Undesirable behaviour can be discouraged by withholding rewards.

"Behaviour is a difficult subject matter, not because it is inaccessible, but because it is extremely complex. Since it is a process, rather than a thing, it cannot easily be held still for observation. It is changing, fluid, and (lasts only briefly), and for this reason it makes great technical demands upon the ingenuity and energy of the scientist. But there is nothing essentially insoluble about the problems which arise from this fact."
B.F. Skinner, *Science and Human Behaviour*

During adolescence, the period in a person's life from puberty (sexual maturity) to adulthood, the approval of **peers** also becomes very important. A peer is anyone of the same age or ability as you. Teenagers usually behave in a way that will gain the approval of their peers. Peer approval can have positive or negative effects on the individual, depending on the values of the peer group. Regardless of the effect, peer group approval is the reward, and it will encourage the teenager to act in a certain way.

The ideas of Skinner have been criticized and rejected by many educational psychologists. Nevertheless he has had a great influence on the methods used to teach students in the classroom.

CLINICAL PSYCHOLOGY

Almost one-third of psychologists are clinical psychologists. They use various forms of psychological treatment, or **psychotherapy**, to treat people who are having personal problems or are suffering from behaviour disorders. The problems can range from depression, to alcoholism, through to marital difficulties. Clinical psychologists sometimes work in hospitals, often along with a team composed of psychiatrists, physicians, social workers and other professionals. They are also employed in schools or homes for the retarded, in prisons, in social welfare agencies, and in child guidance centres.

Sigmund Freud (1856–1939) was a psychologist who developed a system of theory and treatment for mental disorders called **psychoanalysis**. In psychoanalysis, patients are asked to recall experiences from early childhood. As they talk about their experiences, they express hidden fears, anxieties, and other problems that they have suppressed. The psychologist and the patient can then explore these areas in greater depth, and work toward overcoming them, or dealing with them in a positive manner.

Freud believed that the human mind is made up of the unconscious self and the conscious self. The **unconscious self** is that part of the mind of which we are unaware. It is composed of the id and the superego. The **id** is the animal part of the human mind, and is made up of the basic biological needs and urges for food, water, warmth, elimination, sex, and aggression. These urges seek to be satisfied. The **superego** controls the id. It could also be called the conscience.

Sigmund Freud with his daughter, Mathilda.

The **conscious self** is the part of our mind of which we are aware. Freud calls it the **ego**. The ego tries to balance the desires of the id with the controls of the superego. Since society requires that we function in an approved manner, the ego uses the energy of the id and the control of the superego to achieve appropriate goals such as education and a career.

Today, many psychologists and psychiatrists reject part or all of Freud's theories and ideas. However, Freud made an enormous contribution to the growth of psychology as a science. He made us aware of the complicated nature of the human mind, and the important role our early experiences have had in shaping us.

Progress Check

1. What is meant by motivation?
2. What methods do psychologists use?
3. Why is punishment not the best way to obtain desired behaviour?
4. What role do clinical psychologists play?
5. According to Freud, what is the unconscious self? How does it affect behaviour?

1.6 Anthropology

The study of anthropology focuses on both cultural and physical aspects of human development. Anthropologists usually specialize in either one area, although there is overlapping of the two areas. For our purposes we will discuss them separately as cultural anthropology and physical anthropology.

CULTURAL ANTHROPOLOGY

Cultural anthropology is the study of human cultures, the way in which humans live, throughout the world and throughout time. The cultural anthropologist investigates human behaviour in many different cultures, both non-literate and literate, without making value judgements. All cultures are objectively studied and accepted as being equally valid and valuable.

To study a culture, cultural anthropologists usually live with the people they are studying. They eat the same food and use the same tools and utensils. In this way the anthropologists are more likely to be accepted by the people and learn about the way in which their society is organized, they way they do things, their language, and their beliefs.

One of the most important cultural anthropologists was Franz Boas (1858–1942). His study of the Pacific Northwest Indians, conducted at the beginning of the twentieth century, strongly influenced generations of cultural anthropologists. Many anthropologists of the time believed that the cultures of the developed countries were superior to those of the underdeveloped countries. Boas believed that every culture was special and just as valuable as any other. "The more I see of people's customs" he said, "the more I realize we have no right to look down on them."

Boas demanded that his students use scientific methods in the study and examination of human cultures. One of his most famous students was Margaret Mead (1901–1978). She made anthropology understandable to ordinary people. Her early studies were of the raising of children in Samoa, India, New Guinea, and North America. One of her theories was that the way boys and girls behave and think is determined by the role their culture expects of their sex.

Margaret Mead on a field trip to Bali, 1957.

Margaret Mead also did a great deal to establish anthropology as a scientific discipline. She insisted that in order to study and understand a culture, one should live with the people for a long time as one of them. This meant learning their language and customs, worshipping their gods, and eating their foods.

Physical Anthropology

Physical anthropologists are interested in human evolution and in the variations and similarities in physical appearance among groups of people. By examining fossil skeletons, physical anthropologists have traced human evolution over the past few million years. The fossil record, although incomplete, has provided physical anthropologists with evidence of how humans have developed physically, and how they interacted with their environment.

Physical anthropologists are also interested in where human life originated and how human populations spread throughout the world. Human migration has, since the beginning, been affected by how well individuals are adapted to different environments. Climate, altitude, and disease are important features of the environment to which humans, in certain cases, must be adapted if they are going to survive. For example, larger lung size and a barrel chest (large thorax, jutting ribs and a raised rib cage) are adaptations for high altitudes where the air is oxygen thin. Black skin pigmentation is a shield against strong sunlight. Blood pigments that are indigestible to the mosquito that carries a disease called malaria are an adaptation to areas where malaria is prevalent. By tracing adaptations such as these, physical anthropologists try to trace the spread of populations across the globe.

Charles Darwin

A theory central to the study of physical anthropology is the **theory of evolution**, an explanation of how living organisms have developed over time. Charles Darwin (1809–1882), a naturalist who lived in the last century, is credited with originating this theory. His two books, *The Origin of Species* (1859) and *The Descent of Man* (1871), began an intellectual revolution that is still being felt today.

In 1831, at the age of twenty-two, Charles Robert Darwin shipped out aboard the H.M.S. *Beagle* on a scientific voyage around the world. He was the ship's naturalist, charged with making notes of all the plant and animal life encountered on the voyage. It was the Galapagos Islands off the eastern coast of South America that intrigued him most of all.

On these islands he observed fourteen species of small land birds that he knew to be finches. He reasoned that these birds had descended from a single species of finch that had come from the mainland long ago. The modern species of finches differed from each other mainly in size and shape of bill and in the habitat they occupied. Some of the species were seed eating, ground or cactus dwelling, and had short, blunt bills. The other species were insect eating, tree dwelling, and had longer, thinner bills. One species had even developed the behavioural trait of extracting insects from trees with a cactus spine held in its bill. Darwin's belief that the finch species had arisen from an earlier ancestral finch species, ran contrary to the belief of the time. People at that time thought that all species were created specially and that species were **immutable**, or unchangeable.

Darwin realized that organisms produce more offspring than survive to maturity. Yet populations normally remain relatively constant in size. There must therefore be an ongoing struggle for existence. He had observed the variations that appear in nature and that some of these variations were inherited. He therefore reasoned that those organisms that

Charles Darwin

inherit variations that increase their chances of survival will be the ones to reproduce and pass on the favourable variations. In other words, **natural selection** will occur.

Natural selection is the pressure that preserves the best adapted members of a population. Over long periods of time, populations tend to become better adapted to their environment, that is, they evolve. Figure 1.3 gives a summary of Darwin's stages in the natural selection process.

Figure 1.3

- Organisms produce a far greater number of reproductive cells than ever result in mature individuals.
- The numbers of individuals in a species remain more or less constant.
- Therefore, there must be a high death rate.
- The individuals in a species are not identical, but are unique in all characteristics.
- Therefore, some offspring will succeed better and others less well in the competition for survival, and the parents of the next generation will naturally be selected from among those members of the species that have been able to adapt more effectively to the conditions of their environment.
- Hereditary resemblance between parent and offspring is a fact.
- Therefore, by gradual change, subsequent generations will maintain and improve on the adaptive qualities of their parents.

Fossil Discoveries

Charles Darwin greatly influenced physical anthropology by stimulating the search for fossil evidence of early humans. In 1856, just three years before Darwin's first book was published, a partially preserved skeleton of a human was found in a limestone quarry cave in the Neanderthal gorge near Dusseldorf in Germany. The skull had features similar to modern day humans, but was more primitive, with heavy bony ridges over the eyes and a sharply receding chin. The brain capacity of this human was similar to ours. A few people recognized the importance of the find, but most dismissed it as a modern human who had been deformed through disease. Darwin's *Origin of Species* caused many people to reconsider the significance of the find and to change their attitude.

Many more Neanderthal remains were discovered in Europe between 1856 and 1910, and found with them were the remains of long extinct animals such as cave bears, woolly rhinoceroses, and mammoths, as well as a variety of flint tools. It was generally agreed that they had lived between 50 000 and 30 000 years earlier, during the last glaciation, when the earth was covered with ice. Neanderthal sites were later discovered in Africa and Asia as well.

Workers building a railway in southern France discovered five skeletons in a rock shelter. They were similar to modern humans, with high foreheads, prominent chins, and small faces. Many more fossil remains of this type of human have been found since and they have been named Cro-Magnon man after the location where they were originally discovered.

In 1891–1892, the discovery of a different kind of fossil was made by a young Dutch physician, Dr. Eugene Dubois (1858–1940), on the island of Java. He found the jaw, skull, and bones of a creature he felt was more advanced than an ape, but not quite human.

In the 1920's a series of exciting discoveries was made near Beijing (Peking) in China. Anthropologists uncovered many skulls, teeth, and bones of early humans, along with their hearths and some crude tools made of quartz and greenstone from another area. Also found with these fossils were the bones of sabertooth tigers and giant hyenas.

The Olduvai Gorge in East Africa has been the site of many exciting fossil finds. Louis Leakey and his wife Mary spent their lives searching there for evidence of our origins. Their search began in 1931, but the first really important find was not made until 1959 when Mary Leakey spotted part of a skull and two premolars stuck in the side of a cliff. A rockslide had exposed them. After days of careful excavation the Leakeys had gathered enough bone fragments to reconstruct a skull. The age of this fossil is thought to be about 1 750 000 years.

Human fossil finds continue to be made, and each discovery means more information to evaluate. The excitement of finding another possible clue to the development of the human race is what keeps physical anthropologists looking and digging for years at a time.

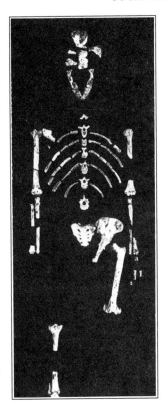

"Lucy", a three million year old skeleton, was found in Ethiopia in 1974. She has provoked as many new questions as she has answered old questions for anthropologists.

PROGRESS CHECK

1. How does the work of cultural and physical anthropologists differ?
2. What did Boas mean when he said "the more I see of people's customs, the more I realize we have no right to look down on them"?
3. Explain what Mead meant when she said the behaviour of boys and girls was determined by sex roles. Do you believe this to be true?
4. Why should a cultural anthropologist live within a culture for a long time in order to study it?
5. What observations did Darwin make while visiting the Galapagos Islands?
6. Why would Darwin's theory of evolution shock many people?
7. How did the discovery of Neanderthal remains help to support Darwin's theory?

OVERVIEW

As our store of knowledge increases, we rely more and more on experts to interpret its meaning. It seems that there are no simple answers when it comes to human thought and behaviour. Sociologists, psychologists, and anthropologists are continually making us aware of previously unknown or unthought of matters. Further, their discoveries and interpretations seem to raise even more questions and controversies. Sometimes it seems that for every expert in a field, there is a different opinion.

One thing we all agree on — the world is rapidly changing. Technology is shaping our thoughts, actions, lifestyles, and concerns. The world seems to be shrinking because of instant communication and fast transportation. Life in many respects is made easier by all the modern conveniences available to us in Canadian society.

On the other hand, life is more complicated in certain respects. Individuals appear to be more distant from each other. They do not have the same closeness and sense of community that people once had before all the conveniences were introduced. More people appear to be suffering, bored, unhappy, or even mentally ill. There are great differences in standards of living, both among Canadians and among the countries of the world. Will our growing knowledge and awareness help us to improve conditions for others and bring people closer together?

KEY WORDS

Define the following terms, and use each in a sentence that shows its meaning.

disciplines
cultures
environment
theories
legends
symbol
immortality
Age of Enlightenment
Industrial Revolution
science

psychology
psychiatry
anthropology
non-literate
motivation
operant conditioning
peers
psychotherapy
psychoanalysis
unconscious self

natural sciences
social sciences
scientific method
empirical
hypothesis
perceptions
behavioural sciences
sociology

id
superego
conscious self
ego
theory of evolution
immutable
natural selection
social institutions

KEY PERSONALITIES

Give at least one reason for learning more about each of the following.

Christopher Columbus
Nicolas Copernicus
Isaac Newton
Voltaire
James Watt
August Comte
Emile Durkheim
Karl Marx

Wilhelm Wundt
B.F. Skinner
Sigmund Freud
Franz Boas
Margaret Mead
Charles Darwin
Eugene Dubois
Louis and Mary Leakey

DEVELOPING YOUR SKILLS

FOCUS AND ORGANIZE

1. (a) Create an organizer that summarizes the discoveries that were made about the physical world and human societies.
   (b) What questions should we ask about the changes that took place in human thinking and societies as a result of these discoveries?
2. (a) Design an organizer that lists the major personalities associated with sociology, psychology, and anthropology, and that summarizes their contributions to these social sciences.
   (b) What questions can we ask about the effects that their theories and conclusions had upon human thinking?

LOCATE AND RECORD

1. Divide the class into groups and do library research on one of these or on other religions of the world — Christianity, Judaism, Islam, Buddhism, Confucianism, or the religions of the aboriginal peoples. Each group should research the particular concept of creation as outlined

by the religion. What similarities in belief become apparent?
2. Divide the class into large groups. Group One are psychologists, Group Two are sociologists, Group Three are cultural anthropologists, and Group Four are physical anthropologists. The tasks of each group are to:
   (a) Research what the particular group of social scientists study.
   (b) Share your research with other members of the class. Once this group research is completed, break into groups of four students to represent each social science group, and relate your research to the members of the group. Be sure to take notes so that you will have a record of all four presentations. Discuss how your discipline interacts with the others, how it differs in focus, and what contributions it has made to the other three disciplines.
3. Conduct a survey of your fellow students using the question: How much enjoyment do you receive from the following? Have them rank their enjoyment through "not at all", "a little", "a great deal".

friendship	television
music	grandparents
boyfriend/girlfriend	brother(s)
dating	sister(s)
stereo equipment	part-time job
mother	school
father	youth group
sports	church/synagogue

When you have recorded all the responses, record the data. Which subject area was most popular? least popular? Speculate why some areas are enjoyed more than others.

EVALUATE AND ASSESS

The following paragraph describes one psychologist's view of behaviour control. Do you agree with his beliefs? Assess his opinions using an example of a classroom situation with which you are familiar.

B.F. Skinner believed that the threat of punishment is not the best way to obtain the desired behaviour. A teacher might keep order in a classroom through threatening punishment, but as soon as the teacher leaves the room, undesirable behaviour often results. Order can only be achieved when the teacher is present. According to Skinner, a much better method is to reinforce the desired behaviour with rewards such as positive remarks and praise. Withholding these rewards should discourage the undesirable behaviour. In such an environment students will more likely develop self-control, good study habits, and a desire to learn. As the student experiences the satisfaction from studying and learning, the rewards become less important. Eventually the presence of the teacher no longer determines the behaviour of the students. Self-control or internal control has replaced environmental control.

SYNTHESIZE AND CONCLUDE

1. Divide the class into groups and research the four most important social problems in your community. You can research your assignment by reading newspapers, magazines, books, or by watching television and making notes. After completing the research, offer suggestions that explain why these are major problems and what your possible solutions to these problems are. Compare your findings and conclusions with the rest of the class. Are they different or similar?

2. The sociologist, Peter L. Berger, writes in his book, *An Invitation to Sociology*, that people in society are like puppets, players, actors, and prisoners. Prepare a chart showing these four categories down the left-hand side, followed by three columns, A, B, and C, to be filled in. In column A describe the characteristics of puppets, players, actors, and prisoners. Indicate in column B situations in which people behave in a way that has been described by the characteristics of each of the four categories. In column C describe how people resemble puppets, players, actors, or prisoners in the situations you have outlined.

APPLY

1. B.F. Skinner, one of the best known American psychologists, has said that human beings are motivated by two factors: seeking rewards and avoiding punish-

ment. All our behaviour, he claims, can be explained by examining how these two factors work. He called this process conditioning. Write a brief report based on at least ten of the headings listed below. Indicate how we have been conditioned, that is, what rewards we gained and the possible punishments we avoided in each case.

Early Life:
eating with a knife and fork
being polite
going to school
arriving at school on time
studying hard at school
obeying rules
having certain friends

Later Life:
dating a certain person
choosing a certain career
getting along with others
buying a certain car
fulfilling our responsibilities at work
living in a certain neighbourhood

2. Speculate about the kind of world we would live in today if the world had little or no technology. What kind of person might you be and what lifestyle might you have if you lived in such a world?
3. Psychologists study motivation — why people act the way they do. For each behaviour listed, give at least two hypotheses that explain why people act this way. People may do things for obvious reasons and also for less obvious, more hidden reasons. For example, we buy clothes for modesty and warmth, but we also buy them to look more attractive. Draw a chart similar to the one below; list the obvious reason first and the hidden reason second.

Behaviour	Obvious Reason	Hidden Reason
ride a motorcycle		
join a club		
play a sport		
go to a concert or dance		
eat a certain food		
listen to certain music		
buy a new car		
date a certain person		

COMMUNICATE

1. Who have been some of the great sociologists, psychologists, and anthropologists? Through individual research projects, write an essay on the career of one personality. Emphasize the contributions that this person has made to one of social science disciplines.
2. Write a short essay or debate and discuss one of the following topics:

- Society is merely a reflection of individual attitudes, values, and behaviours.
- Homo sapiens will eventually evolve into a higher and more developed life form.
- Marx was right in his belief that the economic system determines the nature of that society.
- Increased knowledge has made humans more caring and aware of others.
- Technology has lessened the quality of life.
- Schools should not use marks to make students work.

CHAPTER 2

On Being Human

How do we fit into the animal kingdom? How have heredity and environment contributed to our development?

INTRODUCTION Most of us, when we visit a zoo, are fascinated by the monkeys, chimpanzees, and gorillas. We are amused by the fact that these animals are so similar to us in their behaviour. Sometimes, we stop to wonder how we are different from other animals, and how we are similar. What are the special characteristics that make us human? Is what we call human nature something that we are born with, or something that is learned?

Social scientists, too, are fascinated by such questions. They look for answers in **genetics**, the science of inheritance, in behavioural studies, and in the past. In this chapter we will approach the question of what makes us human. We will examine the influences of environment and **heredity**, the transmitting of characteristics from parents to offspring. We will also trace the evolutionary development of humans.

2.1 *The Animal Kingdom*

We all recognize the fact that we are animals, but at the same time different from the other animals. When we observe chimpanzees and gorillas we can spot many similarities to humans both in their structure and in their behaviour. We can also see many differences. For instance chimpanzees and gorillas are much hairier than humans and they walk with their hands trailing on the ground.

CLASSIFICATION OF ANIMALS

Biologists have classified all the plants and animals according to similarities and differences in their physical structures. The main divisions in the classification system are Kingdom, Phylum, Class, Order, Family, Genus, and Species. Biologists place humans in the Animal Kingdom, Phylum Chordata (having a backbone), Class Mammalia, Order Primates, Family Hominidae, Genus Homo, and Species sapiens. Figure 2.1 shows this classification.

Figure 2.1

Kingdom	*Animalia*	
Phylum	Chordata	frogs, snakes, birds, bats, cats, whales, humans
Class	Mammalia	bats, cats, whales, opossums, kangaroos, humans
Order	Primates	lemurs, monkey, apes, humans
Family	Hominidae	Australopithecine, humans
Genus	Homo	Neanderthal man, humans
Species	Homo sapiens	humans

You may have heard humans, apes, and monkeys being referred to as **primates**. This term refers to the Order to which humans, monkeys, and apes belong. The Order Primates is subdivided into ten families. The one to which humans belong is the Family Hominidae. Early humans and modern humans are often called hominids by anthropologists. What distinguishes the hominids from the rest of the primates? As humans we have a much larger braincase, our teeth are placed in rounded arches (teeth are more or less parallel in other animals), our canine teeth, or eyeteeth, are relatively small and similar in size to our other teeth (in other primates the canines are large), and our big toes are not opposable (we cannot grasp things with our toes the way other primates can).

Within the Family Hominidae there is only one species living today, that is **Homo sapiens**, the scientific name for modern humans. *Homo* means "man" and *sapiens* means "wise" or "intelligent". We are a very successful species, in that we inhabit all the land surfaces of the earth.

Human Groupings

Just the same as other species, all humans do not look alike. We have superficial differences such as variations in eye, skin, and hair colouring, different body builds, and different facial features. In fact, with the exception of identical twins, no two humans look exactly alike. On the other hand, as a species, we look remarkably alike. We owe the differences and similarities of our species to heredity.

In the past, especially in the nineteenth century, anthropologists and sociologists tried to classify human beings into races. They defined a race as a group consisting of people who shared a number of inherited physical features such as hair and skin colouring, hair type, and eye shape.

This concept, that humans could be separated into races, led some people to believe that certain races were superior to others. A great deal of human misery can be attributed to this idea of racial superiority. The most devastating example of racism in this century was the Holocaust, in which 12 million Jews, Slavs, and Gypsies were systematically and deliberately murdered because they were considered to be inferior races.

The concept of race is no longer considered scientific. Firstly, races do not fit neatly into categories, no matter how hard people try. Colours of skin, for example, blend and change gradually throughout the world. Secondly, research in genetics has shown that the observable anatomical traits commonly associated with a so-called race (skin colour, hair type, etc.) are inherited independently of each other, not in a group. Neither are these traits strictly confined to a particular population. They may occur more frequently within a certain population, but they are not exclusive to any one population.

Ashley Montagu, a well-known American anthropologist, believes that "ethnic groups" is a more valid term to use when describing groupings of people. An **ethnic group** is one in which members share common cultural characteristics or traditions. Members of an ethnic group may be of the same nationality or part of a minority group within a country. Physical characteristics play no part in determining whether or not a person is part of an ethnic group.

What do these extracts say about "races"?

"For the layman, as for others, the term "race" closes the door on understanding. The phrase "ethnic group" opens it, or at the very least, leaves it ajar."
Ashley Montagu

"In short, what we call races are subpopulations of a single overall human population. Unlike the evolutionary course undertaken by most other animals, man has not been divided into increasingly isolated populations which then differentiated into new species ... This is not to say that the subpopulations of [man] have been immune to the action of the evolutionary forces, particularly natural selection. But as we have stated previously, man's major adaptation has been to the cultural niche — an environment common to all the subpopulations within our species. Thus every population has an equal capacity for surviving in that

niche regardless of the existing genetic differences between these populations."

J. F. Downs and H.K. Bleibtreu, *Human Variation*

"People still argue that the backward nations have not progressed as far as we have, and that this is a result of certain racial differences. But for most of the past two hundred years, the peoples of Africa and South America, for example, were prevented by European colonists from receiving education and social advancement. Of course, there is the argument that Europeans and North Americans, who today have a relatively high level of technological development, were always cultural leaders but, in fact, the very opposite is true: civilization came late in Europe and was received very slowly by the barbarian peoples who were the ancestors of our famous contemporaries. Europe for most of history has been a backwater that has only developed by adopting advances made in other parts of the world.

... How many times will it have to be reiterated that human beings are not "races" or for that matter the simple principle that all men, by virtue of their humanity, have a right ... to fulfill themselves. None of the findings of physical or cultural anthropology ... can in any way affect this principle, this is an ethical one — an ethical principle which happens in every way to be supported by the findings of science."

Ashley Monagu, *Race, Science and Humanity*

Progress Check

1. What are some differences between humans and other types of primates? What are the similarities?
2. What is meant by the term "heredity"?
3. What term do social scientists now use to replace the term race? What is their objection to the concept of race?

2.2 Heredity and Environment

HEREDITY

You have probably noticed that when a new baby is born into a family, relatives enjoy speculating about whom the baby resembles. Someone will claim the baby has the mother's nose; someone else will say the eyes are the father's; and some other person will be certain that the baby is the "spitting image" of the grandfather when he was a baby.

Family groups are linked by the inheritance of certain common physical characteristics. This process is called heredity, and the mechanism by which it operates is the **gene**. Each cell of your body contains many genes. They are lengths of DNA (deoxyribonucleic acid) responsible for passing on inherited traits from one generation to the next. When a human egg is fertilized by a sperm cell, the resulting fertilized egg contains a set of genetic material from the mother and a set from the father. For this reason children have some physical characteristics of both parents.

Instinctive Behaviour

Besides physical characteristics, there are also certain behavioural traits that are inherited. **Instinct** is a complex, organized behaviour pattern common to all members of a species. What sets instinct apart from other types of behaviour is that it is inherited, not learned.

Social insects such as bees instinctively know their **roles**, or expected behaviours, within the hive. They are genetically programmed to behave as workers, drones, or queens. The web-building behaviour of spiders is also instinctive. Have you ever noticed that spider web designs are characteristic of the particular spider species doing the building? Much of the behaviour of fish is instinctive as well. Instinct determines how they will behave to meet their needs for food, safety, and reproduction. Some fish species instinctively migrate long distances to spawn in a particular stream.

Among the mammals, the more complex the mammal and the better developed its brain, the less its behaviour is determined by instinct. Instead, learning becomes more important for survival.

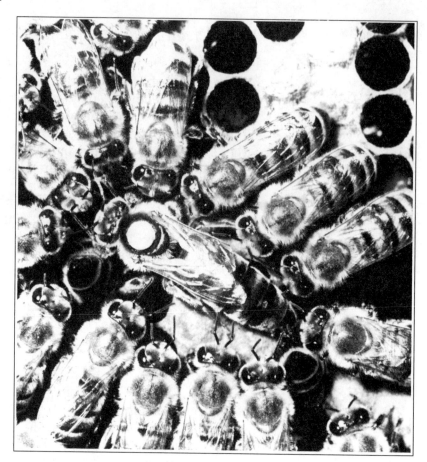

The queen bee holds court — she is constantly attended by a crowd of workers that are ready to feed her as soon as she stretches out her tongue.

By definition, instinctive behaviour is unlearned behaviour. Psychologists used to feel that some human behaviour was instinctive, but they had difficulty finding examples of human behaviour in which no learning was involved. The behaviour of lower organisms is dominated by instinct, but human behaviour is dominated by learning and reason.

Inherited Diseases

Unfortunately, many diseases are also inherited through genes. One such disease is hemophilia, in which a person's blood clots poorly, causing small injuries to bleed excessively. Diabetes, a disease caused by a build-up of sugar in the blood due to a lack of insulin, is another genetically linked disease. As research in genetics progresses, scientists hope to find ways to eliminate genetic diseases.

Genetic Adaptations to the Environment

Sometimes there are **mutations** among the offspring of plants and animals. A mutation is a change that occurs in the genetic material of an individual that can be passed on to future generations. Many mutations are changes for the worse and the individual will die without having a chance to reproduce. A two-headed calf and a fish with a malformed heart are examples of mutations that are unlikely to be passed on. It is unlikely the calf or fish will survive long enough to reproduce. Other mutations are not at all dangerous, and affect the individual only slightly.

A very small number of mutations are positive for particular environments, and give the individual an advantage. Because of this advantage, the individual will be more likely to survive and reproduce. In time, its offspring may replace others of the species that do not have the particular genetic advantage. This process, as we learned in Chapter 1, is called natural selection. It is a process by which genetic improvements can change species of plants and animals, and in time new species may **evolve**, or develop.

It has been suggested, for instance, that people whose ancestors lived in very cold climates tend to have greater bulk in their bodies than in their arms and legs. Perhaps this was a successful adaptation to enable the individual to conserve heat. People who live in the tropics tend to have darker skin than people who live in northern latitudes. Darker-skinned people have more melanin, a skin pigment that protects against the sun's ultraviolet rays.

Sometimes, environments change and become too hostile for some of the forms of life that live there. The environmental changes may occur too quickly for adaptations to occur. If the species is unable to adapt either genetically or behaviourally to suit the new environment, or it is unable to migrate to more favourable areas, it may become extinct. Scientists believe that dinosaurs became extinct about 60 to 70 million years ago, perhaps because of a drastic change in the environment.

In recent times, many creatures have become extinct because of the activities of humans. We are the only species to change our environment in major ways, adapting the environment to suit our needs. We cut down forests, divert rivers, move hillsides, tunnel through mountains, and build cities. Humans have radically changed the world's oceans, rivers, air, climate, and surface.

Residential housing, Burnaby, British Columbia

As a result of human activities, many organisms are faced with the need to adapt to the drastically altered and often polluted environments we have created. The **habitats**, or natural environments of many organisms have been destroyed to make way for cities and farms. Some organisms that are not able to adapt to the new conditions become extinct. The raccoon is one animal that seems to have been able to adapt from life in a woodland to life in a city. Other organisms, whose populations are much reduced by habitat destruction, are on the brink of extinction. Additional pressures such as over-hunting have pushed some organisms into extinction, for example the passenger pigeon, the great auk, and the dodo bird. Many organizations are working toward preserving those organisms that are threatened with extinction. With the help of concerned individuals and organizations, creatures like the whooping crane and the blue whale may be successful in their struggle to survive as a species in a changing world.

ENVIRONMENT

More important to human survival than inherited behaviour or inherited physical adaptations, is our ability to reason and learn. We have learned to build shelters to suit the climate in which we live. We have discovered how to make and use fire for cooking and heating. Our bodies are protected from the elements by appropriate clothing. We have learned to grow crops and raise animals, and to transport them from one area to another. Through reasoning and learning we have adapted our behaviour to the environment in which we live. This has given Homo sapiens a survival advantage.

The Influence of Environment

To a social scientist, environment means more than natural forces such as climate. Environment consists of the external forces and conditions that surround an individual and influence the individual's activities. As humans, our environment includes the society and culture in which we live, as well as the place in which we live.

Have you ever wondered whether the famous people you admire could have accomplished what they did had they had a different upbringing? To what extent did their family and education and where they grew up contribute to their success? To what extent did they inherit their particular abilities? This question, called the nature versus nurture question, has intrigued scientists for a long time. It asks: how much of what we are and what we can do is determined by genetics (nature) and how much is determined by our environment (nurture)?

One of the ways in which scientists have studied the nature/nurture question has been to compare pairs of identical and fraternal twins. Identical twins originate from the same egg, and share the same genetic makeup. Fraternal twins develop from two separate eggs, and are no more or less similar genetically than any other brothers and sisters. Sometimes a study will show that the identical twins are more similar with regard to a particular behaviour or personality trait than the fraternal twins. The trait is then thought to be more likely the result of genetic factors than environmental.

Another method of studying the nature/nurture question is to interview or test adopted children. If they are more similar to their biological parents than to their adoptive parents, then the behavioural trait being examined is thought to be genetically determined. If they are more similar to their adoptive parents, then the trait is thought to be environmentally determined.

Most of the studies that have been carried out indicate that physical characteristics are more likely to be affected by heredity; intelligence tends to be affected by both environment and heredity; education and achievement tend to be more affected by environment; and personality is most likely to be affected by environment.

In just about all aspects of a person's life in which behaviour plays a role, there is an interaction between heredity and environment. It becomes a question of the degree to which heredity and environment interact in any particular situation rather than whether a behavioural trait is the result of only heredity or only environment.

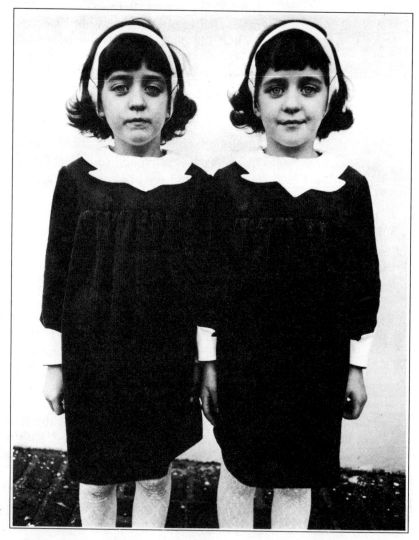

Heredity can explain why twins have similar characteristics, even when raised in separate, different environments.

CASE STUDY

Obesity — Nature or Nurture?

The case histories of 540 adults who had been adopted at birth were studied by a team of medical researchers from Pennsylvania, Texas, and Denmark. Danish subjects were used because Denmark has a registry of all adoptions granted between 1924 and 1947 that included the identities of the biological and adoptive parents.

The question these doctors were trying to answer was whether obesity might be determined by heredity. Their study led to a report in 1986 that found "a clear relation" between the body mass of the adopted adults and their biological parents, not their adoptive parents. Obese children usually have overweight parents. The doctors concluded that "genetic influences are important determinants of body fatness. Childhood/family environment alone has little or no effect" on body mass.

An American medical study involved 4 000 pairs of male twins. This study compared identical twins and fraternal twins. The doctors conducting the study concluded that, for those twins who were overweight, it was twice as likely for both twins to be overweight if the twins were identical than if they were fraternal. This research supported the results of the Danish study. Both studies seem to indicate that genes are one factor that determines body mass.

If the results of these studies are accurate, what techniques can be developed to prevent obesity in people, or to use body mass reduction plans more effectively?

PROGRESS CHECK

1. Why is instinctive behaviour less important and less developed in the more intelligent animals?
2. How can the environment affect the physical appearances of human beings?
3. How have human beings changed the environment? What have been some of the results?
4. What has given Homo sapiens survival advantages?
5. Explain the "nature-nurture" question. What studies have been done, and what were the conclusions? Do you agree with the results?

2.3 Human Needs

BASIC NEEDS

All animals have certain biological needs concerned with survival that must be satisfied. These basic needs — food, water, oxygen, elimination, and reproduction — humans share with other animals. When your stomach is empty, it sends a message to your brain, and your brain responds with a signal to search for food. The impulse to satisfy a need is called a **drive**. The drive for food will persist until your body is satisfied.

HIGHER NEEDS

Human needs are more complex than those of other living creatures. Besides the basic biological needs we also have strong psychological and social needs. We have a need for the companionship of other humans. Relationships with others provide safety, security, love, and affection. We are members of more than one social group — our family, our circle of friends, clubs, societies. Human societies extend from the family group to complex international organizations such as the United Nations.

A psychologist called Abraham Maslow developed a theory about human needs. He proposed that human beings have a **hierarchy of needs** (see Figure 2-2), a certain order in which our needs are arranged. When lower level, more basic needs are satisfied, a person will try to satisfy higher needs.

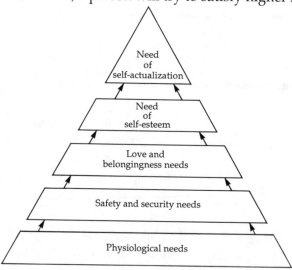

Figure 2.2
Maslow's Hierarchy of Needs

The lowest level of needs are **physiological** needs, relating to the normal functioning of the body. These needs include oxygen, water, and food. If these needs are not filled, people cannot start meeting higher, or more sophisticated needs. (If, for instance, you are very hungry, you will have difficulty concentrating on a textbook until you have eaten.) Even physiological needs have a hierarchy of order. If people are deprived of oxygen and water at the same time, they will first try to satisfy their drive for oxygen, then their drive for water.

Once the physiological needs are satisfied, individuals can think about the next level — the need for safety. Humans will seek an environment where there is safety and security. Once this is done, they will try to achieve a sense of belonging by developing friendships and love relationships. When these needs are met, they will feel pressure to achieve self-esteem and the respect of others. This may be accomplished by obtaining an education, holding a satisfying job, and perhaps raising a family.

Once all these needs are met, people can start to develop and expand their own potential and interests. This highest human need is called **self-actualization**. At this stage, individuals turn from satisfying personal needs to considering and working on problems outside their own environment. Self-actualization cannot begin until the needs lower on the hierarchy are met in the order described.

Maslow thought that only extraordinary people would achieve this final and highest level of development. Maslow would probably have thought that Nellie McClung, who worked to extend the rights of Canadian women was a self-actualized person. Certainly, he thought that Albert Einstein and Albert Schweitzer were self-actualized people.

Einstein riding a bicycle near a friend's home, 1933

Maslow believed that people develop differently and that their success or failure in meeting the higher needs will depend on their personalities, their past experiences, and the culture in which they are raised. He felt that mental illness could be caused by unsatisfied needs. Children who do not receive enough love, according to Maslow, may experience difficulty as adults in achieving warm and caring relationships with others. These difficulties could lead to serious mental and social problems in some situations.

Progress Check

1. How are needs and drives related?
2. According to Maslow, what can be the consequences of needs not being properly satisfied?
3. What is your opinion of Maslow's hierarchy of needs? At what stage are you on this hierarchy?

2.4 Learning to be Human

Culture

It's another school morning and you climb out of bed, brush your teeth, wash, and dress. As you enter the kitchen to eat breakfast, the radio announcer gives the time and then begins reading the news. Your little brother is arguing about whether he should have to wear his jacket. Your parents are preparing for their working day as you grab your notes and the textbooks you need and head out the door. You call over your shoulder that you won't be home until after dinner time that night because you will be working after school.

In other homes in your community, similar scenes are taking place. The clothes that other people are putting on are similar to those of your family, the foods that they are eating for breakfast are similar, the daily activities for which they are preparing are similar, and the values and beliefs that shape the way they think and behave are similar. The way of life that is shared by members of your society is called a culture.

Culture is the common bond that ensures an order and predictability to the way people interact. It permits society to run smoothly. One of the most important roles of the family is to keep alive the culture of the society by passing it on to new family members.

SOCIALIZATION

Newborn infants have no knowledge of the language, customs, beliefs, or values of the society in which they find themselves. As they grow up they must be taught how to become a functioning member of their society. The process through which they learn their society's culture is called **socialization**. Once they have adopted the values and social roles, and absorbed the special knowledge of their culture, they are said to be **socialized**.

Feral Children

Try to imagine yourself growing up in a world without any parents or other human contacts. There are some known examples of children who were lost or abandoned by their parents and subsequently raised by wild animals. These children are referred to as feral children.

In 1920, Reverend Singh, a missionary in India, discovered two female children who were living with a family of wolves. It was observed after they were captured that they had the behaviour patterns of wolves. They ran about on all four limbs, ate raw meat, howled and growled and ate directly from dishes with their mouths. The younger feral child died within a year and never acquired human behaviour. But Kamala, the older, lived for another nine years. Eventually she learned to eat cooked food, to wear human clothes, to understand simple language, and to express emotions. But Kamala never learned to run except on all fours.

Other reports of feral children have since been reported. In 1980, it was learned that a mother in Portugal had locked up her infant child in a chicken coop while she worked. When her nine year old daughter was finally discovered, she flapped her arms like wings and made sounds like a chicken.

In 1986, a seven-year-old boy was discovered in Uganda, Africa, who acted like a monkey. He moved through the trees and made sounds like a monkey. He had learned these behaviours from the animals around him.

The cases of feral children seem to show that proper environmental influences are important in the development of human behaviour.

Isolates

The importance of environmental influences to the development of human behaviour is reinforced by cases of children who were raised in isolation from other human beings. These children, called isolates, were provided with some food and shelter, but were never, or rarely, spoken to or held by another human being. They were forced to live in a dark room where there were no toys or playmates. There was little or no stimulation or activity in their environment.

Two famous isolate cases were those of Anna and Isabelle. When Anna was found in 1938 in Pennsylvania, she was five years old. She had been born out of wedlock and the father of Anna's mother had been furious. He did not want to see Anna around the house. Anna's mother confined her to one room. Her mother brought her milk but otherwise paid very little attention to her. She was rarely bathed or cuddled. As a consequence, when Anna was found, she had few signs of human nature.

Anna was placed in a new home and attended a school for the mentally handicapped. By 1941, the school reported that Anna was toilet-trained, and could use a spoon when eating. She had also learned to brush her teeth and dress herself, but could not fasten her clothes. She had begun to speak but had only advanced to the level of a two year old. Anna died of jaundice in 1942.

In the case of Isabelle, some circumstances are similar while others are different. Isabelle was six years old when she was discovered in 1938. Like Anna, Isabelle was an illegitimate child but her deaf, mute mother had been shut away in a dark room with her. The two often communicated by means of hand gestures, and Isabelle received more care and attention than Anna. But when she was found, she reacted with fear and hostility toward strangers, especially toward men. She made strange croaking noises and her actions were similar to those of a deaf child.

Over a period of two years, Isabelle went through an intensive program of rehabilitation. By the time she was eight and a half, she had reached the educational level of a normal child. Unlike Anna she was full of energy and very cheerful.

Sociologists believe that the early close physical contact that Isabelle had with her mother enabled her to develop into a normal child. Part of this contact involved some simple type of communication. Without these experiences, Isabelle would probably not have acquired human behaviour.

The cases of feral and isolate children seem to indicate that what we call human nature is not something with which we are born. It is learned and acquired after birth. Not only are social contact and imitation important, but also love and attention from others are necessary for the proper development of human nature. The qualities we associate with being human seem to be learned, not innate. However, the ability to learn these human qualities may be inherited.

The environment into which a child is born will strongly influence the child's development. The cultural environment will determine the language, attitudes, values, and some types of behaviour the child will learn. The quality of care the child receives, and the presence of loving and affectionate relationships will help to determine whether the child is properly socialized. A child growing up in an environment where there is proper care and adequate love will have a good chance of developing normally and fitting into social situations.

PROGRESS CHECK

1. Why is culture important?
2. Why would some people say that we are not human at birth?
3. How did the feral children acquire their behaviours?
4. Why did Isabelle develop into a normal human being? Compare her case to Anna's.

5. What do ferals and isolates have in common? How are they different?
6. Why would social scientists study feral children and social isolates?
7. How will the environment affect children's development?

CASE STUDY

Victor
The Wild Boy

Victor, "the wild boy of Aveyron", was discovered in France in 1797. He was seen running naked through the woods. Now and then he was seen digging up potatoes or searching for acorns and roots. Eventually he was captured by hunters and examined by the authorities who said he was about 11 or 12 years old. Although his body bore scars, he did not appear to have any serious physical deformity. What was quite noticeable, however, was that he could not speak any language. He communicated only with cries and grunts. Philippe Pinel, one of the first psychiatrists to examine the boy, concluded that Victor was mentally handicapped and could never be educated. Pinel said the boy "was not an idiot because he was abandoned in the woods, he was abandoned in the woods because he was an idiot".

One man, Jean-Marc-Gaspard Itard, did not agree with this assessment. Itard, who was the resident physician at the school for deaf mutes where the boy was sent, wanted to train him to become human. One of the first things he did was to give the boy a name: Victor. Itard noted that Victor rejected clothing even on the coldest days. He put his hand in a fire. He did not discriminate between objects, and reached indifferently for various objects, even those reflected in a mirror. He did not stare at any one object, and he did not sneeze or cry. Victor recognized food by smell, not sight, and he preferred uncooked food. He appeared to have no particular taste for candy. He showed no emotion, attempted no speech, and now and then would run on all fours.

Some notable changes came after three months of training. Victor started to use eating utensils, wore clothing, and used the toilet. When he sneezed for the first time, he became so frightened he threw himself onto his bed. Just as important he began to show emotion, feelings such as gratitude, remorse, and pleasure when he pleased others. But there were limits to how far Victor could progress. One feeling he never appeared to express was pity and after a period of six years he had still not learned to speak even with

Itard's specialized program. Dr. Harlan Lane, a psychologist, who has done extensive research on this case, concluded that Victor's long period of isolation deprived him of the opportunity to imitate others. Eventually some aspects of Victor's human nature were recovered through training and interaction with people. He learned to read and write simple sentences but he never learned to speak even by the time of his death at more than 40 years of age.

1. What criticism can be made of Philippe Pinel's assessment of Victor?
2. What non-human characteristics did Itard observe in Victor?
3. What human characteristics did Victor develop?
4. According to Dr. Lane, why was Victor's development limited?

2.5   Our Human Ancestors

Now that we have looked at how socialization helps us to become human, let us go back in time to see how we developed as humans and to speculate about the socialization of early humans.

About 15 million years ago, new, somewhat human-like creatures were making an appearance in Africa. They were noticeably different from the existing apes, and also different from modern humans. But they were more human than ape in their structure and intelligence. They were the first **hominids**, or human-like creatures and our earliest human ancestors.

These new creatures began to move out of the forests and to explore the open grasslands for food. They were not hunters; they did not know how to make special tools for killing. They were foragers, eating whatever they could gather that was edible. Scientists speculate that an inadequate food supply in the forests may have been the pressure that contributed to their evolution.

EARLY HUMAN CULTURES

The early hominids had genetic characteristics that set them apart from other animals. They stood erect; they walked on two legs, not four; they could use their hands to manipulate things. In addition, their brains were larger than those of other primates.

How did these early hominids evolve? Some anthropologists believe that a change in climate meant that areas of land that had been forested became more open. Because of this change there was pressure to seek food on the open savannas, or grasslands. Natural selection would favour individuals who were adventuresome or restless enough to seek new habitats, who had the intelligence to adapt their behaviour to changing conditions, and who were physically capable of surviving in the open.

As early hominids spent more and more time on the ground, they began to eat a wider range of food. It was necessary for them to travel long distances for food and water, especially during dry periods. At first they would have returned to the woodlands to sleep, but when they had moved too far away, groves of trees in the savanna offered protection from predators. When trees were not available, they probably found shelter at night on the ledges of cliffs.

It is likely that our earliest ancestors had some form of social organization, but we know very little about their habits. In order to travel through the open savanna, where they were constantly exposed to predators, they probably formed groups for safety. This would require more organization and discipline than had been necessary in the forests. Individuals needed to be posted in front, behind, or at the sides of the group to look out for predators. The babies and very small children had to be carried. In an emergency, such as an attack by a leopard, each individual needed to know what to do.

Chimpanzees, our closest living relatives today, use primitive tools. They use sticks to extract termites from underground nests, and chewed up leaves to sponge water from the forks of trees. It is likely that the early hominids were able to use simple tools as well. They probably used sticks and pointed rocks to dig up roots and shoots of plants, and to extract water from certain plants. These tools may have also doubled as weapons to defend themselves against predators.

Eventually our ancestors developed skills in making specialized tools to be used for hunting, cleaning hides, and preparing food. Such tools have been found in Africa at sites that are more than two million years old. The early hominids living then are named Australopithecus. The fossils of these hominids were first discovered by Louis and Mary Leakey in the 1950's.

The first evidence of the use of fire by hominids dates back about one million years and comes from a cave in Escale, France. Excavations there turned up traces of charcoal and ash, stones that had been cracked by fire, and several hearth areas. As more campsites of early hominids are discovered it is likely that we will find that fires were used much farther back in our prehistory. The ability to use fire meant that people could cook, keep warm, and ward off predators.

In a cave near Beijing, China the remains of an early hominid living site was discovered by archaeologists. The hearths, bones and tools found there showed that early humans used fire and tools, and hunted animals for food at least 750 000 years ago. The hominids found there (Peking Man) and in sites in Africa, Europe, and Indonesia (Java Man) have been named Homo erectus.

About 250 000 years ago, Homo sapiens began replacing the less intelligent Homo erectus. Such a change came very gradually — perhaps over 100 000 years. Archaeologists class these early Homo sapiens as different from their ancestors, because of differently shaped skulls, but their culture seems to have been similar.

The remains of more recent Homo sapiens, from about 35 000 years ago, were first found in France at a location called Cro-Magnon. These people, called the Cro-Magnons, were physically very similar to modern humans. They were highly skilled toolmakers and hunters, and their culture, or way of life, included art and traditions. They buried their dead, and they decorated the walls of their caves in southern France and northern Spain with beautiful paintings.

Living at the same time as the Cro-Magnons were the Neanderthals, named after the place in Germany where they were first found. In appearance they were very primitive, with massive jaws, heavy bony ridges over the eyes, and short stocky bodies. The Neanderthals looked more like Homo erectus than Homo sapiens. They had much larger brains than Homo erectus, however, and for this reason they are considered to be a type of Homo sapiens. It is the Neanderthals that most people picture when the term "cavemen" is mentioned. Yet, one cannot assume the adjective "primitive" to be entirely correct. Their skeletons had been placed in a special position and were found with tools, the remains of food, and in one case, flowers. Obviously they had been buried by others of their group, and it would seem, from the objects buried with them, that they believed in an afterlife.

It appears that the Neanderthals were replaced in a relatively short time by the Cro-Magnons. A number of theories try to explain how the replacement came about. Some anthropologists feel that the Cro-Magnons were better organized for hunting big game. They tended to hunt in larger groups and this, together with their superior weapons, made them better able to compete for game. Others suggest that the Neanderthals may have been killed off by the Cro-Magons or absorbed by the Cro-Magnon tribes. Whatever the case, the Neanderthals disappeared about 30 000 years ago.

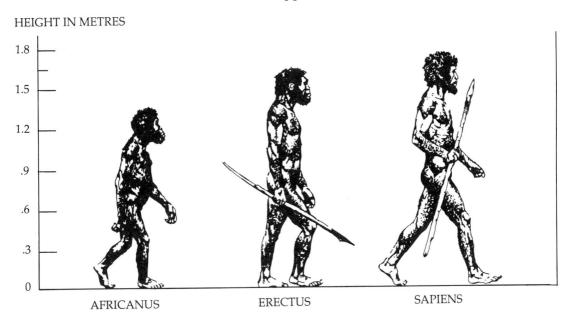

If fossils could be "fleshed-out," the individuals might or might not look like this.

How did our Cro-Magnon ancestors live? At first they lived in bands of a few families, but eventually the bands became associated with other bands into tribes. The tribes were linked by marriage and shared traditions, and at times they would meet to socialize and to hunt on a large scale. Hunting trips might take them great distances from their homes as they pursued herds of large game. For hunting as well as other uses they made a great variety of high quality tools and weapons.

The homes of the Cro-Magnons included natural caves, tents made from the skins of animals, and shelters built from rocks and other materials. The world's earliest known art was produced by the Cro-Magnons. Cave paintings mainly of game animals have been found deep inside caves in France and Spain. They are located so far from any natural light that it is suggested that sacred or special ceremonies might have taken place there. Carvings made from stone, ivory, and bone, and figurines made from clay and powdered bone have been found in Cro-Magnon sites. The Cro-Magnons also wore the first known jewellery. Their clothes were decorated with coloured beads and they wore bracelets of ivory and necklaces of teeth and fish vertebrae.

Like the Neanderthals, the Cro-Magnons buried their dead. Individuals have been found in carefully marked graves with clothes, jewellery, food, and other personal possessions. These are unmistakeable indications that early Homo sapiens believed in an afterlife and practised rituals.

With the melting and retreat of the last glaciers, life became somewhat easier for the Homo sapiens. Fish, birds, and game animals became more plentiful and were available year round. These improved conditions may have made it possible to establish permanent settlements and to have larger families. The result was a markedly increased population.

FROM HUNTERS TO FARMERS

About 10 000 years ago, Homo sapiens devised ways of raising their own food by farming and keeping domestic animals like chickens and goats. This important step meant that people had a more reliable food supply. As a result populations again increased.

Farming communities developed, that were often able to produce more food than was actually needed. A food surplus meant that not everyone had to farm, and members of communities were able to specialize by becoming tool-makers and potters. The goods produced by such artisans were traded for those produced by other communities. Gradually, writing developed, perhaps to record business transactions. At the same time religion became more complex and organized.

By 3500 B.C. urban communities or cities were established in the fertile river valleys of the Nile and the Indus. The first known city was built at Sumer, in Mesopotamia. In a highly developed culture and society, or **civilization**, such as the one at Sumer, complex language, writing, government, art, architecture, trade, and religion flourished. Other early civilizations include the Egyptian, Hebrew, Minoan, Mycenaean, Greek, Roman, and Chinese. Our present civilization has evolved from these earlier ones, and we owe a great debt to them for our customs, knowledge, values, and language.

PROGRESS CHECK

1. What physical characteristics did the hominids possess that other animals lacked?
2. What discoveries and inventions gave hominids a survival advantage?
3. Why did Homo sapiens replace the other hominids?
4. What evidence is there that early Homo sapiens had a well-developed culture?

OVERVIEW

Our awareness of the origins and development of animals and hominids is growing. Through observations and discoveries, experts are improving our understanding about the different influences that help to create certain physical, cultural, and personality characteristics in human societies and individuals. These understandings draw us closer together into the human family because they help us to understand each other better. We are learning more about our roots and where we originated. This helps us to have a greater control over our lives because we understand better why we think, act, and feel the way we do. As far as we know, only humans create a culture that is passed on from one generation to the next. We develop traditions, beliefs, and values, and accumulate knowledge. Our skill with language also sets us apart from other animals, but there have been some studies with other primates that indicate that, if taught sign language, they can express themselves. Our culture encourages us to study, to question, and to classify what we find around us, just as you are doing in this course.

KEY WORDS

Define the following terms, and use each in a sentence that shows its meaning.

genetics   roles   physiological
heredity   mutations   self-actualization

primates   evolve   socialization
Homo sapiens habitat   socialized
ethnic group drive   hominids
gene   hierarchy of needs civilization
instinct

KEY PERSONALITIES

Give at least one reason for learning more about each of the following.

Ashley Montagu   Anna
Abraham Maslow   Isabelle
Reverend Singh   Victor
Albert Einstein   Jean-Marc-Gaspard Itard
Kamala

DEVELOPING YOUR SKILLS

FOCUS AND ORGANIZE

1. (a) Develop an organizer that lists your personal characteristics which have been determined by heredity and those that have been a result of environment.
   (b) What questions can we ask about these characteristics?
2. Divide the class into groups and have a brainstorming session to develop a list of behaviours and qualities that fall under four headings: animals, ferals, isolates, and normal human beings. Four members of each group can write down the suggested ideas on separate sheets of paper for each of the above headings. On separate sheets of paper list the similarities and differences among the four categories. Can it be said that all four share more similarities than differences? Can ferals and isolates be referred to as human beings? Can Homo sapiens be defined as just intelligent animals?
3. By using a chalkboard timeline diagram, trace the development of hominids. List their physical and cultural characteristics. Why did earlier types of hominids disappear?

LOCATE AND RECORD

When anthropologists study a group of people they record their behaviour, material possessions, and beliefs. By referring to information in this chapter and other sources, write a brief report entitled "Cro-Magnon Culture" that includes the following topics: family, religion, physical features, funerals, housing, art, kinship, migration patterns, body adornment, weapons, and hunting.

EVALUATE AND ASSESS

What is the point-of-view of each author on the subject of human beings and human culture? Assess their accuracy.

"... man has another mental quality which the animal lacks. He is aware of himself, of his past and of his future, which is death; of his smallness and powerlessness; he is aware of others as others — as friends, enemies, or as strangers. Man transcends all other life because he is for the first time, life aware of itself."
Erich Fromm, *The Art of Loving*

"It is now generally recognized that intelligence tests do not in themselves enable us to differentiate safely between what is due to innate capacity and what is the result of environmental influences, training and education. Wherever it has been possible to make allowances for differences in environmental opportunities, the tests have shown essential similarity in mental characters among all human groups."
Otto Klineberg, *Readings in Anthropology*

"Differences in culture do not arise because different peoples have different inherited capabilities, but because they are brought up differently. We learn to speak, think, and act the way we do because of our daily associations, and when these change, our habits of speaking, thinking, and acting also change. Children have no culturally based ways of behaving at birth, they only acquire these as they grow up and as a result of a long and complicated process of learning."
Ralph L. Beals and Harry Hoijer, *An Introduction to Anthropology*

SYNTHESIZE AND CONCLUDE

1. Divide the class into groups and brainstorm the cultural similarities and differences among members of the group. List your conclusions in an organiser and rank their importance. Number one would be the most important. After you are finished, determine whether there are more

similarities than differences. Discuss whether class members could work and function together in a classroom if there were no similarities among them.
2. In a short essay, analyze the relationship among the environment, genes, and mutations. What creatures are most likely to survive in different environments?

APPLY

1. Speculate on what would have happened to you if you had been placed in the same position as the ferals and isolates described in this chapter.
2. According to Maslow, individuals may develop psychological problems if certain needs are not met. Indicate what psychological problems people might experience if certain needs were not satisfied in their earlier development and growth.
3. Draw a chart with the following two headings to answer this question: What needs do humans and other animals share?

    Shared Needs   *Distinctive Human Needs*

4. Below is a list of cultural traits that Homo sapiens developed. For each trait, indicate why this trait probably developed.

Development	*Reasons*
families	protection, child rearing
tools/weapons	
cave paintings	
burial of the dead	
farming/animal domestication	

COMMUNICATE

1. According to Abraham Maslow, all people have a hierarchy of needs. Interview five people and ask them to list their needs in order. Eliminate basic or survival needs such as hunger and thirst and focus only on "higher order needs". Ask the respondants what the reasons were for their rankings. Write a brief report on your findings and share them with the class. The following chart gives some of those higher order needs and gives a brief description for each.

Group	Need	Description
SAFETY	Security	you are protected from danger
	Stability	life is steady and not chaotic or confusing
	Order	life is organized and everything is in place
BELONGING	Affection	you give and receive friendship and love
LOVE	Affiliation	you are associated or united with a group or cause
	Identification	you recognize who you are, you "know yourself"
ESTEEM	Prestige	honour, fame
	Success	accomplishments, prosperity
	Self-Respect	liking oneself and feeling important
SELF-ACTUALIZATION	Self-Fulfilment	having the knowledge that one is the best one can be or has developed to one's fullest potential

2. Write an organizer or short essay, or debate one of the following topics.

- There is no individual free will — people merely act and make decisions based on the manner in which they were raised.
- People who experience mental and emotional problems did not receive enough love and attention when they were younger.
- Genes predetermine the lives of human beings.
- Scientists should stop trying to find ways to change genes — it could be the beginning of trying to establish a "super-race" of people.
- Environmentalists are wrong to try and protect the environment from human development — if creatures can-

not adapt to change, they should be permitted to become extinct.
- Smart people will succeed in life even if they are born into a poor environment.
- People are born humans — they do not have to learn to become humans.

3. Write a 1000 word essay (minimum) which speculates about the evolution of Homo sapiens in the next several thousand years. What physical characteristics might develop? What effect would changes have on individuals and on human societies? In your essay, comment on the desirability of speeding up changes by artificially altering human genes to produce, for instance, healthier, stronger, more intelligent human beings. Are there possible dangers associated with this type of genetic engineering. Comment on whether governments should forbid genetic engineering research. Should nature be allowed to follow its own direction?

Class members can summarize the conclusions of their essays and present them to the class for discussion and analysis.

UNIT 2

SOCIAL BEHAVIOUR

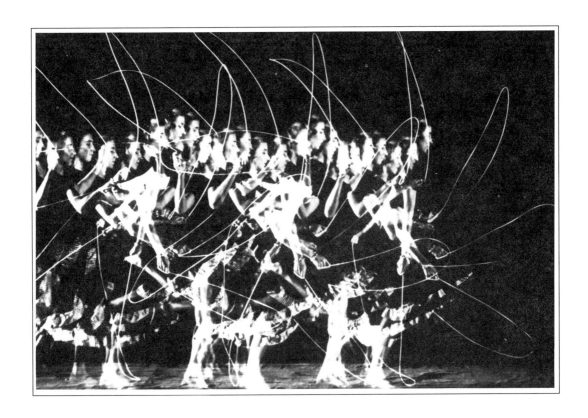

CHAPTER

3

Socialization

What is socialization, and why is it important to our development?

INTRODUCTION All of us have seen toddlers happily digging in the sand or swinging and climbing in a playground. In twenty years time those same little people will be adults. They will be pursuing careers, working at jobs, and making decisions that will affect your community. How do these carefree children turn into independent, responsible, and capable adults?

The answer to this question consists of one word: socialization. In this chapter we shall explore how socialization takes place, its importance, and the roles we are socialized to fill. Through socialization we acquire the knowledge, skills, values, and appropriate behaviours to function in society. Socialization links us to other people and to our community.

3.1 *How Are People Socialized?*

We come into the world helpless infants, and it is some time before we can do things for ourselves. By the end of the first year of life most humans have learned to walk, but they need constant care. Many other animals are mature adults capable of looking after themselves by the end of their first year. Humans take longer to mature than any other animal.

Other animals have instincts that help them to behave in ways that will meet their survival needs. Humans do not have those instincts. We do not instinctively know how to find, grow, or kill food. Nor can we instinctively build a shelter that will protect us from the elements and from predators. What we do have is a large, well-developed brain. Rather than depend on instincts, we depend on learning and reasoning to find ways to meet our needs. We learn and reason throughout our lives.

The socialization process is a learning process, and it is begun by parents. They set examples and teach the type of behaviours they expect from their children. They also teach basic skills such as communication, cooperation and sharing with others, eating in an acceptable manner, and control over impulses.

This family is picking cranberries on the Nova Scotian south shore.

The socializing process begun by parents is carried on by teachers, peers, brothers, sisters, co-workers, and members of the community. They all help to shape us into functioning members of society.

We are socialized to fill the roles that we are expected to play. Right now you are filling several social roles. You are a member of your family, your school community, your peer group, possibly a member of a special interest group, and possibly, if you have a part-time job, a member of a working group. To function successfully in these different roles, you had to be socialized.

The socialization process you undergo from birth is preparing you to become an adult member of society. Perhaps as an adult you will care for and socialize your own children. They, in turn, will have learned from your example how to socialize their own children. In this way our culture is passed from generation to generation.

The Need for Support

The family is the first, as well as the most important agent or institution in the socialization process. Families provide children with an environment that promotes development. The loving attention, encouragement, and stimulation children experience within the family help them to develop into productive members of society.

A study conducted by René Spitz in the 1940's clarified the need children have for loving care. He studied 91 children in an orphanage, all of whom were three years of age or under. Included in this group were 45 babies under 18 months of age who were cared for by just six nurses. They were only able to give minimum care to the babies, so the babies were not cuddled or played with. They were left to lie in their cots with only the ceiling to look at and the cries of other babies to listen to. After two years, 21 of the original 91 children still lived in the orphanage. Of the 21 children still at the orphanage, all were severely retarded in their physical development, mental abilities, and basic skills. Many were unable to walk or to eat with a spoon, and only two could say more than five words.

Studies in which young rhesus monkeys were deprived of their mothers and the company of other monkeys from birth, reinforce some of the orphanage observations. The deprived monkeys did not mature normally, and the females, as adults, treated their babies either with indifference or abuse.

One conclusion psychologists have reached from studies such as these is that babies require more than just feeding, washing, and changing. They need loving contact and stimulation to develop healthily. If it is not possible for a child to be socialized within a family environment, then the child should be played with, talked to, and looked after in a warm loving way by whomever is providing care.

CASE STUDY

Harlow's Monkey Studies

*An important psychiatric study was conducted by Harry F. Harlow, an American psychologist at the University of Wisconsin Primate Laboratory. He separated baby rhesus monkeys from their mothers six to twelve hours after their birth. They were placed with **surrogate**, or substitute, mothers made of wire or of cloth. The wire mothers were equipped with a nipple and feeding bottle, whereas the cloth mothers had no feeding apparatus.*

Harlow wanted to find out which of the mothers the monkeys would stay with the longest. Would their need for food compel the babies to stay with the wire mother, or would their need for comfort make them want to stay with the soft, cloth mother? Harlow found that the babies preferred the cloth mother, and stayed with it up to 23 hours a day. They moved to the wire mothers only when they were hungry.

The babies raised with surrogate mothers and no other monkeys seemed to be normal in infancy. However, when they were adults Harlow observed some very strange behaviour patterns. Often they moved in circles, or stared vacantly while clutching themselves and rocking back and forth. When they were introduced to a group of monkeys, they were withdrawn, and covered their faces with their paws, and hid from the others. They did not know the proper mating behaviour of monkeys, and the females who eventually became mothers did not know how to behave toward their babies. These "motherless mothers", as Harlow called them, either physically abused their babies or were indifferent to them. The indifferent mothers did not nurse, cuddle, or protect their young; the abusive mothers often injured their young so severely that they died.

Despite their mothers' indifference or abuse, the second-generation baby monkeys never stopped trying to be close to their mothers. Occasionally, after months of effort, some of the babies managed to change their mothers' behaviour from rejection of them to acceptance.

1. What did Harlow conclude from his experiments with raising monkeys with surrogate mothers?
2. Do you think that similar results would occur if humans had been used in these experiments instead of monkeys?
3. Do Harlow's studies throw any light on possible causes of child abuse? If so, what might they indicate?

3.2 Socialization and Norms

Babies are naturally **egocentric**, that is, they believe themselves to be the centre of the world. They expect their needs to be met immediately and fully. Gradually they are taught that others have needs as well, and that they may have to wait for what they want.

Children also learn at an early age that not all behaviour will meet with approval. They will be taught certain rules of behaviour. If they break the rules, children will learn that there will be consequences such as being sent to another room, being denied something they were looking forward to, criticism, or mild physical punishment. In this way, children are taught to observe the rules of the society they will be entering.

People in a society share common values, beliefs, dress, patterns of speech, and manners. Children need to learn these **norms**, the standard models of behaviour of their society.

Some norms are much more important than others, and they may be written down as laws. Other norms are simply standard etiquette such as saying "please" and "thank you". Sociologists have classified norms into three basic types — folkways, mores, and laws.

Folkways include etiquette and basic good manners. Standing up to give an elderly person your seat on a crowded bus is an example. Dressing up in your good clothes to attend a wedding or funeral is another example of folkways. Keeping yourself clean is a folkway that helps the general hygiene of your society. Failure to observe folkways will not result in severe punishment. **Ostracism**, or exclusion from the group, ridicule, and negative gossip are some of the penalties someone might expect for disobeying folkways.

Mores are important rules of behaviour in a society — they are the moral standards of a society. Many of the mores are rules that tell us what not to do. Some mores of Canadian society are: do not cheat, do not steal, do not destroy public or private property, do not drive after you have been drinking, do not physically harm others, and do not kill. The penalties for disobeying mores are much more severe than for disobeying folkways. People who disobey mores are endangering the stability of the society, and they must be punished accordingly. Sanctions or punishment could range from being expelled from school for cheating on an exam to life imprisonment for killing someone.

Celebrating a birthday is a folkway of our society.

Laws are norms that have been formally stated and that are enforced. They specify the standard of behaviour and the punishment for violating it.

WHY DO WE CONFORM?

The socialization process teaches people the norms of their society. But what makes people observe these norms? Probably the most important reason is the strong need people have to be accepted by others and to belong to the group. If people did not conform to the norms of their society they would be ostracized and possibly jailed.

As individuals we feel the pressure exerted by the rest of society to **conform**, to act or behave in agreement with the actions of others. The pressure that influences us to conform comes from several groups. These sources of influence can be parents, teachers, and peers. Within your peer group you most likely have a code of behaviour and dress. You, along with the other members of your group, try not to violate this unwritten code. If you do, you risk having the other members ridicule you, gossip about you, or boycott you from the activities of the group.

Social influence, such as the influence of our peer group, can be very strong when the group is large. It is much more difficult to resist the influence of a large number of people than it is to resist one or two. Suppose that in a peer group of thirty people, twenty-eight believed it was essential to get married by age twenty. The remaining two people felt it was important to wait until age thirty. Which group do you think would influence you more strongly?

It is even more difficult to resist the combined influence of more than one group. If all the people you associate with — parents, teachers, friends, co-workers, team members, and students — do not smoke and frown on others smoking, it is unlikely that you would begin to smoke.

With society always undergoing change, social influence constantly shapes our behaviour so that we can adapt to new social conditions. Smoking used to be thought of as physically harmless and as a sociable activity. Now, because of advances in medicine, we know that it is very harmful to our bodies to smoke or to inhale second-hand smoke. Smoking is no longer a norm; nonsmoking has become a norm. In fact, in some areas of society nonsmoking is considered such an important norm it is being written down as a law.

Progress Check

1. Why is it important that children learn the rules and absorb the beliefs of their society? How does this learning process take place?
2. What will happen to people who break norms?
3. How does society influence people to follow certain patterns of behaviour? Give examples.
4. How are you influenced by your peer groups?

3.3 Learning and Conditioning

We learn to fit into our society, and this learning process is begun soon after birth. But why do we want to learn and how do we learn? In this section we will look at some theories of learning developed through experimentation.

CLASSICAL CONDITIONING

A Russian doctor, Ivan Petrovich Pavlov (1849–1936), while working on digestion in dogs, made an extremely important contribution to our knowledge of learning.

He found that dogs salivated (produced saliva in their mouths) while eating as well as at the sight of food. The food was a stimulus, and the salivation was the response to the stimulus. Pavlov experimented by using stimuli that had nothing to do with eating. He tried ringing a bell or turning on a light a few seconds before presenting dogs with their food. After a few times of doing this, the dogs began to salivate as soon as they heard the bell or saw the light in expectation of the food that would follow. In other words, the bell or light was the stimulus that brought forth the response of salivating. This process, by which the dogs learned to salivate when presented with a stimulus that had nothing to do with eating, is called **classical conditioning**.

Another experiment in classical conditioning, this time with a human, was performed in 1920 by psychologists John B. Watson and Rosalie Rayner. It should be noted that this same experiment could not be performed today because it would be considered unethical.

Watson and Rayner introduced an 11-month-old infant named Albert to a white rat. Albert showed no fear of the rat, but as he reached out to touch it, a loud noise frightened him. The psychologists had hit a steel bar close behind Albert's head. When Albert had recovered from his fright and once again reached out to touch the rat, the noise was produced again. This time Albert fell forward and covered his head at the loud noise. One week later the rat was again shown to Albert. He reached for the rat, but withdrew his hand before touching the rat.

That day the rat was shown to Albert several times, paired with the loud noise of the bar being struck. Soon Albert's response to the rat was to crawl away from it as fast as he could, crying at the same time, even if the bar was not struck. Albert had been conditioned to fear the rat. This fear carried over to objects that resembled the rat. Albert cried in response to a ball of absorbent cotton, a Santa Claus mask, and a fur coat.

Operant Conditioning

There is another kind of learning that takes place by trial and error. Take the case of a hungry stray cat. The cat investigates a garbage can and finds food in it. It is very likely that the next time the cat is hungry it will try looking in a garbage can. In a different case, each time a hungry stray cat investigates a garbage can it is chased by a vicious dog. Quite likely the cat will learn not to investigate the garbage can. The cat learned the responses that have positive or rewarding consequences, and those that have negative or punishing consequences while operating or working in its environment. Such learning is called operant conditioning.

Edward Lee Thorndike (1874–1948) was an American psychologist who worked at Columbia University at about the same time as Pavlov was performing his experiments. Thorndike studied the way chicks, cats, dogs, and monkeys learned in different situations.

In one of his well-known experiments he put a cat in a box that had a door opening from the inside by a mechanism such as a pedal, string, or button. A piece of food was placed outside the box. The cat could smell the food on the outside. After struggling to get out of the box for a while, the cat eventually tripped the mechanism, more or less by accident, and escaped from the box to eat the food. On the next trial the cat again struggled to get out of the box. This time it confined its efforts to the part of the box where the release mechanism was located, and escape was faster. Eventually the cat learned to trip the release very quickly. Operant conditioning had taken place. This particular type of operant conditioning, in which an animal learns a response in order to get a reward, is called **reward training**.

In other experiments, animals have learned a particular response in order to avoid a punishment. Rats have been trained to turn a wheel in their cage to avoid an electric shock. This type of operant conditioning is called **active avoidance** because the animal learns a response in order to avoid punishment.

Learning not to make a particular response in order to avoid punishment is called **passive avoidance**. Animals can quickly learn to avoid stepping down off a platform in their cage if they receive an electric shock every time they do.

One other type of operant conditioning is called **omission training**. In omission training, a reward is withheld if the animal responds in a certain way. The animal quickly learns not to make the response that will result in the reward (food) being withheld. This kind of conditioning is not studied much since it is felt to be unimportant.

LEARNING TO FIT IN

Humans also learn through trial and error and through punishment for inappropriate behaviour and rewards for appropriate behaviour. When a person is rewarded for certain behaviour, psychologists call the reward a **positive reinforcer** because it reinforces the behaviour. A punishment is called a **negative reinforcer** because it does not reinforce the behaviour. A negative reinforcer has a negative effect on the behaviour it is punishing.

Positive reinforcers for small children who are being taught acceptable behaviour can be hugs, kisses, praise, and smiles of encouragement. Through trial and error the baby learns what behaviour earns rewards. Throwing food from a highchair is great fun for a toddler, but it does not win affectionate responses or praise. On the other hand, first attempts to eat with a spoon will meet with praise, and the behaviour will be reinforced.

A small girl playing ball with her father will be praised when she catches or throws well. The positive reinforcement of the father's praise will tend to encourage the child to enjoy a game of ball and to improve her skills. She will have learned from her father that ball playing is acceptable behaviour.

Suppose instead that the child was playing with matches. The reaction on the part of the father would be quite different from his reaction to playing with the ball. The father's reaction might be to frown, take the matches away, and possibly scold the child for playing with them. The child will have learned that playing with matches does not please her father, and that it is not desirable behaviour. This learning has resulted from negative reinforcement.

Almost everything we do has been learned through some form of conditioning. We know how to walk, eat with knives and forks, put on our clothes, and cook our meals through conditioning. We also know how to talk, what words to use or not to use, when to smile or not smile, and with whom to associate.

PROGRESS CHECK

1. What did Pavlov's experiments with dogs reveal? Do you think human beings could be conditioned in a similar manner?
2. Do you think that humans like Albert should be used in experiments by psychologists? Explain.
3. What were the results of Edward Lee Thorndike's experiments with cats? Is human behaviour shaped in a similar manner?

CASE STUDY

Piaget's Theory

Jean Piaget (1896–1980), a Swiss psychologist, contributed greatly to our knowledge of how children learn, how they perceive the world, and how their intelligence develops. He showed us that a child's conception of the world is different from an adult's. Children pass through a number of stages of mental development and at any one stage will understand things in a certain way. They can be stimulated and encouraged to explore their environment and to think about what they find. However, they cannot learn certain "truths" until they reach the stage of development in which that understanding is achieved.

Piaget described four major stages of mental development in children: the sensorimotor stage, the preoperational stage, the concrete operations stage, and the formal operations stage. He maintained that all children must go through these stages, not one can be skipped. At each stage the child achieves a certain level of mental abilities. As new information is taken in, it is interpreted. The interpretation will agree with the child's current stage of mental organization. Sometimes the information has to be changed or distorted to make it agree with their ability to understand it.

As children take in more information about their environment their minds are stretched and their mental abilities grow. Their views of the world become more and more accurate. Periodically they will make a breakthrough in understanding, and will move on to the next stage of mental development.

As you read the descriptions that follow of the stages of development, keep in mind that the ages indicated for each stage are only approximate. Each child is different, however each child passes through all the stages in the order they are given.

Sensorimotor Stage — Birth to Two Years

In the sensorimotor stage of development the child learns through the senses (sensory) and through actions (motor). At birth, infants have no sense of separateness from the universe. Instead, they feel that they are the centre of the universe. During this stage they learn that objects exist apart from them. The most important achievement in this stage is the realization that an object exists even when it is taken away. At first an infant will not look for a toy when it is removed from sight. If the toy cannot be seen, it does not exist for the infant. But toward the end of this stage the infant will look for it after it has vanished.

Preoperational Stage — Two to Seven Years

Children who are in the preoperational stage are able to imagine objects and people when they are not present. They are no longer limited to learning through the senses and through physical actions. Now they can talk about objects, draw pictures of them, and imagine the objects are present in their play.

This stage is called preoperational to indicate that children are not yet able to understand certain basic mental operations. It is not possible for them to focus on two dimensions such as height and width at the same time. A child shown two identical glasses containing the same amount of water will agree that the amount of water is the same for each. However, if the contents of one of the glasses is poured into a tall narrow container, the child will say that there is more water in the tall container. If the experiment is repeated and the water from one of the glasses is poured into a short wide container, the child will say that there is less water in that container.

In the preoperational stage children believe that objects that move are alive and have thoughts and feelings. A stone rolling down a slope is alive and perhaps running away. Moving clouds may be playing chase with each other.

Another feature of children at this stage of development is that they have an absolute idea of right and wrong, and there is no in-between. They believe that people should be punished for doing something wrong, regardless of the reasons. If a boy broke 12 glasses while helping his mother set the table, while another boy broke 1 glass while sneaking some jam, preoperational children would say that the boy who broke 12 glasses should be punished more severely.

Concrete Operations Stage — Seven to Eleven Years

Children in the concrete operations stage can picture an object in their mind, and mentally perform an action with the object. In this way they can try out an operation in their mind to see if it would work or what the results would be. Then they can perform the action physically. An example is a chess game where the players mentally move chess pieces and judge the relative merits of different moves before they make them physically.

When presented with the problem of the water being poured from one container to another, children in this stage have no trouble recognizing that the volume of water remains the same no matter what shape the container.

In this stage children develop the ability to rank or classify objects and people in more than one way. They can see, for instance, that their friend's mother can also be a doctor, the wife of their friend's father, a member of the school board, and a tennis player.

An important development of this stage is that children become less egocentric. They are better able to consider another person's point of view.

Formal Operations Stage — Eleven Years and Older

The formal operations stage is the stage at which logical abstract thinking becomes possible. Once children can think logically in abstract terms, then they have the potential of creative reasoning. Children can now have a conversation that involves ideas and abstract thoughts.

The formal operations stage is the highest level of mental development. Psychologists feel that by the age of 20, mental development is for the most part established. It will still be possible to learn new things, of course, but the mental processes that deal with the information will not develop further.

Piaget's views on childhood development have made us aware of the ways in which children look at and understand the world. In some cases, this awareness has affected our methods of treating children and educating them. His theory has also challenged some traditional types of adult perceptions.

1. Outline the stages that children must go through in their development.
2. Write down some of your own observations about childhood behaviour. Do they support Piaget's theory?

3.4 The Importance of Roles

In order to live and function in society, we must all become social actors. William Shakespeare put it this way some 400 years ago:

> "All the world's a stage
> And all the men and women merely players:
> They have their exits and the entrances;
> And one man in his time plays many parts, ..."

As players or actors we each play roles and act in a certain manner depending on the situation. We have been socialized to play these roles, and we know the behaviour that is expected in each role.

At this very moment, you are playing the role of a student. The expected behaviours of this role are to attend school, to be on time, to read, to take notes, to participate in class, to observe school rules, and so on. After school, you may have a job in a supermarket as a cashier or clerk. In this role you are expected to be helpful and courteous to customers, to handle the goods and money efficiently and conscientiously, and to follow orders given by your boss or senior employees. Other roles can include those of daughter, son, brother, sister, friend, team member. Each role demands certain kinds of behaviours.

Most people perform their roles automatically, and do not think about the changes in behaviours they must make. A new role, however, may cause you to stop and consider what is expected of you. Going for a job interview, or starting a new job may be a time to ask those with experience how you should behave.

Roles make it easier for people to relate to each other. We know what to expect from others and what is expected of us. Social interaction depends on the ability of people to play their own roles. It also depends on their ability to understand others and to predict how other people will act according to their roles. Understanding others and predicting their behaviour requires that we use imagination. We must be able to mentally take the other person's role.

Learning roles, because it is an essential step in socialization, is begun at an early age. Sociologist George Herbert Mead studied the way children develop the ability to see things from another person's point of view. He described three stages in the development of this ability.

The first stage is the **preparatory stage** in which children imitate the behaviours of the people around them. They may talk on the phone to imaginary people, pretend to read the newspaper, and walk around in their parents' shoes or hats. Although they are imitating these behaviours, they do not understand the meanings of them.

In the next stage, the **play stage**, children pretend to be someone else. They take the roles of doctors, teachers, and bus drivers and practise their roles. By playing these roles children gain skills in taking roles and in understanding how other people think and feel. They also begin to realize that they are separate and unique individuals. This stage takes place around the ages of five or six.

By the time children have reached the **game stage**, they are capable of understanding a set of roles that belong to a system. Mead explained this stage in terms of a baseball game. Children at the game stage can understand how the team functions because they know the rules and the different roles the players will take. With this knowledge children can play different roles on the team and predict what the other players will do.

By being able to take on a set of roles instead of just one, children can understand how others will respond to their own behaviour. They can then control their impulses and regulate their behaviour to conform with what is acceptable to others.

ROLE MODELS

During the preparatory and play stages of development, children imitate the behaviours of the closest and most trusted people around them. These people are referred to by sociologists as significant others, or **role models**. They have an important influence on the behaviour of young people.

By imitating and learning roles, and playing these roles when interacting with others, children develop their own identity, their sense of self. As they play each role, they will get feedback from others. Sometimes the feedback will be positive and will tell them that they have been accepted in a certain role, for example, the role of leader. Sometimes the feedback will be negative, indicating perhaps that they have not yet learned the role sufficiently well, or that they do not fit that role, for example the role of class "brain".

ASCRIBED ROLES, ACHIEVED ROLES, AND LIFE CHANCES

There are some types of roles over which we have no control — we are born to play them. These are called **ascribed roles** or **ascribed status**. Infants are born male or female, and they are born into a particular family. They also have no choice when they will be born, so they will have no control over their age at any point in their lives. Gender, environment, and age will determine to a great extent the roles people play as children and as adults.

Environmental conditions such as the interests, abilities, values, social class, and religion of the parents influence the roles of the children. Gender dictates how children are socialized and how people will interact with them. The age group to which people belong also influences the roles they play at any stage in their lives. An appropriate role for a teenager to play is not necessarily appropriate in middle age. A parent is expected to play a more responsible role than a child.

The culture into which infants are born will often determine the language(s) they use, the customs they observe, the religion they adopt, the values they hold, and the social class to which they belong. All of these conditions can influence the opportunities, or **life chances**, they will have during their lifetime.

Certainly though, people have control over much of what they do in life. They do not remain under the total influence of their parents for long. They meet and are influenced by people outside the home. They participate in new activities and absorb new ideas.

Through their own efforts people can obtain certain levels of education and income, professions, skills, families of their own, and life-styles. They may also change these conditions at different points in their lives. For example, many people, once they retire, decide to go to university to earn a degree, or they decide to take up a new career or a new hobby. The roles associated with these conditions are roles over which people have control. They are not roles they were born to play; they are **achieved roles** or **achieved status**.

Role Sets

The network of roles people play in their relationships with others in a particular situation are called **role sets** by sociologists. At school your role set probably includes the roles of friend to some students, classmate to others, pupil in your relationship with your teachers and school staff, and team member with a coach. Perhaps you are the neighbour of someone on staff or a sibling of one of the students — these are two more roles you might play. Because the roles you play in a role set are related to each other, the expectations and the behaviours are also related. A role set, then, is easier to play than a number of unrelated roles, each with different expectations and behaviours.

Fronts and Settings

People usually project a particular image or **front** in each of their roles. Employers might have an air of authority and confidence when interacting with employees, and an attitude of understanding and helpfulness with customers. The different fronts are adopted to project the desired image for each role.

When in a new situation, we often imitate the kind of front others use who are experienced in that situation. For example, in a new job as a clerk in a clothing shop you might imitate the front projected by an experienced clerk.

Sometimes the type of surroundings and the props people have are important requirements to establish an appropriate front for the roles they are playing. These **settings** can include the neighbourhood in which they live, the way their houses are decorated, the cars they drive, and the clothes and accessories they wear. People will frequently consciously create the setting necessary to project a desired front. The setting makes it easier for them to relate appropriately to others in the particular role they are playing.

ROLE CONFLICT, ROLE STRAIN, AND ROLE AMBIGUITY

Sometimes the different roles played by people conflict with each other. This **role conflict** results from trying to satisfy different expectations that are not compatible. As a student you are expected to do your homework and perform to the best of your ability in school. You may also feel the need to earn money from a part-time job. Among your peers there may be pressures to stay out late and to neglect your school and family duties. These role conflicts may result in **role strain**, the tension people feel when they are unable to satisfy the expectations of others.

Sometimes role strain is the result of not having a clear idea about the types of roles you want to pursue in life. You may be undecided about the kind of career, the type of relationships, the level of education, or the commitments you want to have in life. This indecision is referred to as **role ambiguity**. Adolescence is a time in human development when role ambiguity can easily occur, and with it anxiety and uncertainty. It is a difficult period in life because there is a changeover from child to adult behaviour. With this changeover there is pressure from several groups to adopt certain roles and behaviours. There are also many decisions a person must make at this point in life that will have long-range consequences.

PROGRESS CHECK

1. Make a list of all the roles you perform in any given week.
2. What functions do roles perform? When do roles change?
3. According to George Mead, how do children learn roles?
4. List personal examples of role conflict and role ambiguity.

3.5 Gender Roles

The different roles assigned to men and women on the basis of sex are called **gender roles**. These roles consist of behaviour patterns that a society feels are appropriate for each sex. Social scientists are still trying to work out the basis for these different roles. To what degree does biology influence gender roles? To what degree are gender roles based on tradition? How do little girls learn to be feminine and little boys masculine?

Gender Expectations

From an early age, boys tend to follow the behaviour patterns of their father and girls the behaviour patterns of their mother. To a small extent, the learned role a boy or girl lives up to may be based on the biological differences between males and females. Mainly though, the gender role will be based on the expectations of the adults who influence the child.

According to Canadian sociologist Metta Spencer, gender roles exaggerate whatever small biological differences do exist. Our expectations of what is female and what is male behaviour are based on these exaggerated differences. The expectations become **stereotypes** which children learn and try to follow. A stereotype is an exaggerated image of certain characteristics that a group or an individual may possess.

Sometimes the stereotypes children learn run contrary to their actual experience. A kindergarten class insisted that only men could be bus drivers, even though the bus on which many of them came to school was driven by a woman. A little girl, whose mother was a doctor, maintained that men were doctors and women nurses.

Stephen Richer conducted a study of four year olds in an Ontario kindergarten in 1979. He discovered that sex role stereotyping was already well established at that age. Boys and girls tended to sit with members of their own sex, and they played traditional male and female games according to their sex. Boys played with cars, trucks, and building blocks. Girls played with doll houses, baby dolls, and dressing-up clothes. The boys' behaviour tended to be more aggressive than the girls.

Richer's findings were supported by a similar study conducted by the federal Department of Labour in 1985. Several hundred children in grades 1, 3, 6, and 8 were interviewed to determine the level of sexual stereotyping of their career goals. It was found that 93% of the boys wanted traditional male occupations such as doctor, forest ranger, or bus driver. Only 43% of the girls chose similar occupations. Most of the girls were "preoccupied with marriage and motherhood". Sexual stereotypes were obvious when not one boy chose to be a flight attendant or librarian, while none of the girls wanted to be a construction worker, plumber, or stockbroker.

These two studies show that sex role stereotyping plays an important part in the socialization of children. It is clear that sexual stereotyping is learned, and is an important influence on children's attitudes and interests in later life.

Career Choices

Avis Glaz, a vice-principal, studied 1 167 Ontario teenaged girls in 1980. She found that many of the girls wanted to perform such non-traditional jobs as firefighters and police officers when they left school. Most of the girls, however, expected to fill the traditional female jobs of child-care workers, nurses, and teachers. What they wanted to do was quite different from what they expected to do. On the other hand, the career dreams of boys were similar to what they expected to do on leaving school.

Ms. Glaz also found that those girls who believed in traditional female roles had lower career expectations than those who believed that people should be restricted in career choice only on the basis of their abilities, not on the basis of their sex. The teenaged girls who were tradition minded chose low-paying, low-status work. The girls whose thinking was more liberated chose higher-paying, higher-status work.

Glaz suggested that female stereotyping might be overcome by bringing successful career women into the classroom to talk to students.

Traditionally, Canadian females have been socialized to accept less important jobs and less pay than males, and to accept the greater share of housework and child raising responsibilities. For those women who pursue a career, the dual load can be exhausting. Men have traditionally been socialized to be the authority figures and breadwinners in the family. Lately these roles are being re-examined by men and women alike. With more and more women in the workforce out of necessity, a new balance must be found.

Socialization and Stereotypes

"The threat of failing to behave like a member of one's own sex is used to enforce a thousand details of nursery routine and cleanliness, ways of sitting or relaxing, ideas of sportsmanship and fair play, patterns of expressing emotions, and a multitude of other points in which we recognize socially defined sex-differences, such as limits of personal vanity, interest in clothes, or interest in current events ..."

Margaret Mead, *Male and Female*

One way in which we are socialized is by the reading materials we encounter. It is often the case that in children's story books all the active, interesting activities are performed by males, and the passive roles assigned to females. The message received by little girls is that their role is to stay home and admire the adventuresome males. These days many writers are aware of this subtle stereotyping, and they are making an effort to portray females as active, adventuresome, leaders.

Another way in which reading materials reinforce stereotypes is through the use of "he", "his", and "him" when talking about either sex. Publishers now try to avoid this kind of sexist language by using words referring to both sexes. The use of non-sexist language is especially important when writing or speaking about career expectations and goals. In this way, young people will be socialized to expect that men and women will be given equal treatment and opportunities.

One stereotype has been very harmful to some males. Traditionally "being a man" meant not crying in public, being tough and aggressive, and hiding inner feelings and emotions. Living up to this stereotype has made it difficult for many men to develop close caring friendships, and to

WORLD OPINION

A sampling of what people in other countries think:

JAPAN

Recent surveys in Canada show that sexual stereotyping is still very evident among Canadian youth, with one study finding "sex the most significant variable in determining responses" on career choices among children aged 6 to 14.

In Japan, with a reputation as a bastion of male supremacy, a recent survey shows some evidence of a liberalizing of that view. Respondents were asked:

Do you agree or disagree with the opinion that men should work to earn income and women stay home and look after the family?

	1984 %	1985 %
Agree	72	60
Disagree	20	27
Other/no answer	8	13

In many offices, women are assuming managerial and leadership positions. Do you think this is a good trend?

	1984	1985
A good thing	61	65
Not a good thing	30	23
Other/no answer	9	12

Would you like to have a female boss?

	1984	1985
Yes, I would	26	23
No, I would not	60	54
Other/no answer	14	23

If you had a choice, would you like to be born a man or a woman?

	Male		Female	
	1984 %	1985 %	1984 %	1985 %
Man	90	90	42	39
Woman	7	5	53	54
Other/no answer	3	5	5	7

(Mainichi Shimbun, Dec. 4-5, 1985 National adult sample of 2,388.)

Figure 3.1

respond to their children with warmth and love. Women, on the other hand, have been characterized as living by their emotions and being incapable of making rational decisions.

Some of these stereotypes still exist, but they are being questioned more and more. With modern male and female roles no longer as clearly defined as they once were, there is more overlap in what is expected from both sexes.

PROGRESS CHECK

1. Do you agree with the views of Metta Spencer? Explain.
2. Outline the conclusions of Stephen Richer in his study of 4-year-old children.
3. Analyze the results of the Avis Glaz study. Do you agree with these findings?
4. How can sexual stereotyping be avoided?

3.6 *Parenting*

Parents assume the main responsibility for the socialization of young children. They teach their children to walk and talk, and to behave as members of society. Often parents are judged by how their children behave, and how well they succeed in school and in social situations.

Today's parents tend to be better educated, better off financially, and older than parents in the past. They also have fewer children. With smaller families, parents are often able to give their children more attention. This may be one reason why children from small families tend to do better in school and tend to be more motivated to achieve.

There are many more books on the market today that deal with parenting. It is no longer felt to be something that comes naturally. Instead, many people realize that there are skills involved in parenting that can be learned and used. The experts often disagree, and the styles of parenting promoted by the experts change over time. Experts do agree that important requirements for good parenting are the ability and the desire to provide children with sufficient love, attention, and stimulation.

Parenting is currently undergoing change. Single parent families are on the increase. In families having both parents, the incidence of both parents working is increasing. For

Figure 3.2
The following newspaper clippings are from 1986 papers; the Gallup Poll was taken in 1984.

Most think ideal family has two kids

Most Canadians feel the ideal family should have no more than two children, according to a Gallup poll released today.

Of 1,059 adults interviewed in May, 61 per cent felt families should have two children or less. Thirty-nine per cent felt families should have more than three children.

That is a drop from a year ago, when 56 per cent interviewed perceived families should have two children or less and 44 per cent thought a family of three or more was ideal.

But it is a return to 1982 poll results which showed that 63 per cent thought two or less to be an ideal figure while 37 felt a family should have over three children.

Four per cent of Canadians view one or no children as an ideal number today. This compares with 2 per cent a year ago.

The decrease appears to occur fairly generally across age, sex, education and income groups.

Such polls are accurate within 4 percentage points, 19 times out of 20.

Two children ideal number to most Canadians, poll says

Most Canadians feel that two children is the ideal number to have, according to a Gallup poll released today.

Last month, Gallup asked Canadians the following question: "What do you think is the ideal number of children for a family to have?" Fifty-eight per cent said two.

This is a marked contrast from 1945 — the beginning of the post-war baby boom — when 60 per cent felt four or more children was the ideal.

By 1970, with increasing inflation and unemployment, only 33 per cent of Canadians favored large families.

Today, only 13 per cent would opt for four or more children. Twenty-six per cent this year said three children would be ideal, and just 3 per cent said one child or none at all.

GALLUP POLL
"What do you think is the ideal number of children for a family to have?"

Ideal Number of Children

	2 or less	Three	4 or more
1984	61%	26%	13
1983	56	27	17
1982	63	25	12
1980	59	27	14
1974	52	24	24
1970	34	33	33
1959	22	23	55
1945	17	23	60

Ideal No. of Children (Pct. Saying 4 or more)

	1984	1983
National	13%	17%
Sex		
Male	11	14
Female	15	20
Age		
18 to 29 years	12	13
30 to 49 years	9	14
50 years and over	18	25
Education		
Public school	25	28
High School	10	15
University	14	16
Income		
Under $20,000	17	22
$20,000-$29,999	14	18
$30,000 & over	9	10

many families two incomes are necessary for raising children. It is estimated that Canadians spend an average of $100 000 (1984 dollars) to raise a child to the age of 18. Daycare facilities are in many cases taking over the responsibility of looking after young children whose parents work.

With more and more young children being cared for in daycare centres, social scientists are studying the social and emotional effects on the children. Most are in agreement that very young children benefit from having loving attention from one main person. This kind of attention is not always available in daycare facilities. As children grow older, they enjoy the companionship of other children. At this stage in their socialization, daycare may be a stimulating experience and may help them develop social skills.

Women perform the bulk of the work involved with raising children. Many Canadian women, however, have a career outside the home. Sometimes career women suffer from role strain, the tension felt from performing roles that conflict. They may feel that they are unable to spend enough time with their children because of the responsibilities of their job. Alternatively, they may feel they cannot concentrate on furthering their career because of the time they need to spend at home nursing sick children or running the household. They may also sense that their husbands, relatives, or friends feel that they are not being "good" mothers by going out to work.

Whether parents work or not, parenting is not an easy task. However, the joys of being close to children, of guiding them, and of watching them develop their potentials, balance the responsibilities involved in their care.

What do these statements say about parenting?

"At one time people learned how to raise children from members of their own family. Now parents are turning more and more to professionals."
Esther Lefevre, clinical psychologist, Montreal

"Working women today tend to look upon having a baby as another professional endeavour. They want to know a lot more about what kinds of activities are best for their children."
Alison Gopnik, psychologist, University of Toronto

"Parenting used to be thought of as an innate skill. Now people see it as something that can be improved."
Roy Ferguson, child care specialist, University of Victoria

"Women who work carry a burden of guilt because we are not with our children as much as we could be ... so we want to make sure we give our kids the best."
Eva Czigler, CBC producer and parent

"Parents are constantly asking me whether they have done something wrong because their kid is not toilet-trained. I tell them to forget about it. I have never met an adult who was not toilet trained."
Terence Creighton, psychologist, University of Calgary

"What is important is a sense of predictability. A child needs to know that when he cries he will be taken care of and when he is hungry he will be fed ... My message to parents is very simple .. I tell them that because the facts are unreliable and the science of child develop-

ment is still immature, they should treat every expert skeptically. That includes myself, of course. Don't believe any of us. Use your common sense."
Jerome Kagan, child psychologist,
Harvard University, Boston

"When I was a kid everybody knew how a kid was supposed to be raised. Nowadays nobody is sure ... values are so widely different from one place to another."
   Dr. Thomas Miller, child psychologist, Vancouver, B.C.

"People talk about spending "quality time" (instead of long periods) with their kids, but that's just a justification. A child's earliest experiences are terribly important, and a parent needs to be there. It's a hard, tiring job ... but you have to do it. It should be a work of love."
Dr. Harold Breen, Ottawa psychologist

"If a child lives with criticism he learns to condemn.
If a child lives with hostility he learns to fight.
If a child lives with ridicule he learns to be shy.
If a child lives with shame he learns to feel guilty.
If a child lives with tolerance he learns to be patient.
If a child lives with encouragement he learns
   confidence.
If a child lives with fairness he learns justice.
If a child lives with security he learns to have faith.
If a child lives with approval he learns to like himself.
If a child lives with acceptance and friendship he
   learns to find love in the world."
Author Unknown

PROGRESS CHECK

1. What do experts agree are important requirements for good parenting?
2. What are some of the social changes affecting parenting today?
3. Discuss role conflict for career women who are mothers. Can men who have a career and are fathers have the same role conflict?

OVERVIEW

The socialization process largely determines the kinds of persons we become. We know that children require love and attention to develop into healthy, normal human beings. Parents teach their children many basic skills that society expects from its members in order that they can function in and contribute to that society. Rewards and sanctions are given for both desired and undesirable behaviours to teach people right from wrong.

Children go through specific stages and these determine their level of skills and understandings of the world around them at certain periods in their development. Gradually children learn to perform different roles in particular situations. They also learn to expect different behaviours from others in specific roles. Over time, children are socialized to interact with others in a wide variety of activities.

The gender of children has been important in determining the types of behaviours, interests, and skills that society encourages in the socialization process. In recent years this has become less important as parents, schools, and the rest of society become more convinced and aware of the equal potentials and abilities of all children.

We are now in the process of learning to understand the importance of proper daycare facilities, learning materials, and role models in the socialization process. The socialization of children involves a multitude of influences and is extremely complex. But we do know that if children are to become fully developed human beings, they must be given a positive, loving, and stimulating environment.

KEY WORDS

Define the following terms and use them in a sentence that shows their meaning.

surrogate	active avoidance	life chances
egocentric	passive avoidance	achieved role/status
norm	omission training	role set
folkways	positive reinforcer	front
ostracism	negative reinforcer	setting
mores	preparatory stage	role conflict
laws	play stage	role strain
conform	game stage	role ambiguity
classical conditioning	role model	gender roles
reward training	ascribed role/status	stereotype

KEY PERSONALITIES

Give at least one reason for learning more about each of the following.

René Spitz
Harry F. Harlow
Ivan Pavlov
John B. Watson and Rosalie Rayner
Edward Lee Thorndike

Jean Piaget
George Herbert Mead
Metta Spencer
Stephen Richer
Avis Glaz

DEVELOPING YOUR SKILLS

FOCUS _____

What questions should students ask about the following topics: socialization, norms, the learning process in children, roles, and parenting?

ORGANIZE _____

1. Design an organizer that compares the importance of instincts in insects, animals, and human beings.
2. After reading the section on gender roles, develop an organizer that identifies male and female stereotypes and the possible reasons for their use.

LOCATE AND RECORD _____

1. Using information in the textbook, design a plan to compare the studies of René Spitz and Harry Harlow on hu-

man and animal behaviours. Your comparisons can be made under the headings "cause" and "effect".

2. Develop a four-column organizer to identify the characteristics associated with Piaget's different stages of mental development in children. Add additional characteristics that are associated with each stage of a child's mental development by referring to additional articles and textbooks on Piaget's theory.

3. Try to remember points in your life when you experienced role strain. Develop a chart that lists these experiences and outlines their causes (role conflict, role ambiguity ...). In a separate column write down the consequences of this role strain and determine how these situations resolved themselves.

4. Research Report — Interview a social worker or official of the Children's Aid Society. Ask these questions:
   (a) What effects does the loss of biological parents have on a child?
   (b) Are children raised in institutions such as orphanages less successful in school than children raised in normal families?
   (c) Are children from single-parent homes less successful in school than those from two-parent homes?

   Prepare for the interview by researching these questions in your school library. Look up as many sources as possible for each question. Combine information from your reading with information from the interview to write your final report.

EVALUATE AND ASSESS

1. George Herbert Mead observed that children go through three different stages of development and each stage determines the level of their abilities. Divide the class into groups and discuss his theory through your own experiences and observations. Talk to your parents, friends, and such experts as child care workers, and ask for their opinions. Assess whether there is enough evidence to support Mead's theory.

2. Most of us accept many of our parents' values, but at the same time we reject some of their values. List the values you accept and those you reject. Try to determine the

reasons for your acceptance or rejection. Discuss your list with members of the class. If you have a brother or sister, ask them to make a similar list and compare the two.

SYNTHESIZE AND CONCLUDE

1. Form groups of four to six students and have each group develop an organizer that lists various institutions and people who help to socialize individuals. Beside each, list examples of its social influences. In a third column assess the importance of the effects of the influences of each in the socialization process in relation to each other.
2. Compare the experiments of Pavlov, Watson and Rayner, and Thorndike. What relationships exist in the three separate experiments?
3. Compare the different roles that you have in your everyday life. How are the behaviours in these roles similar and different? How do these behaviours make it easier to relate to others?
4. Prepare a comparative study that answers the question: What are the benefits and drawbacks of daycare? Use your resource centre or another library for information. Present your findings to at least one other person in the class.
5. Interview two boys and two girls between the ages of 6 and 8. Ask them these questions:
   (a) What job would you like when you grow up?
   (b) Should girls and boys do the same kind of work or should their jobs be different?

   After compiling your data, analyze the information by answering these questions.
   – Did the boys and girls differ in their job choice?
   – Did boys and girls feel they should do the same type of work? How many did not feel this way?

   Combine and analyze your information with the rest of the class. What pattern emerges? Why might this be the case? As a follow up, complete a similar survey with grade 9 students in your school. Are there any differences in views between them and the grades 1 to 3 students you interviewed?
6. Piaget said that children pass through several mental stages of learning. As their minds develop, they are able to

learn different concepts. If their minds have not developed past a certain stage they will not be able to grasp some concepts. Verify Piaget's theory by conducting your own research into how children think.
(a) Have at least 5 four- to six-year-olds watch the following experiment. Have two identical tall glass containers filled with the same amount of water. Make sure the children note that both contain the same amount of water. Then pour the contents of one of the containers into a short wide container. Ask the children which container has more water. Record their responses.
(b) Ask the same children to decide which child in the following two scenarios deserves the greater punishment. One child was helping to wash the dishes and broke twelve glasses. The other child was sneaking candy from the kitchen counter and knocked two glasses onto the floor.

APPLY

1. Divide the class into small groups and discuss how stereotypes can influence the socialization process. Do the students in your group use stereotypes in their conversations and thinking? To test the opinions of those who maintain they do not use stereotypes, examine the want ads of local newspapers — do the students automatically assume that certain jobs are for females and others for males? Discuss how stereotypes could be eliminated and speculate about the possible outcomes if this goal was achieved.
2. Speculate on what would happen if specific folkways, mores, and laws did not exist in society. The class can be divided into three groups. Each group can list examples of folkways, mores, or laws and then make conclusions about the kinds of situations that would exist if these norms were absent. Each group can share its findings with the rest of the class.
3. Summarize Harlow's experiments with monkeys by completing the following sentences.
   (a) Monkeys who were exposed to both the wire and cloth mothers ...
   (b) Monkeys who were raised in isolation ...
   (c) Harlow's experiment tells us that ...

4. Examine a catalogue for a major department store selling toys. List the types of toys offered and the gender to which they are likely to appeal. What stereotypes do these toys reinforce? Follow this up by visiting a daycare centre and interviewing five boys and five girls. Ask them what toys they prefer and record their responses on a chart similar to the following.

Survey Question: What toys do you play with?

Child (male or female)	Toys Preferred	Gender Usually Associated With Toy Male/Female/Either

– What pattern emerged from your survey?
– Do boys usually play with one kind of toy while girls play with another?
– Do you think there is a relationship between the toys these children play with and their behaviour?

5. Analyze the following hypothesis:
"Each of us tries to project a different image to different people. We may even project different images to different parents or to different siblings or friends." To test your hypothesis ask five students what image they seek to project in the company of the following people.

    (a) Parents: Mother
                            Father
    (b) Teachers (3 different teachers)
    (c) Siblings: Brothers
                           Sisters
    (d) Friends: Female
                          Male
    (e) Boss at work
    (f) Coach
    (g) Other people (specify)

After interviewing five people analyze your results. What is the pattern? Is the hypothesis correct? Why would we present different people with different images or fronts?

COMMUNICATE

1. Divide the class into groups of two students each. Each pair of students can discuss the importance of the dif-

ferent role models at the various stages in their lives. Each student will then make a chart that lists the earliest stages to the latest stages in their lives and the role models and their influences on the students. Students can discuss whether or not their role models had a positive and/or negative influence.

2. Write an organizer or short essay, or debate one of the following.

- We are human when we are born and don't have to "become human" later.
- Psychologists cannot accurately apply the results of their experiments with animals to human behaviour.
- Children should not be told to act like boys or girls but should be permitted to develop their own personalities.
- The teenage years are the toughest period in a person's development because of the lack of clarity of roles at that time of life.
- Child rearing experts cannot tell us any more about children than parents or grandparents can.
- Free universal daycare is a right for all children and parents.
- Parents should stop trying to provide their children with stimulating environments and let them play on their own.
- Married couples should not have children until their 30's.
- The single life is superior to married life.
- Women who are married and have children should spend most of their time at home.
- In the future children will be increasingly socialized according to their individual abilities and personalities; less importance will be placed on their gender.

CHAPTER 4

Personality

*What is personality?
What are some of the theories
about how personality develops?*

INTRODUCTION In the previous chapter we discussed how we are socialized to behave like other humans in our society. We imitate certain behaviours and are rewarded; we try different behaviours and are rewarded for some and punished for others. Thus through conditioning, we learn how to interact with others in acceptable ways. This process shapes us so that we can fit into our society.

While it is possible to see the similarities in human behaviour, it is also evident that there are many differences in the way individual humans behave. Within your own family you can see how family members react differently in the same situation. In this chapter we will explore the theories that account for the development of these differences, and how the differences affect our approach to life.

4.1 Who Are You? Describing Personality

Everyone uses the term "personality" but what does it mean? Often personality is discussed as if it were a characteristic that individuals either have or do not have — "she has personality", or as a characteristic that can be measured — "he has lots of personality".

The scientific meaning of the term personality is much more complex and definitions vary. For our discussion, we will define **personality** as: the set of characteristics that makes each of us unique. These characteristics are relatively permanent — they do not vary from day to day. All of us have a set of relatively permanent characteristics that makes us unique, and therefore all of us have a personality. No one has more or less personality than anyone else — just a different set of characteristics. Our personality shapes the way in which we act, think, and feel about events, ideas, and other people. Our personality also shapes the way others respond to us. For example, people will usually respond in a positive manner to someone with a friendly, outgoing personality.

DESCRIBING PERSONALITY IN TERMS OF TRAITS

Another word for the relatively permanent characteristics that make up our personality is **traits**. Traits are expressed as types of behaviour. When we say that a person has the trait of honesty, for example, we mean that the person behaves fairly consistently in an honest way. The person expresses this trait by telling the truth, and by not stealing or cheating.

One way of describing a personality is by naming a person's traits. Psychologists Gordon Allport and Henry Odbert prepared a list from the dictionary of all the adjectives that distinguish one person's behaviour from another's. Their list of adjectives numbered 17 953.

Another psychologist, Warren Norman, managed to double Allport and Odbert's list. Then he set out to simplify the list. The first step was to organize the list into pairs of opposite adjectives, for example, good natured and irritable. Then he experimented by asking people to describe their friends using the list of paired words. When he analyzed the results, he found that five basic traits were identified by their descriptions. Another study was conducted to see if people would use the same basic traits to describe other people they had just met. The identical five basic traits were again used. Table 4.1 lists the five basic traits and the pairs of opposite adjectives that make up the five.

Another psychologist who has studied personality traits and has attempted to classify them is Raymond Cattell. He gathered data on how people describe the personalities of others. Analyzing this data, he arrived at 35 groups of traits, or trait clusters. From the 35 trait clusters he isolated 16 that were most basic. With the 16 basic trait clusters he developed a personality questionnaire for self-testing, called

TABLE 4.1 Five Basic Traits and Pairs of Opposite Traits Used by People When Rating Others

Adjectives		Basic Traits
talkative frank sociable	silent secretive reclusive	Extroverted
mild, gentle good-natured not jealous	headstrong irritable jealous	Agreeable
fussy, tidy responsible scrupulous	careless undependable unscrupulous	Conscientious
poised composed calm	nervous, tense excitable anxious	Emotionally Stable

polished
imaginative
artistically
  sensitive
	crude
	simple
	artistically
	  insensitive
		Cultured

the Sixteen Personality Factor Questionnaire. Cattell has had many groups of people answer the questionnaire. A combined personality profile for a group of airline pilots shows that pilots, on average, are emotionally stable, tough minded, controlled, and relaxed. See Figure 4.1.

Describing Personality by Type

Many people have tried to simplify the ways of describing personality. There have been several methods suggested for fitting people into personality categories, or **personality types**.

As early as the second century, there was a theory of personality types proposed by a Greek doctor, named Galen, who practised medicine in Rome. He described four personality types, based on the bodily fluid people of each type had in excess. Descriptions of these four types were: cheerful and optimistic, depressed and pessimistic, quick-tempered and irritable, and calm and uninvolved.

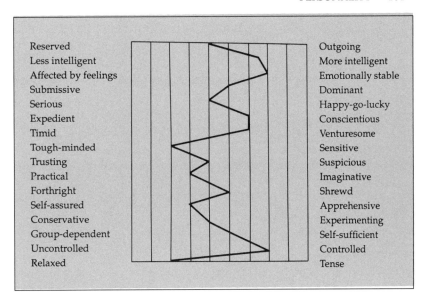

Figure 4.1

Carl Jung, in this century, proposed two main personality types, the **introvert** and the **extrovert**. He saw the extrovert as being outgoing, vigorous, and taking action. The introvert, on the other hand, tends to think about and analyze things, and is less inclined to take action. One problem with Jung's personality types is that most people fall somewhere between extroversion and introversion. Another problem is that there are many aspects of personality that are not taken into account by just two types.

A more recent and more workable system of personality types was devised by Hans Eysenck, a German-born psychologist living in England. He combined Galen's four personality types with Jung's two types, and added two more — stable and unstable types. The following diagram shows how Eysenck categorized personality types. See if you can place yourself on Eysenck's chart.

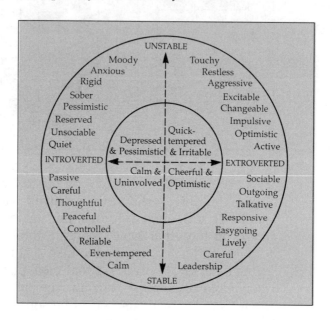

Figure 4.2

Another attempt at classifying personality was made by psychologist W. H. Sheldon. He classified people according to three different body types — **endomorph**, **mesomorph**, and **ectomorph**. Then he related the three body types to certain personalities. The endomorph has a round soft body, enjoys eating and drinking, is sociable, tends to be conventional, and needs to rely on others during periods of distress. The mesomorph is muscular and enjoys strenuous exercise. Mesomorphs are outgoing, adventurous, and direct. The ectomorph has a thin angular body and is shy, sensitive, nervous, and given to worry. When troubled, ectomorphs seek quiet and solitude. Few people fit these types precisely, so Sheldon developed a scale for classifying the degree of endomorph, mesomorph, and ectomorph in each person.

Have you ever watched other students in the cafeteria at lunchtime? Some of them spend most of their lunch hour socializing and gossiping, others bring books to read. In class, some students become intensely involved in certain lessons, others are withdrawn or shy, and some may be "turned off" completely.

However crude our attempts to fit people into personality types, it is natural for humans to want to classify personalities. In the next section we will talk about ways in which we attempt to assess a person's personality.

PROGRESS CHECK

1. What personality traits do you have? Compare them to those listed in Table 4.1.
2. What extroverted and what introverted personality traits do you have?
3. Whom do you most closely resemble, an endomorph, a mesomorph, or an ectomorph? Explain.
4. What personality traits do you find most attractive in others? Do your friends have these traits?

4.2 Assessing Personality

At times in your life you may be asked to rate someone else's personality — the person might be under consideration for a job or a promotion. On the other hand, you might be asked to complete a questionnaire that will give others an idea of your personality.

Often a number of people will be considered for one job. It is not possible to try each one out in the job, so management needs some way of predicting which candidate will be best able to handle the responsibilities. Along with an aptitude test which tests potential ability, a job applicant might be given a personality test. The personality traits measured by the test relate to such things as how well a person will fit in with co-workers, how likely the person is to provide effective leadership or to achieve success, or how well the person is likely to perform under stress.

Very often the personality assessments are specially developed to measure certain traits thought to be relevant to a particular situation. A social agency requiring employees to work with mentally handicapped children might be looking for traits such as patience, affection, and responsibility. If you were training astronauts for a space program, what traits would you look for? If you were placing babies for adoption, what traits would you want prospective adoptive parents to have?

There are over 500 personality tests of various kinds in use currently. The testing techniques used most often to describe an individual's behaviour are: ratings, inventories, interviews, projective tests, and behavioural assessment.

Personality Rating

Personality rating consists of assigning a score or rank to a person for each trait given. Suppose you were asked to rate your teachers on how well they get along with students. One of the traits being rated might be thoughtfulness toward students. The scale on which you would rate each teacher might look like this:

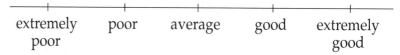

extremely poor   poor   average   good   extremely good

You would indicate your rating on the scale by checking one of the descriptions.

For a rating to be of value, the person doing the rating must know the person being rated reasonably well, and must be able to remain objective, that is, not be influenced by personal feelings.

Personality Inventory

Another kind of personality test is the inventory. It consists of lists of questions to which the person doing it answers "yes", "no", or "don't know". Sometimes the questions are worded so that they require a "true", "false", or "cannot say".

The most commonly used personality inventory is the Minnesota Multiphasic Personality Inventory. Its purpose is to diagnose those people suffering from psychological problems. It has been found useful in screening job applicants, as well as in counselling. The inventory consists of 550 statements regarding behaviour that require a "true", "false", or "cannot say" answer. From a person's answers to the inventory, psychologists are able to measure such things as anxiety, self-control, defensiveness, and depression. When this test is scored on a computer, an individual's personality profile can be compared with other profiles stored in the memory of the computer.

Another well known personality inventory is the California Psychological Inventory. It was developed for general use, in particular for use with high school and college students. It is said to be able to predict such things as achievement in school and college, leadership, managerial ability, and juvenile delinquency.

Precautions are taken to prevent people from cheating on personality inventories and giving answers that will make them look better than they are. "Lie" questions are built into the inventory. A lie question might be "I read the newspaper editorials every night." A person having a high "Lie" score is faking answers to look better.

INTERVIEWS

The value of a personality assessment made on the basis of an interview depends on the skills of the interviewer. Many people routinely make personality assessments who are not formally trained to do so. A test was run in which 12 managers, who were used to making assessments, interviewed 57 applicants for a job. The managers then ranked the applicants in order of suitability for the position, based on what they had learned about each applicant's personality in the interviews. When the rankings were compared by a psychologist, they were found to be very inconsistent. One applicant, for example, was ranked most suitable and least suitable by two different managers.

Sometimes an interviewer has the results of a personality rating and a personality inventory available at the interview. Then the interviewer can clear up any areas of the previous tests that might have been unclear, or that might have yielded inconsistent results. The person being interviewed can also give more complete answers than were possible on the written tests.

PROJECTIVE TESTS

Another technique used by psychologists to assess personality is the projective test. In this type of testing the person is called upon to use imagination to answer questions. The psychologist may ask the person to make up a story about a picture, tell what an inkblot brings to mind, finish a sentence, or create a picture in response to an idea.

What is brought out in projective testing is mainly aspects of personality that are usually hidden. Very often the aspects that are revealed are those a person feels are undesirable. We tend to project our undesirable traits onto others. For example, a person who is not loyal to friends, and gossips about them, may feel that everyone is disloyal. In a projective test a person's undesirable traits are projected onto the picture or inkblot. What is seen, then, is a projection of that person's inner thoughts. The psychologist, by evaluating the responses, learns about the person's inner feelings and desires. Of course, if the psychologist interprets the person's responses incorrectly, the assessment of personality will be inaccurate.

BEHAVIOURAL ASSESSMENT

The behavioural method of personality assessment looks at how a person behaves in a specific situation. That behaviour is then assumed to be an accurate reflection of how the person would act in similar situations.

One way in which behavioural assessment is done is to set up a situation in which the person being tested can cheat or steal. The circumstances are such that the person feels no one will know of the cheating or stealing. It is assumed that a person who will cheat or steal in this kind of test will also cheat or steal in day-to-day living.

RESPONSES AND INTERPRETATION

Different people, depending on their personalities and motives, respond to test questions in different ways. People who are very nervous about the test may not be able to concentrate and may have difficulty understanding and answering the questions. Other people may tend to answer "yes" to most questions because agreeing with others is one of their personality traits. Other people who commonly disagree with others may answer "no" to most questions. People who want to be seen in a good light — job applicants for example — may answer questions in a way they perceive to be most socially acceptable. Other people — potential military draftees for example — may wish to do poorly on the test, and will answer questions in a way that will make them appear socially maladjusted.

Some of the tests can detect these different attitudes. The Minnesota Multiphasic Personality Inventory has questions that detect whether people are choosing answers carelessly, are choosing answers to look bad, or are lying.

Interpreting the personality test data is done in two ways.

Psychologists can personally assess the personality and behaviour of an individual, as revealed by the test data, by using their knowledge and judgment. Personal assessments are used most commonly for projective tests. Secondly, psychologists or other trained people administering tests can assess the test data using prepared tables. These are tables of statistics that have been assembled from the responses of many people with many different personality traits. The psychologist interprets the responses of the individual by comparing them to the prepared tables.

With computerized tables of statistics now available, the job is much simpler. An individual's responses are fed into the computer, compared to the tables already in the computer's memory, and the data are analyzed. The computer then provides a personality profile, a summary of test scores, and a lengthy interpretation of the test scores.

Progress Check

1. What do personality tests try to achieve?
2. What do personality ratings try to determine? Do personality ratings on a close friend with regard to performance in school, sports, maturity levels, and getting along with others. Are your ratings objective, or are they biased?
3. What is the purpose of personality inventories? Do you think that the results of such tests can be accurate?
4. What problems can occur when trying to make a personality assessment through interviews? How can these problems be overcome?
5. How are projective tests used to assess personality?
6. Explain the behavioural assessment of personality and comment on how effective it might be. Are there any possible dangers in using this type of personality assessment?

4.3 *The Family and Personality*

One of the main influences on personality is socialization. It is the process of socialization that shapes our tastes, goals, attitudes, values, and behaviour, and the process begins within the family when we are infants. The way we view ourselves and others is greatly affected by the relationships we develop with our parents and siblings as we are growing up.

BIRTH ORDER

The influence of birth order (whether an individual is the youngest, oldest, middle, or only child in a family) on personality development was studied by psychologists Rudolf Dreikurs and Alfred Adler. Both observed that the youngest, middle, and oldest child were likely to have very different personalities. All three children will be reared under different circumstances depending on their position in the family.

The first born is an only child and the centre of attention until a second child is born. With the birth of the second child, the first born may experience a degree of jealousy because it is necessary to share the parents' love and attention with the new baby. Older children may feel it is necessary to compete with the second born for recognition. The competition can take many forms such as striving to do well in school. A first-born child, or an only child, tends to be more serious, more conforming, and more likely to go to college or university. They are likely to be oriented toward adults and activities of which adults approve, such as homework.

Older children may influence the personalities of younger children. Often younger children copy the interests, behaviours and attitudes of their older siblings, especially if they are the same sex.

The second child does not experience being an only child — from birth there is another sibling in the family. Usually the second child is more active in order to keep up with or compete with the older child. Second-born children are often risk takers with a high energy level. They tend to be more cheerful and less serious about school than the first born.

Middle children may feel neglected and uncertain within the family group. Since they have neither the privileges of the youngest nor the rights of the oldest, middle children may develop traits or skills that will bring power and attention. They may, for example, work hard to excel at sports.

Youngest children can become discouraged and feel inferior because they are the weakest and smallest. Some youngest children become over-achievers in order to catch up with their older siblings. Sometimes too, they may be pampered and become overly-dependent on others.

Of course, not every oldest, middle, or youngest child will automatically fit these patterns. How children are treated by their parents and siblings, and how they perceive themselves in relationship to other members of the family will be most important in the shaping of their personalities.

From the moment they are born, children act, think, and feel in response to how they perceive their world. What actually happens is not always as important as the way a person perceives the situation. A youngest child may feel neglected when in reality that child receives as much attention as the other children. The child's personality will be influenced by the perceived neglect, however, not by the actual attention received.

Dreikurs points out that children want to belong, to be accepted, and to feel fulfilled within the family unit. If they feel they are not, they may develop behavioural problems. Their negative behaviour is intended to attract attention or to overcome the power of older siblings or parents. Sometimes it is meant to get even with or to punish those responsible for their lack of acceptance within the family.

In Adler's opinion, children are born feeling inferior because of their dependence on adults, their weakness and their small size. They strive to overcome this feeling of inferiority and to develop superiority. The behavioural patterns they use to reach this goal eventually become their characteristic personality traits. Those people who do not manage to overcome their feeling of inferiority are said to have an **inferiority complex**. They feel that they are not as good as others in certain respects. If such feelings persist into adulthood, they limit the full development of a person's potential.

INFLUENCE OF GENDER

A sociologist named Orville Brim investigated the influence of gender on personality. He found that in a family consisting of only boys or only girls, the stereotyped "male" and "female" traits are reinforced. The boys tended to be more ambitious and competitive, and the girls tended to be more obedient and caring.

In a family consisting of boys and one girl, the girl often showed more "male" traits, and was frequently called a "tomboy" because of her interests. A single boy in a family of girls often exhibited more of the so-called "female" traits. The influences of brothers and sisters on each other's behaviour and personality can be important.

Karen Horney (1885–1952) pointed out some of the ways in which parents may distort the personality of a young girl, for example, by discouraging independence or academic achievement, or by showing favouritism to a brother.

FAMILY SIZE

The size of the family can also influence the personalities of family members. Two sociologists, James Bossard and Eleanor Boll, found personality differences between small families of two children and large families of six or more children.

In the small families studied, the children were more dependent on their parents. Very often they were not required to do household chores in order that they might concentrate on school work and engage in social activities outside the home. Other research indicates that children from small families tend to do better in school and to demonstrate greater motivation to succeed.

On the other hand, the children of large families, studied by Bossard and Boll, were expected to perform specific household chores and share in the household responsibilities. Consequently they did not have as much time to devote to school work and social activities. They tended to rely more on their siblings and were not so dependent on their parents.

Personality is a complex subject because it takes into account all the ways we act and feel — our attitudes, interests, values, beliefs, habits, and goals. Our personality is not fixed at birth. It is shaped by our experiences at home, at school, and at play, and it is influenced by the people with whom we come in close contact. In the very important early years when the foundations of personality are being laid, children are mostly at home. The family is therefore a major influence in personality development.

PROGRESS CHECK

1. Why do first-born children or only children tend to do better in school? Do you agree?
2. What are some personality traits often associated with middle or youngest children in the family?
3. According to Adler and Dreikurs, what are some of the behavioural problems that children might develop if they are not properly socialized?

4. How can the sex of family members affect personality development?
5. Why is personality such a complicated subject?

4.4 *Some Theories of Personality Development*

Psychologists have been studying personality for close to one hundred years now. During this time many theories have been proposed to explain the nature of personality and how personality develops. In section 4.1 we discussed some of the theories of describing and classifying personalities. In this section we will deal with the theories of personality development.

FREUD'S PSYCHOSEXUAL THEORY OF PERSONALITY DEVELOPMENT

Undoubtedly the first important theoretical study of personality was made by Freud. According to his theory, the personality is composed of three parts: the id, the ego, and the superego.

The id is the part of our mind which is already developed at birth. It is made up of all the basic biological needs and drives, and it wants immediate gratification of these needs and drives. The conscious part of our mind, the ego, is next to develop. It begins its development during the second six months of life when we start to distinguish between ourself and the outer world. The ego is the rational part of our mind and is involved with learning, remembering, and reasoning. The superego, together with the id, forms our unconscious mind. It is thought to develop between the ages of three and six when we learn the values and moral standards of our society. The superego has two functions. It is our conscience, telling us what is right and wrong: acceptable and unacceptable. It also provides us with an image of what we could be ideally.

How do the three parts of the personality interact? The superego is constantly in conflict with the id, trying to stifle the pleasure-seeking and aggressive impulses of the id. The ego must find a balance between the pleasure-seeking id and the conscience of the superego. Difficulties in resolving the different demands and achieving a balance can result in anxiety and sometimes abnormal behaviour.

When anxiety is severe, the ego tries to reduce it by resorting to what are called **defence mechanisms**. These are strategies that the mind adopts to cope with the situation, but the person is unaware of their purpose. A few of the main defence mechanisms used by people to combat anxiety are repression (banning from memory unacceptable impulses, thoughts, or events that cause anxiety, for example, forgetting the details of a fatal car crash); denial (rejecting painful reality, for example, the death of a loved one, and denying that it has happened); displacement (transferring unacceptable feelings about a person or situation to someone else considered less powerful, for example, transferring anger that should be directed at a boss to a co-worker instead); projection (blocking unacceptable feelings or thoughts by attributing them to someone else, such as a miser believing that everyone else is stingy); and rationalization (making poor behaviour appear to be good by giving it an acceptable reason, for example, stealing something from a store and reasoning that the store has overcharged in the past and that the stolen article makes up for it). An understanding of how the defence mechanisms operate helps to explain personality and behaviour.

To review then, we have discussed the structure and the parts of personality, when the parts developed, the interactions of the parts, the results of interaction, and some strategies used to reduce conflict produced by the interaction. The id, ego, and superego, according to Freud, make up the structure of the personality. Sometimes the ego cannot reconcile, or meet, the different demands of the id, superego, and reality. The result is anxiety. When anxiety is extreme, the ego resorts to the use of defence mechanisms. The next question is how personality develops.

Freud proposed a theory of personality development that is strongly tied to the biological development of the individual, and to the striving of the id for pleasure. One part of the body is a source of sensual pleasure for the individual at each stage of personality development. The sensual pleasure provided by that body part relieves tensions that have built up. If, at any stage, pleasure is denied, frustration results. At each stage certain personality traits emerge. They are based on the type of balance that is achieved between pleasure and frustration.

Freud's First Stage
The first stage of development Freud called the oral stage. It lasts from birth to approximately 18 months and coincides

with the period during which infants are preoccupied with sucking, licking, biting, gumming, and chewing anything they can put in their mouths. They gain pleasure, apart from satisfying hunger, from these oral activities.

If the oral drive is not allowed to be satisfied, as for example, if the mother does not allow the baby to nurse long enough or frequently enough, the infant will become frustrated. The frustration can be expressed as distressed crying, shallow breathing, rigid muscles, or lethargy. Freud suggested that if babies build up a great deal of anxiety when deprived of sufficient oral pleasure, they may become overly dependent on their mothers. Later, these individuals could develop dependencies on the people around them, a type of personality Freud called oral-dependent.

Some psychologists feel that an unsatisfied oral drive during this period will be expressed in later life by overeating or excessive smoking. These activities are thought to compensate for lack of early oral satisfaction.

Freud's Second Stage
The second stage of personality development, Freud called the anal stage. It is in this stage, between the ages of eighteen months and three years, that the child gains pleasure from the elimination of body wastes. This pleasure can be frustrated by the parents who attempt to bring the process under control by toilet training, sometimes before the child has the necessary muscle control. The child, however, can exert control by resisting toilet training or by retaining wastes and offering them at the proper times with the expectation of parental praise.

Freud believed that a child's experiences during toilet training greatly affects their adjustment as adults. Excessively strict toilet training, according to Freud, could result in feelings of dislike for authority figures in later life, extreme neatness, or strictness in budgeting time and money.

Freud's Third Stage
In the third stage of development, called the phallic stage, the child's pleasure centres around the sex organs. This stage, when children become aware of their genitals, occurs between the ages of two and six approximately. Their sexual awareness may appear sometimes as masturbation and sometimes as a need for close contact with the parent of the opposite sex.

Freud proposed that during the phallic stage a little boy develops the **Oedipus complex**, a sexual attachment to his mother. This complex is named after a character from Greek mythology, Oedipus, the king of Thebes, who killed his father and married his mother without realizing their true identities.

In Freud's view, a little boy directs his sexual desires toward his mother. However, he feels his father will retaliate and harm his sex organs. Because of this fear, and because of his guilty feelings, the little boy tries to become just like his father. His feelings for his mother he converts to loving affection. In this way the little boy resolves his Oedipus complex and at the same time develops a conscience — the superego.

Sometimes during this period boys who are overanxious may become obsessed with their masculinity. Later in life this anxiety may express itself as a preoccupation with so-called male physical activities, intense vanity, and exhibitionism. If major problems are encountered during the phallic stage, the boy may become self-centred and may be unable to have mature relations with the opposite sex.

A comparable complex, called the **Electra complex**, develops in girls during this stage. Electra, again a character from Greek mythology, helps her brother kill their mother and her lover as retaliation for the murder of their father, Agamemnon. The little girl, in Freud's view, wishes to have her father to herself, and resents competition from her mother. As with the boy, the little girl's rebellious thoughts produce guilt and fear. She therefore chooses to identify with her mother and as a result develops a conscience (superego).

Freud's Fourth Stage
The fourth stage of development is called the latency period, so named because it is a period in which little sexuality is observed. It occurs between the age of six and the onset of puberty. During this time boys seek the companionship of other boys, and girls the companionship of girls. Their energies and attention are turned toward school and play, and they are less concerned with their bodies.

Freud's Fifth Stage
The fifth and last stage of development, called the genital stage, begins with the sexual maturity of an individual at puberty (around the age of twelve). At this stage a major

struggle takes place between the id, which has increased sexual energy, and the superego, which attempts to enforce social rules controlling sexual behaviour. It is necessary for the individual to control this sexual energy in order to participate in normal adult male-female relationships.

During adolescence there is a great deal of socialization with the opposite sex usually leading to dating and eventually marriage. On the other hand, during this period adolescents may often experience feelings of inadequacy, confusion, or guilt related to their sexuality.

Critics of Freud's theory of personality feel that it concentrates too heavily on sexual development. Others point out that the women whom Freud studied and on whom he based his ideas were all turn-of-the-century upper class Viennese women, many of whom were suffering from mental problems. Another basic criticism is that Freud relates personality development to childhood only, and fails to account for changes that take place in adulthood. Psychologists today recognize that changes continue to take place after maturity. New experiences may give us more (or less) self-confidence. New relationships may help us to overcome earlier problems and disappointments.

CARL JUNG'S THEORY OF PERSONALITY

Like Freud, Jung believed that personality has different components, or parts. He maintained that there are two sides to everyone: the persona and the shadow. The **persona** is the public mask we wear as we play our roles in society. It is the way we want people to see us. The **shadow** is that part of our mind of which even we are not completely aware. It is there that we keep those parts of our personality that we wish hidden from others, for example, aggressive and sexual drives. The story of Dr. Jekyl and Mr. Hyde demonstrates the dual nature of the personality. Dr. Jekyl was a mild-mannered, gentle man who is preoccupied with his work during the day. By night, he was transformed into the evil Mr. Hyde who roamed the streets brutalizing people.

Frederic March in the movie roles of Dr. Jekyl and Mr. Hyde

Jung believed that personality development is directed toward maturity, achievement, and self-actualization. Freud believed the aim of personality development was to resolve the conflicts that arose during each stage.

PROGRESS CHECK

1. According to Freud, how do the id, the ego, and the superego interact with each other?
2. How do people cope with extreme anxiety?
3. Outline Freud's stages of development in children. Give your opinions about his theory.
4. According to Freud, what personality traits can emerge during the socialization process of children?
5. Do you agree with Carl Jung's theory of personality? Give reasons for your answer.

ERIK ERIKSON'S THEORY OF PERSONALITY DEVELOPMENT

Psychologist Erik Erikson, like Freud, concluded that personality develops through various stages. He describes eight stages, which start in early childhood and continue through to old age. At each stage, according to Erikson, personalities can develop in two opposite directions.

Erikson observed that children pass through the developmental stages at different rates — some faster, some slower. Certain abilities are learned during each stage. If these abilities are not developed, individuals will encounter difficulties later in life in situations where the skills are required.

TABLE 4.2 Eight Personality Stages

Stage 1	Year One	Basic trust *versus* basic mistrust
Stage 2	Years Two-Three	Autonomy *versus* doubt
Stage 3	Years Four-Five	Initiative *versus* guilt
Stage 4	Years Six-Eleven	Industry *versus* initiative
Stage 5	Years Twelve-Eighteen	Identity *versus* role confusion
Stage 6	Young Adulthood	Intimacy *versus* isolation
Stage 7	Middle Age	Generativity *versus* stagnation
Stage 8	Old Age	Integrity *versus* despair

Stage 1

During the first stage, which coincides with the first year of life, children develop either a sense of trust or of mistrust depending on how secure they feel, especially in relation to their mothers. If they can predict with some certainty that their parents will look after all their needs, they will develop a sense of trust. If they receive sufficient attention and warmth, they will also know that they are loved. Alternatively, if their world is unpredictable and their parents' love is uncertain, children may develop a sense of mistrust.

Stage 2

In the second stage, which occurs between ages two and three, children struggle between a feeling of **autonomy**, or independence, and feelings of shame or doubt. This is a stage when many skills are being mastered: walking, climbing, talking, exploring, dressing themselves, and many others. They are learning independence by exploring and mastering their environment. Encouragement from parents, when each new step toward independence is taken, will send them on to greater efforts. At the same time they are learning control of impulses and drives, and this control gives them a further sense of independence and accomplishment.

Parents need to allow their children the necessary freedom (within safe bounds) to develop their skills and to experience their growing independence in this stage. If parents are too restrictive, over-protective, and critical during this time, children may develop a sense of doubt, shame, and uncertainty about themselves and their abilities. These children may doubt their abilities in later life and may have difficulty operating independently.

Stage 3

The third stage, between the ages of four and five, involves a struggle between initiative and guilt. It is a time in which children develop initiative in their activities and their thinking. Through exploring, planning, and setting goals, they develop a better understanding of their environment and a sense of purpose. They also work on improving language skills and ask many questions about the world around them and their place in it. "Why" is a word they use very frequently. Sometimes parents restrict their children's physical and verbal explorations too severely during this period, or downgrade their children's accomplishments. As a result, these children may doubt the value of setting goals and doubt their ability to reach them. They may feel guilty for having ambitions and become shy in their dealings with others.

Stage 4

The fourth stage occurs when children first attend school at ages six to eleven, a time when their world has expanded beyond the family. There is a conflict at that time between industry (effort) and inferiority. To succeed at school, children must learn to complete assigned tasks and to do them well. Praise from parents and teachers for their accomplishments will reinforce a sense of success through industry. If they do not learn to produce acceptable work or if their work is criticized they may feel inadequate and inferior in comparison with the other children.

Erikson stressed the importance of family life in preparing children for school. Children from families that stressed language writing and reading skills are better prepared for school life and generally experience greater success. They have the satisfaction of completing their assignments and receiving good marks, they are encouraged by parents and teachers, and they are respected by their friends. Consequently they develop a greater desire for learning and academic success.

School presents a challenge to children to do well. Those who meet the challenge and do well are able to pass through this stage successfully. Those who experience failure, may feel unworthy and inferior. They may have negative emotions about school and about learning in general.

Stage 5
The fifth stage corresponds with puberty and the conflict between identity and role confusion. In this stage young people must cope with the physical changes their bodies are undergoing, and with the challenge of discovering and shaping their identities. They must answer the question "Who am I?" In this search for identity they may model themselves after parents, friends, teachers, or media heros. Sometimes they may be uncertain about which role models or sets of behaviour are best to imitate, and about what they want to do with their lives.

Adolescents, because they are generally physically mature are expected to act in an adult manner. However they are usually not emotionally or mentally mature, not having gone through the life experiences that help to bring about maturity.

Some teenagers find this stage difficult. They are under pressure to "find" themselves and to fit into expected roles. They are also expected to channel their energies towards gaining an education and establishing a career. Sometimes teenagers find it difficult or impossible to meet the expectations of others. They may be regarded as socialization failures because they have not achieved the goals set for them by society, and because they have not established acceptable identities and personalities.

Stage 6
The sixth stage, early adulthood, is a period of struggle between intimacy and isolation. Young adults, once they have established a sense of identity, seek intimate relationships with others. Some of the relationships may develop into

deep friendships and others into love affairs and marriage. Within these relationships, people learn to share and to consider the needs and feelings of others. Close relationships can also end in rejection, disappointment, and bad feeling. For some people the fear of being hurt may cause them to shun intimacy. Without intimate relationships, however, people may become withdrawn and isolated, or very self-centred.

Stage 7
In the stage that marks mid adulthood there is a struggle between **generativity**, or full productivity, and **stagnation**, or lacking any development. Erikson uses the term generativity to extend the idea of productivity to include raising a family, working productively and creatively, and caring for those beyond the family circle, even future generations, in a selfless way. Through generativity, people develop a sense of achievement and self-worth.

The opposite is represented by those who are not concerned with the needs of others, and whose lives have stagnated, that is, there is no development. They may be dissatisfied and feel that they are not achieving anything in life that is meaningful. This dissatisfaction may lead to despair, envy of others, and anger over missed opportunities.

Stage 8

The final stage, which comes at about the time of retirement, is a time of stuggle between integrity (completeness) and despair. At this point many people can look back on their lives with a sense of completion and satisfaction. They have met past challenges and crises with integrity, and they have accepted, with few if any regrets, the way that life has worked out for them. Others may feel depressed and unfulfilled because of past failures and mistakes, and missed opportunities. They look back on life with a feeling of despair.

Erikson believed that most people pass through all of these stages with differing degrees of success. How well they develop in each stage will determine how well they succeed in later stages. The successes and/or failures that they experience along the way will mold their attitudes, determine their levels of confidence, and influence how they feel about and relate to themselves and others.

Erikson's theory provides reassurance that personality can be modified by experience at any stage, including those stages that occur after childhood. In this respect it differs greatly from Freud's theory.

Progress Check

1. Outline Erik Erikson's stages of development in the life cycle. What influence does each have on personality?
2. At what stage of development are you in Erikson's life cycle? Do you believe that you have passed through the earlier stages successfully?
3. Compare the theories of Freud and Erikson. In your opinion, which explains personality development better and why?

4.5 *The Sense of Self*

All of us have a sense of who we are — our strengths and weaknesses, our beliefs, the knowledge we have gained, certain things that we are proud of, and other things that we do not like about ourselves. This mental image of what we believe about ourself is called a **self-concept**.

Self-concept develops and changes as we grow older, partly due to new experiences, and partly due to our expanding ability to think in abstract ways and to evaluate. As young children our self-concept is based on simple ideas such as gender, interests, and physical abilities — "I am a girl. I like teddy bears. I can ride a tricycle." When they are a little older and can think in abstract terms, children expand their self-concept to include traits such as friendly, honest, helpful, or lazy.

CAREER PROFILE
ERIK ERIKSON

Erik Erikson who is of Danish and Jewish descent, was born in Germany in 1902, married a Canadian woman named Joan Mowat, and emigrated with her to the United States in the 1930's. There he became the first child psychiatrist in North America.

In the field of social science, he pioneered the view that human personality develops in a series of stages. He stressed the importance of proper development in each stage, but maintained that later development could compensate for improper growth in an earlier stage.

He studied with Freud in Vienna, Austria and was inspired by Freud's works. Unlike Freud, however, Erikson believes that a child's personality is greatly influenced by the social environment.

Erik Erikson has led an interesting and a committed life. During the 1940's and 1950's he opposed the Communist witchhunt in the U.S.A. One way in which he upheld his convictions was by refusing to sign a statement on a job application denying that he had ever been a Communist. Although he was not a Communist, he believed that the government and employers were going too far in demanding such statements from people. During the 1960's, the Eriksons actively supported the anti-war movement, and they are now active in the nuclear disarmament movement.

Erikson has been a strong influence in psychological and psychiatric thought. To continue the study and further the applications of Erikson's work, the Erik H. and Joan M. Erikson Center has been established in Cambridge, Massachusetts.

Erikson is an accomplished author as well. He won a Pulitzer prize for his book *Gandhi's Truth*.

The Eriksons have raised three children, and now make their home in California.

Parents have a strong influence on how children see themselves. A parent who consistently tells a child "You are so scatter-brained" might eventually have a child who thinks "I am scatter-brained". This child will have little confidence when it comes to problem solving and school work. Often parents who negatively influence their child's self-concept are totally unaware that they are doing so. Sometimes they think that they are only making a joke and they do not realize that their child has taken the "joke" to heart.

The late Dr. Haim G. Ginott, a child psychologist, advised parents that constructive criticism means "pointing out how to do what has to be done, entirely omitting negative remarks about the personality of the child". If a child spills a glass of milk, the parent need only say "I see the milk is spilled. Here's another glass of milk and here's the sponge". There is no message in those words to make children feel badly, or to convince them that they are clumsy, and that no matter what they do they will probably spill, drop, or break things.

Encouraging children to have a positive self-concept helps them to approach new situations with confidence in their abilities. A child with a positive self-concept can take initiative, work independently, and take pride in accomplishments. What is more, the child can recover from setbacks and not be crushed by them. Children with negative self-concepts, on the other hand, are certain that they will fail at whatever they tackle. This lack of confidence often means that they will not attempt things, or that they will give up at the slightest problem.

Children tend to compare themselves to others at school, and these comparisons influence their self-concept. A teacher can help children who have negative self-concepts by encouraging them to express themselves and to make decisions and take action. These children need to know that they are valued and accepted as they are.

Judging ourselves by how others see us, the **looking-glass self**, does not end with childhood. What others think of our actions, attitudes, and expectations is very important to our self-concept and our feeling of self-worth. We imagine how we appear to those whose opinions we value, and we adjust our behaviour accordingly. This is part of the socialization process.

> " ... During adolescence the young person develops a true sense of self. While children are aware of themselves, they are not able to put themselves in other people's shoes and to look at themselves from that perspective. Adolescents can do this and do engage in such self-watching to a considerable extent. Indeed, the characteristic self-consciousness of the adolescent results from the very fact that the young person is now very much concerned with how others react to him. This is a concern that is largely absent in childhood."
>
> **Professor David Elkind**

THE CONCEPT OF SELF IN TWO SOCIETIES

North Americans tend to be preoccupied with their sense of self. We are always talking about "ourselves", "himself", "herself", and "themselves". We have many commonly used expressions that relate to self: talking to oneself, proud of oneself, thinking to oneself, and looking to oneself.

William Caudill, an expert on Japanese culture, has found that other societies, Japanese society for example, do not place such great importance on the development of a sense of self in children. Japanese parents do not spend so much time or effort on teaching language to their children as North American parents do. They view their children, not so much as separate beings, but as part of the parents. Japanese

parents assume that they understand what their children require. It is therefore not important that their children learn to express themselves at an early age.

Caudill observed that Japanese parents tend to carry their children around more, and to have more physical contact with their children than North American parents. In North American society, when space permits, children are usually placed in separate rooms as infants. In Japanese society, it is common for children to sleep with their parents until around the age of ten. This tends to reinforce the importance of the family group, as opposed to the individual. In Japan, children are viewed first as members of the family group, and second as individuals. The needs of the group are considered to be more important than the needs of the individual.

Progress Check

1. Why is it important for adults to encourage children to develop their abilities?
2. What is meant by the statement that "people tend to live up to the images that others have of them"? Do you agree?
3. Compare the socialization of Canadian and Japanese children. How do these different socialization processes relate to the development of personality?

Overview

Personality is a difficult subject to understand. Everyone has a unique personality because of the many different influences that each one of us has experienced in our lives. We all have many personality traits — some are positive and others are considered negative by different people around us. We try to understand ourselves and the reactions of others toward us. These concerns take up a considerable amount of our energy and time.

Most people want to be liked and cared for by those whom they consider important in their lives. But sometimes certain personality traits prevent people from getting close to others, or hinder them from achieving their goals. The experts who have been referred to in this chapter have many reasons to explain why people have certain traits. They would agree, though, that desirable behaviour and personality traits permit people to lead happier and more productive lives. People having negative traits or personality disorders have difficulty leading happy and fulfilled lives. Learning to understand the reasons for their behaviour, and taking steps to correct negative and damaging personality traits can change their lives. Often professional help is needed to understand and to make the changes.

Of course it would be a very boring world if everyone had the same personality. Different personality traits are desirable and expected in people. To a great extent they are the result of our past experiences. By becoming aware of how past experiences influence personality development, we may learn to understand ourselves and others better.

KEY WORDS

Define the following terms, and use each in a sentence that shows its meaning.

personality	mesomorph	shadow
traits	ectomorph	autonomy
personality type	defence mechanism	generativity
introvert	Oedipus complex	stagnation
extrovert	Electra complex	self-concept
endomorph	personna	looking-glass self

KEY PERSONALITIES

Give at least one reason for learning more about each of the following.

Gordon Allport & Henry Odbert	Alfred Adler
Warren Norman	Orville Brim
Raymond Cattell	Karen Horney
Galen	James Bossard & Eleanor Boll
Carl Jung	Sigmund Freud
Hans Eysenck	Erik Erikson
W. N. Sheldon	Haim Ginott
Rudolf Dreikurs	William Caudill

DEVELOPING YOUR SKILLS

FOCUS

The focus of this chapter is personality. Students and experts in psychology and sociology want to know more about personality and the different theories of personality development. By examining these theories we can learn more about our own personalities and arrive at conclusions to help explain why people have similar or different personalities.

"Personalities can be extremely important in determining the quality of our relationships with other people and the success we have in our personal and public lives." Divide the class into groups and brainstorm a list of possible questions that can be asked about the subject of personality. Each group should have someone write down the ideas that are suggested by group members. At the conclusion of the brainstorming session the group can choose the best questions. These will be clear and to the point. One question might be "What types of personalities are considered to be most desirable? Least desirable?"

Each question will bring out different points-of-view and information that will reveal something new about the subject of personality. Answering these questions through discussion, reading, research, and writing will lead students to a better understanding of personality and the issues surrounding this complex subject.

ORGANIZE

1. In an organizer describe Freud's stages of biological development and their effects on personality development. In a third column outline the potential negative effect for those who do not successfully complete each stage.
2. Develop an organizer that outlines Erik Erikson's theory of personality development. Describe the abililties that are developed during each stage and the potential problems that can be experienced.

LOCATE AND RECORD

1. Divide the class into groups to investigate Freud's theory about the interactions among the id, the superego, and the ego. Individual students in each group can draw a diagram that clearly outlines their functions and relationships.
2. Choose one expert who is referred to in this chapter and use the index in the textbook to find out as much informa-

tion you can about this person and theories. Go to your library and use the card index to find additional information. Record your findings in your own words.

EVALUATE AND ASSESS

1. By referring to the textbook create an organizer that lists the purposes and goals of personality testing techniques under the following headings: ratings, inventories, interviews, projective tests, and behavioural assessment. In two columns list the positive and negative qualities of these different testing techniques. In a final column identify the testing technique that you think is most effective and the one that is least effective. List the reasons for your choices.

2. Examine the following clues and identify the main ideas or terms to which the clues relate.
   (a) a way of behaving / remains with a person for a long time
   (b) repression / denial / displacement / projection / rationalization
   (c) the way we want people to see us in various roles / the part of the mind that we try to hide
   (d) a mental image we have of ourselves / it changes as we grow older
   (e) judging ourselves by the opinion of others
   (f) a person who is outgoing / a person who keeps things inside

SYNTHESIZE AND CONCLUDE

1. On a chart list your greatest personality strengths (things you do well or personal traits you think are good) and weaknesses (things you do not do well or are problems for you).
   Using a chart indicate which of the factors associated with personality development, that are listed below, have led to your strengths and weaknesses. Explain how this occurred.

   Heredity: characteristics or features with which you were born such as facial features and anatomy

Parents: how they influenced you or have dealt with you

Siblings: how you relate to your siblings, how you feel about your place in the family

Family Values: the things important to your family, such as, mores, sports, religion, success, honesty

Role Models: people after whom you have patterned yourself

Community: schools, clubs, neighbourhoods, peer groups that have influenced you

Circumstances: things that happened to you when you were growing up. Some examples are a divorce, death, or injury in the family; the birth of a new child; a move to a new neighborhood and/or school

Other Factors:

2. Examine the theories of Rudolf Dreikurs, Alfred Adler, Orville Brim, and James Brossard and Eleanor Boll. Apply their theories and observations to your own family situation. Which are most correct and least correct in your situation?

3. Not only do we develop a "looking glass self" based on what others want us to be, but different people and situations bring out different qualities in us. Do certain people and situations bring out good or bad qualities in you? Explain.

4. In some ways our personalities reflect how we feel about ourselves. Do we like who we are, or are we unhappy with our self-concept? What do you like about yourself? What do you not like about yourself? Whould you rather have someone else's identity? Whose? Why?

APPLY

1. Analyze your personality by listing your traits under two columns — positive and negative. In another column compare your present traits to those that you had three years ago. Have they changed, and if so, in what direction? In a final column speculate about the kinds of personality traits that you want to have in five years. Share your conclusions with other class members.

2. Some experts believe that people are attracted to others who have similar personality traits as themselves. List the traits that you find most attractive and appealing in others. Compare these traits with your own personality. Are they similar or different?
3. Use Erik Erikson's theory on personality development and apply it to friends and relatives. Do your observations support Erikson's opinions?

COMMUNICATE

1. Write an essay assessing the theories on personality development in this chapter. In your essay describe the theories that are most accurate and give your reasons.
2. Write a short essay and/or debate one of the following topics.

- People are too preoccupied and concerned about the opinions of others: they should do what they want with their lives.
- Most people are brainwashed by society and have no real individuality.
- The personality of a child is predetermined at birth.
- Personality tests are an invasion of a person's privacy and should be banned.
- The experts who developed personality theories have given us valuable insights and understandings.

CHAPTER

5

Women and Equality

What is prejudice, a stereotype, discrimination? Who were some of the Canadian pioneers in the struggle for equality of women? How equal are women today?

INTRODUCTION In much of history women have been regarded as the property of the male — first of their father, then of their husband. Males often considered females to be the "weaker sex" in need of protection. The price of this protection was that women were not allowed to own property, to vote, to work outside the home, or to be in control of their own lives.

Attitudes toward women on the part of both sexes have changed radically in this century, and continue to change. In this chapter we will be looking at the history of the women's movement in Canada, and examining the current status of Canadian women.

5.1 *Prejudice, Stereotypes, and Discrimination*

As people grow up they develop preferences and dislikes. Many of these preferences and dislikes are absorbed from the society in which they live. Some have to do with types of food, music, dance, and dress styles; others with ways of behaving, goals, values, and religious beliefs. People who share the preferences and dislikes of their society, feel a part of the society.

Individual members of a society or group will be aware of the way their group as a whole feels about other groups. Sometimes they will view groups that have fewer members or less power, **minority groups**, as inferiors or as threats to their security. Such a view may be based on the characteristics or actions of just a few members of the minority group. When people are prejudged on the basis of characteristics thought to be common to all members of their group, we say that the person or group doing the prejudging is **prejudiced**.

Painted on a high school wall in Montreal, JAP refers to "Jewish American Princesses", a derogatory label; in 1974 five Jewish students in this school were beaten up by other students.

When prejudiced people meet an individual from a group they dislike, they do not consider the person as an individual. Instead, they apply the negative feelings they have about the group to that individual. They stereotype, or label the individual. They disregard the person's individuality and expect that the person will conform to their prejudiced image of the group. This image, the stereotype, is a kind of mental picture that exaggerates the characteristics of a typical member of the other group.

Social scientists have found that people who are prejudiced usually ignore any new information that does not fit their stereotyped images. They maintain their negative opinion of a group even when they are presented with evidence indicating that their negative opinion is wrong.

Many people, who deny that they are prejudiced, tell jokes about groups perceived as being different on the basis of religion, ethnic origins, sex, age, social class, or handicap. In these jokes members of the minority group are stereotyped, and seen as being different from the rest of society. Such jokes tend to reinforce stereotypes, and prolong people's misunderstanding of others.

The media is often guilty of reinforcing prejudice and stereotypes. Up until recently, movies and television programs portrayed Indians as blood-thirsty drunken killers who preyed on defenseless, innocent people. This very negative stereotype has been very difficult for Indians to overcome. It has made many Canadians less sympathetic to native peoples' demands for a fairer deal in Canadian society.

In a report to the federal government in 1985, Media Watch, a women's group that monitors images of women in the media, stated:

"Woman continues to be cumulatively portrayed in ways that suggest her place is in the home, her interests do not include issues of the world or public affairs but are limited to interests in other people and her own appearance. Her primary roles are those of caretaker of men and children and sexual object of men ... Women continue to be psychologically and physically abused in the guise of entertainment ... she is usually submissive and has no power ..."

Analyze several television ads to see if similar comments could be made today.

The CBC policy manual for on-air staff states that

"Ill-advised use of stereotypes tends to reinforce prejudice and constitutes an assault on the dignity of the individual..."

Listen to your favourite radio station for comments made between recordings. Did you hear any remarks that suggest stereotypes of individuals?

Why are people prejudiced? Low income, low status occupations, and poor education often contribute to prejudice. Unskilled and poorly paid workers often find themselves competing with newly arrived immigrants for a limited number of jobs. This competition often creates tension and bad feelings, especially when economic conditions are bad and jobs are scarce. At those times there is a sharp increase in prejudice and stereotyping.

People who are frustrated by having little power may use minority groups as **scapegoats**. They blame the minority group for the poor circumstances in which they find themselves. The Jews in Nazi Germany were an ethnic minority treated as scapegoats for the poor economic situation that followed the World War I. **Anti-Semitism**, prejudice and persecution directed against Jews, reached horrific proportions during World War II. Women too were scapegoated by the Nazis. The women's movement, which was very strong in pre-Nazi Germany, was dissolved when Hitler came to power. **Feminists**, people who believe in equal rights for women, were removed from public positions such as teaching jobs. Women were banned from holding any decision-making position, and married women were expected to stay home, leaving paid jobs to men.

Research conducted by social scientists indicates that, typically, many very prejudiced people dislike all minority groups. Furthermore, they were often raised in families that had a high level of prejudice, and part of the socialization process in the family was learning parental prejudices. Insecurity, fearfulness, and a great respect for authority are other characteristics of very prejudiced people. To reduce frustration, anger, and insecurity, they use minority groups as scapegoats.

CASE STUDY

Attitudes Toward Minorities

For the past four years, more than 2 000 Canadians have been interviewed by professional interviewers on behalf of the League for Human Rights of B'nai Brith Canada. The following are the 1986 results of three of the questions asked.

Percentage of respondents who occasionally or rarely have contact with:

Jews:	64.1
Italians:	75.1
Poles:	62.3
Blacks:	75.3

Percentage of respondents who said that _____ have "too much" power in Canada today.

Jews:	16.7
Italians:	11.3
Poles:	3.3
Blacks:	6.4

Percentage of respondents who said they would not vote for a person of _____ descent, if nominated by the party they normally vote for in their riding.

Jewish:	9.4
Italian:	8.2
Polish:	8.0
Black:	8.6

An analysis of opinions within an educational category shows that prejudice declines with increasing education. The highest levels of prejudice were found among people who have eight years or less of schooling.

What is your opinion of the conclusion that there is a relationship between literacy and attitudes?

Sometimes a group that has power, a **dominant group**, treats another group unfairly. They might, for example, prevent the less powerful **subordinate group** from voting in elections, or deprive them of equal access to job opportunities. This process of unequal treatment is called **discrimination**.

Social scientists have found that members of subordinate groups may accept their lower status in society and may even feel shame because of it. Eventually, sufficient numbers within the group may become aware of the discrimination against their members, and feel frustrated or angered by it. Then a few group members, or **activists**, will begin working vigorously toward making others aware of the unfair treatment, and trying to change the political system that allows the discrimination.

Often activists are seen by others in their group and by the larger society as extremists, and their methods and goals are viewed as radical (extreme, drastic). This attitude is especially common among those who expect to lose some power if the activists achieve their goals. At a later period in history "extremists" may be honoured by society for their vision and courage.

PROGRESS CHECK

1. Explain the connection between prejudice and stereotypes.
2. How do jokes reinforce stereotypes about minority groups?
3. Why is the media often accused of reinforcing prejudice and stereotypes?
4. How can income, occupation, education, and economic conditions affect people's attitudes toward minority groups?
5. Why are activists often disapproved of by the larger society?

5.2 Attitudes Toward Women

Although the term minority group implies a group that has fewer members, it has a broader meaning. Sociologist Louis Wirth stated, "We may define a minority as a group of people who, because of their physical or cultural characteristics, are singled out from the others in the society in which they live for differential and unequal treatment, and who therefore regard themselves as objects of collective discrimination."

Fifty-two percent of Canada's population is female. Despite the fact that women are in the majority, women as a group are the largest minority group in Canada to suffer from prejudice and discrimination. To explain how women can be considered a minority group requires a look at male-female relations in the past.

Some social scientists believe that the foundations for modern male-female relations developed thousands of years ago. They speculate that the superior physical strength of the males of primitive societies enabled them to assume the dominant roles.

In early societies life was uncertain and tribal members depended on each other for their survival. The cooperation of everyone was essential. Some social scientists believe that it was easier in these primitive society to assign chores and duties based on sex than on any other factor. Since women were the ones to bear the next generation and to nurse them, their lives were more valuable from a biological point of view than the men's. If women did not produce children, the group would not survive. It therefore made sense for men to be the ones to risk their lives hunting. Caring for children, tending the fires, making clothes, food gathering, and food preparation would be the obvious duties of those who stayed home — the women. Through this division of labour, gender roles were probably developed and passed on with the culture.

In later, more complex societies these traditional gender roles were expanded and refined. Women in agricultural societies had an important role in food production, in addition to child rearing and household management. As societies moved to a more industrialized base, economic production was no longer centred in the household. According to sociologist Richard Hamilton, women lost status with industrialization. Most women no longer had an important role in the economy of the society. Women became dependent on males who held jobs outside the home.

Throughout history most societies have been **patriarchal**, meaning the male is the head of the family. There have been a few societies in which power and prestige were shared by both sexes. However, no true **matriarchy**, a society in which women are dominant over men, has been found.

In this century, the women's movement is challenging traditional patriarchal authority, and focusing attention on the discrimination that is practised in our society. Status and identity of women are two of the issues they are tackling.

In the past in Canada, the status of a woman was based on her husband's status in society. Not only was status dependent on marriage, identity was as well. Traditionally (since about the 17th century) when a woman married she changed her surname from that of her father to that of her husband. In so doing, she was sharing her husband's social identity.

With the growth of political and social awareness on the part of women within this century, these practices are being challenged and changed. In Quebec since 1981, women who marry retain their surname for all legal purposes. Children can be given the surname of either parent, or a combination of the two surnames. In Ontario women now can choose between their husband's name or their own.

140   HUMAN SOCIETY

PROGRESS CHECK

1. Why are women classified as a "minority group" when they often represent a majority?
2. According to some social scientists what was originally responsible for the formation of traditional male and female roles?
3. According to Richard Hamilton, why did women lose status when economies became industrialized?
4. Traditionally what determined a woman's status and identity? Is this still true today?

5.3 The Early Women's Rights Movement in Canada

During the latter part of the 19th century and the early part of the 20th century, the focus of women who wanted social change was on obtaining the vote, or **suffrage**. Women recognized that without a vote, they had little control over their own lives.

In this 1915 photo are members of the Political Equality League which presented the women's case for voting rights to the Manitoba legislature: Mrs. A. V. Thomas, Mrs. F. J. Dickson, (front) Dr. H. Crawford, Mrs. E. Burritt.

By the 1880's, small groups of women, referred to as **suffragettes**, were working to gain women's suffrage. They signed petitions, wrote letters, and sent representatives to persuade the lawmakers to change the laws so that women would have the right to vote. One of the first of these groups was the Women's Christian Temperance Union, the WCTU, formed to fight alcoholism among the working class. They recognized that in order to change the liquor laws they needed the right to vote. Another early women's group was the National Council of Women. It fought for the rights of working class women and children who worked long hours under conditions that were often harsh and dangerous. They also recognized that women's suffrage was necessary before they could change conditions.

The suffragists were regarded by many Canadians as extreme activists, and as such they were ridiculed and criticized. Those who opposed women's suffrage argued that women were inferior to men and could not be trusted with the responsibility of voting. Others claimed that it was God's will that women should be dependent on men, and that men would make the wisest decisions for all. Women's place was in the home, after all. So how could they be expected to be knowledgable about issues outside the home, and be capable of making political decisions?

The following is an excerpt from **Equal Suffrage** *by James Hughes, published in 1910. Hughes did not agree with these reasons why women should not have the vote, but they were the most common reasons given by those opposed to suffrage.*

1. *It is unwomanly to vote.*
2. *Most good women — intelligent, domestic, godly mothers — are opposed to suffrage for women.*
3. *Politics will degrade women.*
4. *Wives might vote against their own husbands and thus destroy the harmony of the home.*
5. *Women are fairly represented by men. Their fathers and brothers vote.*
6. *Women as a sex have no wrongs which male legislators cannot be expected to redress.*

> 7. *The transfer of power from the military to the unmilitary sex involves a change in the character of a nation. It involves, in short, nothing less than national emasculation.*
> 8. *If a woman becomes a man, she must be prepared to resign her privilege as a woman. She cannot expect to have both privilege and equality.*
> 9. *Women are more nervous than men, and the excitement of elections would undermine their constitutions and tend to unbalance them.*
> 10. *Women must bear and nurse children. It is impossible that they should compete with men in occupations which demand complete devotion as well as superior strength of muscle and brain.*

Between 1916 and 1925 women were granted the right to vote in every province of Canada with the exception of Quebec. It was 1940 before Quebec women could vote in provincial elections. The federal government gave the right to vote to women who were related to members of the armed forces in 1917. By 1918 all women of voting age could vote in federal elections. Figure 5.1 shows the exact dates for political equality for women in Canada.

World War I (1914–1918) caused a shortage of workers in vital industries and many women were called upon to work in the factories. They proved equal to the demands made upon them, not only performing the same jobs as men, but performing them just as well as men. In most countries they won the right to vote shortly after the war.

Figure 5.1

Province	Suffrage	Eligible to Hold Office
Manitoba	Jan. 28, 1916	Jan. 28, 1916
Saskatchewan	Mar. 14, 1916	Mar. 14, 1916
Alberta	Apr. 19, 1916	Apr. 19, 1916
British Columbia	Apr. 5, 1917	Apr. 5, 1917
Ontario	Apr. 12, 1917	Apr. 24, 1919
Nova Scotia	Apr. 26, 1918	Apr. 26, 1918
New Brunswick	Apr. 17, 1919	Mar. 9, 1934
Prince Edward Island	May 3, 1922	May 3, 1922
Newfoundland	Apr. 13, 1925	Apr. 13, 1925
Quebec	Apr. 25, 1940	Apr. 25, 1940

Suffragettes were often portrayed as frustrated single women with nothing better to do than make trouble. Gradually, however, more Canadians became aware of the suffrage issue and the arguments supporting it. They also recognized that the stereotype of suffragettes was false when they came into contact with these tireless women.

Emily Stowe (1831–1903) was active in a number of areas. She earned a medical degree in 1876. Partly through her efforts, the University of Toronto began to admit women in 1886. She also worked to obtain the Married Women's Property Act, which gave women partial control over their own property for the first time, and she established the first suffrage organization in Canada called the Toronto Women's Literary Club.

Nellie McClung (1873–1951) is well known in Canada for her successful efforts to obtain voting rights for women in the provincial elections in Alberta, Saskatchewan, and Manitoba. She fought for equal pay for women who performed the same work as men, and was elected to the Alberta provincial legislature in 1921. In addition to her political work on behalf of Canadian women, Nellie McClung wrote sixteen books and was the mother of five children.

Emily Murphy (1868–1933) became the first female judge in Canada in 1916 and worked hard for the recognition of women's rights. In 1927, Emily Murphy, Nellie McClung, and three other women (Henrietta Louise Edwards, Louise McKinney, and Irene Parlby) called the Famous Five, asked the Supreme Court of Canada to rule on the question of whether the word "person" included women. At this time Canadian women were unable to sit in the Senate because they were not considered by law to be persons. In 1928, the Supreme Court ruled that under Canada's constitution, the British North America Act, women were not persons, and therefore not eligible to be appointed to the Senate. The five women disagreed with this ruling, and appealed to a higher court in Britain — the Judicial Committee of the Privy Council. This body, in 1929, overruled the Supreme Court decision and affirmed that Canadian women were indeed legal persons and thus eligible for appointment to the Senate.

To some Canadians living today, the accomplishments of these women may seem minor. At the time, however, they were major victories in the struggle for recognition of the legitimate rights of women. Women today have benefitted from the hard work and dedication of the suffragettes and other activists in the women's movement. To achieve their goals, they had to overcome centuries of socialization to increase public awareness of the injustices in the system.

PROGRESS CHECK

1. How could suffragettes be described as activists?
2. How did World War I affect women's rights in Canada?
3. How did Emily Stowe, Nellie McClung, and Emily Murphy contribute to the struggle for equality?
4. What did the "Famous Five" accomplish?

5.4 Canadian Pioneers in the Struggle for Equality

The following biographical sketches summarize the political accomplishments of a few of the courageous women who followed their vision of equality for women, and made themselves heard in Canadian society.

Irene Parlby (1878–1965) held the position of President of the United Farm Women of Alberta from 1915 to 1919. Two years later, in 1921, she was elected to the Alberta Legislature and was appointed to a cabinet position. She was one of the Famous Five who appealed the Supreme Court decision of 1928 which ruled that Canadian women were not legal persons. In 1930 she was the Canadian representative to the League of Nations.

Louise McKinney (1868–1933) was elected to the Alberta Legislature in 1917, and in so doing was the first woman to be elected to a provincial legislature. She was another member of the Famous Five who appealed the Supreme Court person ruling. Earlier in her career she had worked diligently for women's suffrage. She had also been a strong supporter of the temperance movement, a movement dedicated to ending the consumption of alcohol by bringing in prohibition.

Henrietta Louise Edwards (1849–1933) wrote books on the legal status of women and children in Canada. She fought for mother's allowance for Canadian women, and was a member of the Famous Five. Besides her political involvements she raised three children.

Agnes MacPhail (1890–1954) was the first woman to be elected to the House of Commons and held her seat from 1921 until 1940. Her life was devoted to improving the living and working conditions of ordinary Canadians, including workers, farmers, the blind, and the handicapped. In 1927 she introduced the old age pension for people over 70 years of age ($20 a month). She also worked at prison reform and better health care for Canadians. Being a pacifist, she abhored war and violence.

Cairine Wilson (1885–1962) in 1930, was the first woman to be appointed to the Canadian Senate. As Chairperson of Canada's National Committee on Refugees, she helped many European refugees to settle in Canada at the end of World War II. She was the first Canadian woman delegate to the United Nations, where she served on the International Save the Children Fund. The status of women and the plight of refugee children were two issues to which she devoted considerable time and effort. In addition to her involvement in social and political issues, she raised eight children.

Muriel Fergusson (1899–) graduated as a lawyer in New Brunswick in 1925, and became actively involved in provincial politics. Later she was appointed the first woman Probate Court Judge in New Brunswick and was the first woman from New Brunswick appointed to the Canadian Senate. In 1954 she helped New Brunswick women win the right to serve as jurors, and has supported the struggle for equal pay for work of equal value. Muriel Fergusson stressed that more women must get involved in politics if they are to achieve true equality with men.

Thérèse Casgrain (1896–1981) worked hard to obtain the provincial vote for women in Quebec, finally won in 1940. She also actively supported the rights of Quebec women to enter non-traditional professions such as law, and fought against a law that prevented women in the Quebec civil service from earning more than $1500 a year. In 1970, following an active career in Quebec politics, Thérèse Casgrain was appointed to the Canadian Senate. She had four children.

Pauline McGibbon (1910–) in 1974, was the first woman to be appointed Lieutenant Governor (Ontario) in Canada and in the British Commonwealth. She was also the first female member of the Canadian Club and the first woman to be appointed to serve as Chancellor of the Universities of Toronto and Guelph, and as Governor of

Upper Canada College. Throughout her career she has maintained that "competence has no sex" and she has actively supported equal treatment for women in all areas of Canadian society.

Bertha Wilson (1923–) began to study law at Dalhousie University in 1955 at the age of 32. At that time she was one of only ten females in a class of 200. Her legal career has included being appointed a judge of the Ontario Court of Appeal, and in 1982, being the first woman to be appointed a judge of the Supreme Court of Canada. As one of nine Supreme Court judges, she is serving on the highest court in Canada.

PROGRESS CHECK

Make a chronological list of the accomplishments mentioned in these short biographies.

5.5 Discrimination in the Work Force

Despite the efforts of feminists working to achieve equal rights and equal treatment for women in Canada, there is still a long way to go. Discrimination against women has been practised for so long it often goes unrecognized and it is difficult to eradicate or erase.

PAY EQUITY

One of the most obvious and yet enduring inequalities is that of earning power. In the past, women were financially dependent on men, usually their fathers or husbands. It was assumed that women entering the labour force were either doing so because their families were poverty stricken, or because they wished to earn "pin money" for luxuries. On the other hand, men were believed to be supporting a family, and should therefore earn higher wages. Those women who were supporting a family on their own income formed the poorest segment of society.

In this country most women are no longer financially dependent on men. They earn money to support themselves or their families, or they contribute to the support of their families. Figures for 1983 indicate that women constitute 41% of the work force. It is predicted that by the year 2000, the majority of women — 60 to 70% — will be working outside the home. The financial situation for working women has improved steadily in this century, but **pay equity**, which means paying men and women for equal work of equal value, is still a long way off. Some statistics follow to illustrate the situation.

- In 1985 more than 60% of poor Canadians were female.
- Statistics from the National Council of Welfare for 1985 showed that 9.4% of male-headed families lived in poverty whereas 42.0% of female-headed families lived in poverty.
- For every $1 earned by men, women earned:
  $.53 in 1911
  .60 in 1969
  .66 in 1986
- In December 1984 Statistics Canada reported that men earned an average of $11.93 an hour, while women earned an average of $8.84.
- In 1985, the Canadian Union of Public Employees reported that women with the same skills and education were paid $6 000 to $10 000 less per year than men.

In 1984, Statistics Canada reported that education greatly affects the wage gap between men and women. Women who have eight years or less of schooling earn 57% of what their male counterparts earn. Those women with university or college degrees earn 70% of what their male counterparts earn.

Employers cannot be expected to voluntarily give up the unequal system from which they benefit. It is therefore up to government to initiate changes in pay scales. In fact, many women earning low wages require government assistance in the form of welfare payments, the Guaranteed Income Supplement to the elderly, or subsidization of health care, child care and housing. If these women earned more money, they would be less likely to require assistance while they were part of the workforce and during retirement. The government's best interests therefore, would be served by implementing pay equity.

EQUITY IN HIRING AND PROMOTIONS

One of the reasons that women consistently earn less than men on average is that women are employed mainly in lower-level positions. In addition, the jobs they hold fall into a narrow range of occupations, that, for the most part, do not require the full use of their abilities and potential.

In 1984, the *Royal Commission on Equality in Employment*, headed by Judge Rosalie Abella, reported that 65% of Canadian women filled clerical, sales, and service jobs. These positions, which are among the lowest paid, are the same type as those occupied by Canadian women in 1901.

A study of 11 federally owned companies, conducted by Judge Abella, revealed that only 4% of the 1639 upper-middle managers were female. Female workers were concentrated in the lowest ranked and lowest paid positions in these government-owned companies.

In 1986, the *Canadian Business Magazine* reported that only 2 of Canada's top 500 business executives were women. In a survey of companies operating in Canada, only 1% of the 15 000 directors were women. The figures for Canada's chartered banks show that although 75% of banking employees are female, only 28% are in senior management positions.

Women continue to remain in the lower paying "pink collar" jobs, the so-called "velvet ghetto". Those who make it to management positions are often directed into areas that are considered female, such as public relations and human resources management. This stereotyping results from people considering women to be more sensitive and better able to relate to others.

Within the teaching profession, discrimination is evident also. Below are figures reported by Statistics Canada for 1982–83 for nine provinces (Quebec excluded), the Yukon, and the Northwest Territories.

	Teachers	*Heads of Departments*	*Principals*
Male	66 295	8300	9000
Female	98 309	2361	1394

In Ontario there has been a backward slide. In 1972 16.1% of elementary school principals were female; in 1982 only 12.5% were female. Students, observing this imbalance, perceive that men occupy roles of authority while women play supporting roles.

Women teachers tend to be concentrated in the earlier years of school. This concentration tends to support the stereotyped image of women being the natural care takers of young children. In 1981 the Superior Council of Education in Quebec reported as follows:

Percentage of female teachers
Pre-school   99
Elementary   89
High school   41
University   16

Despite the high percentage of female teachers in the elementary schools (89%), only 35% of the principals and vice-principals were females.

Although there is no doubt that women have been discriminated against in the workforce, this is not the only reason for their poor representation in positions of authority. Traditionally, women have borne the main responsibility for the home and childcare. Consequently many women choose jobs with set hours to enable them to carry on the dual role of businessperson and homemaker. Upper management jobs usually demand extra hours that would conflict with their family duties. As some working women would put it "I need a wife!"

A 1984 survey of Canadian career women commissioned by Isobel Bassett (*The Bassett Report: Career Success and Canadian Women*) points to several reasons for women's lower position in the workforce. Although discrimination is the reason most often cited, other factors relating to socialization are also important, as shown in Figure 5.2.

Factors Career Women Cite for Restricting Their Progress in the Workforce.

Sex discrimination	88%
The tendency to put family ahead of career	84%
Lack of assertiveness and self-confidence	75%
Lack of ambition	73%
A tendency to turn down promotions so as not to be more successful than their husbands	66%
Conservatism and fear of taking risks	66%
Lack of expertise in economic and business principles	64%
Inappropriate educational background for business success	45%
An inability to understand the nuances of teamwork as well as men do	38%

Figure 5.2

Discrimination and the Law

Laws and official government bodies exist in Canada to protect minority groups from discrimination. Each province has a Human Rights Code, and a Human Rights Commission that investigates complaints involving discrimination, punishes offenders, and compensates proven victims of discrimination. Federally, there are also the Canadian Human Rights Act which prohibits discrimination in employment and the federal Human Rights Commission operating in a similar manner to the provincial Human Rights Commissions. The Canadian Charter of Rights and Freedoms guarantees protection from discrimination.

The following are two excerpts from the Canadian Charter of Rights and Freedoms.

"Every individual is equal before and under the law and has the right to the equal protection and equal benefit of the law without discrmination and, in particular, without discrimination based on race, national or ethnic origin, colour, religion, sex, age or mental or physical ability."

Section 15

"... the rights and freedoms in [this Charter] are guaranteed equally to male and female persons."

Section 28

CASE STUDY

Three examples of complaints to Human Rights Commissions illustrate how Canadian society is slowly changing some of its attitudes.

In 1985, when Gladys Kickham applied for a position on the police force, she received the highest overall marks in the job competition. The hiring committee subsequently changed her marks and hired a male candidate instead. In 1986 $10 000 was awarded to Gladys Kickham by a Human Rights Commission in Prince Edward Island.

In 1986 complaints were received by the federal Human Rights Commission concerning sex discrimination by the Canadian Armed Forces. The Canadian Armed Forces has maintained that women are not physically or psychologically able to perform the duties of a combat soldier, a fighter pilot, or a crew member of a fighting ship, submarine, or tank. Georgina Brown of Winnipeg, a licensed commercial pilot, was one of several women who filed a complaint. She claimed that she was discriminated against on the basis of sex when she was not hired as a fighter pilot. Katherine McCrae of Toronto also claimed that, on the basis of sex, she was turned down for a helicopter squadron position, although she was qualified. As a result of these and other complaints the defence headquarters located in Ottawa has stated that by 1987 the Canadian navy and army will begin experimental operations to determine whether women will be admitted to combat roles.

These and other cases have attracted the attention of the media, and have made the Canadian public more aware that attitudes and practices are changing. Changes are observed in many areas of society as women move into jobs that have been traditionally held by males, such as managing banks, driving heavy machinery, and serving on police forces.

EQUITY AND MENTAL HEALTH

Besides the economic benefits that women would experience from pay equity and equal job opportunities, there are also psychological benefits. In order to have a sense of self-worth women need to know that all career doors are open, that they are not limited to jobs with low status, low pay and low satisfaction. They need to know that they will not be treated as inferiors.

An American study directed by doctors from Columbia's College of Physicians and Surgeons in 1954 concluded that 21% of women in their forties were psychologically impaired. The comparable figure for men was 9%. When the study was redone in 1974, the change was dramatic. The figure for women in their forties suffering psychological impairment was down to 8% while men remained the same at 9%. Betty Friedan, American feminist and author, concludes that "women who once would have suffered despair and the feeling that their life was over as they hit forty were finding a new sense of self-worth and opportunities for growth." The two doctors who conducted the study, Drs. Srole and

Fischer, explained "Equality in general is good for mental health. Getting out of poverty, out of dependence, having some control over your own life, some measure of autonomy, independence and mastery of your life is good for people. That's the basic change for women."

PROGRESS CHECK

1. What are some common misconceptions about women who work outside the home?
2. What statistics support the argument that women are not being treated fairly in Canadian society?
3. Why should the government implement pay equity?
4. What are "pink collar jobs"?
5. Why are there so few women in positions of authority?
6. Show how specific laws seek to protect Canadian women.
7. Why is it important to publicize cases involving discrimination?

CASE STUDY

The Use of Sexist Language

Many people have become aware of and are concerned about **sexist** *language. Sexist means discriminating against people because of their gender. Sexist language leads to stereotyping males and females. One way in which language contributes to stereotyping occurs through the use of masculine pronouns — he, him, and his — when supposedly referring to either sex. The message most people pick up from reading or hearing masculine pronouns is that only males are involved.*

Many words that describe occupations and titles contain the word "man", for example, fireman, mailman, policeman, chairman, and businessman. The implication is that only men can perform these jobs.

Some titles are also sexist in nature. Traditionally females have had to reveal their marital status by using the title "Miss" or "Mrs.", whereas all males use the title "Mr." In recent years, many women have adopted the title "Ms" to replace Miss and Mrs. Some government offices and businesses have gone one step further, eliminating titles entirely, and using names only in correspondence.

Another form of sexism in language is found in children's readers, textbooks, and literature. One survey, conducted in 1960, of all third grade readers published since 1930 found that 73% of the stories featured male characters. Where girls or women did appear in the readers, they were portrayed as being timid, inactive, unambitious, uncreative, and morally and intellectually inferior to males. The same stereotypes were found in later surveys.

Publishers of textbooks in recent years have been trying to eliminate sexual bias from their books. They have been using both male and female pronouns, or using the plural pronouns "they", "them", and "their".

Here is a list of some words that are considered sexist, along with the non-sexist terms that can replace them. See how many more you can add to the list.

Sexist term	*Non-sexist term*
fireman	firefighter
showman	performer
cleaning lady	cleaner
office girl	office helper
insurance man	insurance agent or representative
salesman	salesperson or sales representative
policeman	police officer
man hours	labour hours
gentleman's agreement	honourable agreement
forefathers	ancestors
manhole cover	access cover
businessman	business person
chairman	chair or chairperson
anchorman	anchor or anchorperson
housewife	homemaker
man made	synthetic or machine made
stewardess/steward	flight attendant
actress	actor
poetess	poet
saleslady	sales person
bachelor's degree	undergraduate degree

1. How does sexist language reinforce stereotypes?
2. Do you believe there is a need to change words that are considered to be sexist in nature?

5.6 The Socialization of Women

GENDER ROLES — BIOLOGY OR SOCIALIZATION

In 1973 the *Royal Commission on the Status of Women* reported that Canadian women tended to be "dependent, passive, lacking in self-confidence, and often frustrated". On the other hand, men are usually characterized as being independent, aggressive, active, and self-confident. Are these differences innate, or are they learned? Does socialization account for the different traits of men and women, and the different roles they play in society, or is their a biological basis to them?

There are different schools of thought with regard to the importance of socialization in the adoption of female and male roles.

Some social scientists feel that male and female behavioural differences are innate, and caused by hormones. **Hormones** are chemicals that regulate the various processes of the body; they are sometimes called chemical messengers. According to some social scientists, the hormones produced prenatally (before birth) by the sex organs of the **fetus**, or unborn child, affect brain patterns and the later behaviours of each sex. There are two kinds of sex hormones, **androgens** or male hormones, and **estrogens** or female hormones. Both androgens and estrogens are produced by males and females, but in different proportions. According to this school of thought, these hormones are responsible for making males more aggressive and females more passive and caring. Some researchers go so far as to claim that hormones are responsible for the patriarchal organization of most societies.

Other social scientists argue that both prenatal sex hormones and socialization play a part in male-female behaviour. Prenatal sex hormone levels may direct a child toward aggressive or passive traits, but socialization will determine whether these tendencies will develop into male and female traits.

The balance of social scientists believe that male and female behaviours are totally learned, and that prenatal hormones do not play a part. They feel that children are raised to conform to the sex stereotypes their culture upholds.

Early Socialization

Socialization in our culture begins at birth. When girls are wrapped in a pink blanket and boys in a blue one, their parents' expectations for them are based on their sex, and they are treated differently.

The reinforcement of sex-appropriate behaviour by parents tends to ensure that boys will learn to be aggressive, independent, active, and adventuresome; while girls learn to be passive, dependent, verbal, and social. Boys are more often given complicated and challenging games and toys, such as building sets, which develop mental skills. Other toys that are typically given to boys — trains, boats, airplanes, cars, trucks, and heavy equipment — stimulate imaginary adventures in the outside world. Girls, on the other hand, are often given dolls, baking sets, and tea sets with which to "play house". They are expected to stay closer to home, to stay neater and cleaner than boys, and to look "pretty".

Advertising helps to reinforce these stereotypes. Toy doctor kits picture boys using them, while nurse's kits picture girls. Games and activity toys often picture boys using them and girls standing by watching and admiring. Boys are the active ones in commercials who return from a game filthy dirty, requiring mothers to use a superior detergent to get their team uniforms clean.

Girls have also been encouraged to play games such as hopscotch and skipping — fairly non-competitive games.

In picture books and story books, girls are often underrepresented, and where they appear, their roles are often insignificant and stereotyped. Mostly they appear in passive roles, watching and helping. Mothers are mostly confined to the house, cooking, cleaning, and taking care of the baby. The message little girls can receive from their reading is that there are few opportunities open to them besides being wives and mothers and housekeepers.

In the past, women were not expected to have careers outside the home. Caring for their husbands and children was supposed to be sufficient fulfilment. A successfully socialized woman was seen as being satisfied with her home-centred role as wife and mother, and content to be dependent on her husband. Studies, however, have shown that some women who were not employed outside the home felt less happy and less fulfilled than women who were employed. The role of the homemaker was too limiting for those women, and did not allow them to express their talents, their intelligence, or their creativity.

With such a large proportion of women now in the work force, society is changing, and the female role is also in the process of changing. The way in which girls were socialized in the past is no longer valid in today's society. It is not realistic for girls to be socialized to be dependent and passive, or to see marriage and children as their only lifelong goals. Statistics on working women, poverty, and divorce point to the need for girls to be socialized to develop and have confidence in their abilities, and to train for an occupation or career in addition to planning for a family.

Recent studies have shown that girls still have unrealistic expectations and views about their future roles and their place in society. A report released in 1985 by the Canadian Advisory Council on the Status of Women, revealed that the majority of a group of 122 girls between the ages of 15 and 19 believed that they would be happy for the rest of their lives. They did not expect to experience divorce, poverty, or unemployment. They were ignorant of the statistics that show that one in five Canadian women lives in poverty, that 40% of marriages end in separation or divorce, and that 88% of them will have to provide for themselves at some point in their lives. They were unaware the female workers are concentrated in low-paying, low-status jobs.

It is not uncommon for women to be forced into a homeless state.

This study also revealed that the younger girls had higher career ambitions than the older girls. Older female teenagers tended to lower their goals and education ambitions, and to conform to female stereotypes in which competition and success outside the home were undesirable. The report also pointed out that girls who had the lowest ambition levels had mothers who were either unskilled workers with low paying jobs, or homemakers. These girls expressed the desire to marry a good provider and have children at an early age. Those girls who had mothers with satisfying and well-paid careers outside the home, were more ambitious and self-reliant.

Another recent study carried out by the Superior Council of Education in Quebec supported the findings of the Canadian Advisory Council. The results of this study showed that many young women have a **Cinderella complex**, the expectation that they will marry financially successful men who will comfortably support them and their children. Many of these girls lacked ambition and confidence in their abilities. They were likely to avoid courses in mathematics and science.

A 1982 survey of female highschool graduates in Alberta found 76% had no plans for future careers. Furthermore, they too were ignorant of the statistics regarding the high percentage of women in the workforce.

The women's movement has been trying to expand girls' horizons, and make being a female a more desirable role to play. Many educators are now trying to break down the stereotypes that cause girls to shun academic success. Some parents are now encouraging their daughters to train for non-traditional careers, to aim to be doctors instead of nurses, and engineers instead of secretaries.

Sociologist, Lenore Weitzman, sums up the problem in the following way: "As long as women are denied real career options, it is realistic for them not to put all their energy into occupational goals. As long as they lack role models of successful career women, are denied structural supports to aid a career, are told they are neurotic or unfeminine if they are dedicated to an occupation, we cannot expect large numbers of young women to aspire to professional careers."

Concern for the role women will play in the future is worldwide. The year 1975 was designated International Women's Year by the General Assembly of the United Nations, with the aim of "defining a society in which women would participate fully in economic, social, and political life." The period 1976 to 1985 was designated the Decade for Women: Equality, Development, and Peace. A *World Plan of Action* was adopted to make progress in these areas.

The objective of the *Plan* was to outline what should be done to improve the status of women and to stimulate action at all levels of government throughout the world to ensure equal opportunities and rights. The guidelines developed by this international body covered such areas as international cooperation and participation on international forums, family planning, women's rights within the family, education and training, employment and economic roles, political participation, health and nutrition, housing, the special needs of elderly women, and female criminal offenders.

In the area of education and training, the *World Plan of Action* guidelines state:

- Programs, curricula and standards of training should be the same for boys and girls.
- Texts and teaching materials should be rewritten to reflect an image of women in positive and participatory roles.
- Research should be undertaken to identify discriminatory practices in education and training to ensure educational quality.

With regard to employment, the World Plan of Action Guidelines urges that:

- Special efforts are needed to encourage positive attitudes among employers, workers, men, and women in society and to eliminate obstacles based on sex-typed divisions of labour.

Awareness of how traditional socialization of girls puts them at a disadvantage in today's world is growing internationally. In Canada old stereotypes are gradually being discarded. Many girls now are being encouraged to pursue an education that will prepare them for a rewarding career. More and more women are overcoming the traditional expectations of society and enjoying the satisfaction of using their abilities to the fullest. They act as role models to encourage more girls to live up to their potential.

Progress Check

1. Explain the controversy surrounding the effects of hormones on male and female behaviours and potentials. Which theory do you most agree with and why?
2. How are boys and girls socialized to learn "appropriate gender roles"?
3. How do advertising and books reinforce sex stereotypes?
4. What did the 1985 Canadian Advisory Council on the Status of Women reveal about the attitudes of Canadian teenagers? What criticisms can be made of these attitudes?
5. What criticisms are made about the attitudes of some young women who graduate from high school?
6. How can the socialization of young women be improved?

5.7 The Canadian Working Woman

During World War I (1914–1918) and World War II (1939–1945), it was acceptable for married women to work. Wartime child care centres and training programs were set up to enable women to make the transition from home to the workplace. Hundreds of thousands of Canadian women worked to aid the war effort at non-traditional jobs — in factories producing war materials, driving buses and trucks, ferrying bombers across the Atlantic.

With the end of the wars the Canadian government passed laws regulating women's participation in the workforce. Those married working women who were not self-supporting were not allowed to hold government jobs unless no men were available to fill them. Married women generally were expected to give up their jobs to make room for the soldiers returning from overseas. Working women were portrayed by the media in a very unflattering light, and homemakers were held up as models of "real" women. Times and attitudes have changed since then. The results of a Gallup Poll shown below illustrate how people are becoming more accepting of mothers working outside the home:

> Percentage of Canadians who believe that women with young children should not work outside the home:
> 1960   93%
> 1982   52%

A study was conducted by Professor Monica Boyd of Carleton University, called *Canadian Attitudes Toward Women: Thirty Years of Change*. She studied 43 Gallup Polls of Canadian attitudes between 1953 and 1983, and found that,

- by the early 80's fewer Canadians viewed the man as the authority figure in the family, they saw a greater equality between marriage partners.
- in 1971, 58% of Canadians believed that women could run businesses as well as men — by 1985 this percentage had risen to 83%.

From her study results, Professor Boyd concluded that the Canadian public's confidence in women's abilities is increasing.

Changing attitudes toward working women are also reflected in the laws that have been passed to protect their rights as persons. In 1970, pregnant women were granted a 17-week maternity leave without pay, and the right to return to their jobs at the end of their leave. (In Sweden women have a nine-month maternity leave with pay, and in Finland, a year's leave with pay.)

Provinces like Ontario have recently (1974) passed laws which guarantee women the right to pay equity. Now men and women by law must be paid the same wages for performing the same jobs.

In 1981 Canada signed the United Nations Convention on the Elimination of All Forms of Discrimination Against Women, and agreed to work toward ending inequalities between men and women. In that same year, Canadian women joined together to put pressure on the government to include Clause 28 in the Canadian Charter of Rights and Freedoms to guarantee the "rights and freedoms ... equally to male and female persons..." This protection is now part of the Constitution Act, 1982.

Women's groups are currently working to guarantee the principle of pay equity — equal pay for work of equal value. This would mean, for example, that secretaries (usually women) would be paid the same wages as shippers (usually men) if their jobs are judged to be of equal value to the employer. Pay equity legislation for **civil servants**, or government employees, has been passed by the federal government, and the provincial governments of Quebec and Manitoba.

In Ontario, pay equity legislation was passed in 1987 that will apply to 2 000 000 working women. This law will require most employers with ten or more workers to introduce pay equity programs. Jobs will be ranked on the effort required to perform them, the skills demanded, the level of responsibility required, and the working conditions. Employers refusing to obey the new law will face fines of between $2 000 and $25 000. When this new law was proposed, many employers angrily objected and threatened to stop hiring women to avoid the equity issue. Most people, however, have expressed approval of the new law, which they see as an important step in creating a fair situation for working women.

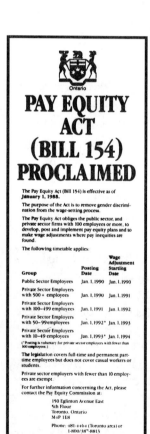

In her 1984 *Royal Commission Report*, Judge Abella recommended that all employers be required by law to introduce employment and pay equity. She believes that governments must lead the way in implementing these changes, and that companies which refuse to introduce equity programs should be denied government contracts. She stated that Canadian society must learn to take female workers seriously, and grant them the same opportunities, respect, and treatment that other workers receive.

Employment equity is the policy of hiring members of minority groups into jobs previously closed to them, provided they have the qualifications. They will then serve as role models for other members of their group. Their encouragement will help to end **gatekeeping** where dominant groups control certain jobs and only permit members of the group to be hired. Employment equity would apply to women, native peoples, the disabled, and members of ethnic groups.

Women can be as technically competent as men given the same training.

Iraqi women learn carpentry in Baghdad. Iraq is one of the most liberal of the Middle East countries with respect to women's rights.

PROGRESS CHECK

1. Why did the Canadian Government pass laws at the end of the two World Wars to prevent married women from working if they were not self-supporting?
2. How have the attitudes of Canadians changed towards women?
3. What laws have been passed to protect Canadian women who work outside the home?
4. What improvements have been suggested to provide additional protection to working women?

CAREER PROFILE
ROSALIE ABELLA

Rosalie Abella began her career as a lawyer in Ontario and she was appointed a Family Court judge in 1976. The federal government, in 1983, asked her to head a Royal Commission to investigate employment equity in eleven Crown Corporations. Her *Royal Commission Report on Equality in Employment* was released in 1984, and its contents surprised many people. She revealed widespread discrimination and inequality in employment among visible minority groups including women, native peoples, and the disabled. Members of minority groups were found concentrated in low-paying, low-status jobs, and were found to experience higher unemployment rates than other Canadians.

To reverse the discrimination, Judge Abella recommended that employment equity be introduced by the government in both the private and the public sectors of the Canadian economy. Included in employment equity are pay equity, government-funded education and training for minority groups, promotions for minorities, access to daycare for working parents, fair pensions, and paid leave for the parents of young children who choose to take care of them in the home. Judge Abella urged the government to work toward achieving employment equity and toward ending discrimination.

Her report has created a new awareness of the unfair treatment of minorities in Canada. As a result, governments and some employers are beginning to implement some of her recommendations.

Rosalie Abella is amazingly energetic. She is the mother of two boys and has been married for 18 years. Her "spare time" is devoted to volunteer work in the arts community, the University of Toronto, the Jewish community (her parents were survivors of the Holocaust), and the legal community. She is the author of four books and 22 articles, and regularly gives speeches to groups in Canada, the U.S.A., and Europe on social issues such as family law, women's rights, and human rights. Presently she is Chairperson of the Ontario Labour Relations Board.

OVERVIEW

In the past one hundred years there has been a slow but steady change in attitude toward Canadian women. People have become increasingly aware of the prejudice and discrimination against women that exist in Canadian society. Traditional attitudes toward women no longer apply in a society that has changed so dramatically. There is now more support than ever before for the fair treatment of women who work both inside and outside the home, and more recognition of their efforts. The principles of equal pay, treatment, and opportunities for everyone are now accepted by more Canadians.

Laws have been passed and government departments have been set up to make and enforce guidelines to protect women and other minority groups from discrimination. But old attitudes and employment practices are slow to change, and prejudice and discrimination still exist throughout Canadian society. Many employers and employees still hold their stereotyped images of women, and do not accept them as equals or potential authority figures. Often these people are unaware of their biased attitudes.

By increasing public awareness of discrimination and stereotyping, and by making and enforcing laws to prevent discrimination, it is hoped that Canadian society will continue to make progress. Ideally the problems and inequalities that exist now will be dealt with and everyone will receive fair treatment in Canadian society. Many men and women feel threatened by the changes in attitudes toward both sexes that are taking place at all levels of society.

KEY WORDS

Define the following terms and use each in a sentence that shows its meaning.

- minority groups
- prejudice
- stereotype
- scapegoat
- anti-Semitism
- feminist
- dominant group
- subordinate group
- discrimination
- activist
- patriarchy
- matriarchy
- suffragettes
- pay equity
- sexist
- hormones
- fetus
- androgens
- estrogens
- Cinderella complex
- paternal
- civil servants
- employment equity
- gatekeeping

WOMEN AND EQUALITY 167

KEY PERSONALITIES

Give at least one reason for learning more about each of the following.

Emily Stowe
Nellie McClung
Emily Murphy
Irene Parlby
Louise McKinney
Henrietta L. Edwards
Agnes McPhail
Cairine Wilson

Muriel Fergusson
Thérèse Casgrain
Pauline McGibbon
Bertha Wilson
Rosalie Abella
Isobel Bassett
Lenore Weitzman
Monica Boyd

FOCUS AND ORGANIZE

1. What questions can we ask about prejudice, stereotypes, and discrimination?
2. What are the five most important causes of discrimination against women? List them in their order of importance.
3. What are the five most important events/recommendations that have occurred or been made that will bring about equality for women? List them in the order of their importance.
4. Create an organizer that lists examples of prejudice, stereotypes, and discrimination.

LOCATE AND RECORD

1. Using the concept of stereotyping make a list of at least five stereotypes associated with women and men. What conclusions can you make about these stereotypes? Are they fair or accurate?
2. Make two lists. One shows the "traditional" female occupations and the second shows the "traditional" male occupations. To what extent have females entered "male" occupations. Use information from the text, newspaper articles (ask for vertical files in library), and books.
3. By referring to the textbook and its biographies and documents, make a list of examples of discrimination against women in Canada. Indicate who was doing the discriminating and the time period when it occurred. After completing the chart indicate whether these types of discrimination still take place today. If not, what reasons can you give for their elimination?

4. Personal Experience Report: Make a list of any examples of prejudice, stereotyping, or discrimination that you have personally experienced at school, at work, or elsewhere because of your gender. (Both males and females can be targets of these negative feelings, labels, and actions.) Make another list that records experiences in which you yourself engaged in these activities. Relate your experiences to the class.

EVALUATE AND ASSESS

Toronto Star, April 10, 1982

GUIDELINES ON AD STEREOTYPING

The Canadian Advertising Advisory Board's committee on sex-role stereotyping has established guidelines for "more realistic portrayals of men and women in advertising." If you think an ad contravenes one of them, the CAAB wants to know about it. According to the guidelines, advertising should:

- Recognize the changing roles of men and women and reflect a broad range of occupations for both.
- Reflect contemporary family structure, showing men, women, and children pitching in equally in household tasks.
- Portray men and women of various ages, backgrounds and appearances actively pursuing a wide range of interests sports, and hobbies.
- Show men and women as equally capable, resourceful, self-confident, intelligent, imaginative, and independent.
- Not exploit women or men purely for attention-getting purposes: their presence should be relevant to the advertised product.
- Use non-sexist language, as in working hours rather than man-hours, synthetic rather than man-made, business executives rather than business men or women.
- Portray both men and women as users, buyers, and decision-makers, both for "big-ticket" items and major services, as well as small items.
- Reflect a realistic balance in the use of women as experts and authorities.

1. By watching television, listening to the radio, and looking at advertisements in newspapers, magazines, and on signboards, assess whether ads used in these types of media are following the Canadian Advertising Advisory Board's guidelines on sex role stereotyping. Why is it important to ensure that ads follow these guidelines? Write a

report with your conclusions by making specific references to ads you have seen.
2. Order the following films and find out more about these topics.

Suffragists: After a Century   produced by the Federation of Women Teachers. A discussion of the progress of the women's movement by such famous Canadian personalities as Rosalie Abella and Laura Sabia.

The Impossible Dream   produced by United Nations / Kratky Films. An animated film about the double workload of a woman performing a full-time job and being a homemaker.

Killing Us Softly: Advertising's Image of Women produced by Cambridge Documentary Films. Explores the image of women as presented by modern advertising. Write a one-page report on one of these films.

SYNTHESIZE AND CONCLUDE

1. Examine the following chart, and give three possible explanations for Canada's low ranking.

Country	Percentage of Men's Wages Earned by Women
Italy	86%
Denmark	86%
Sweden	83%
France	78%
U.S.A.	64%
Canada	64%

2. Examine the following Gallup Poll results from 1985.

Why would people fifty years and over be less likely to say women should have an equal chance for jobs?
Why would people with less education be less willing to give women an equal chance for jobs?
Why did 77% of those surveyed say women should have an equal chance while in 1960 only 23% agreed?

GALLUP POLL RESULTS

"Do you think married women should be given equal opportunity with men to compete for jobs, or do you think employers should give men the first chance?"

	Equal Chance With Men	Give Men First Chance	Qualified	Can't Say
National				
Today	77%	18%	3%	2%
1960	23	70	5	2
By Age - Today				
18-29	84	14	2	-
30-49	81	15	2	2
50+	65	26	5	5
By Sex - Today				
Men	76	19	3	2
Women	77	17	3	3
By Education - Today				
Public School	62	29	2	7
High School	77	18	3	2
University	89	7	3	1

3. Examine the following table compiled by Statistics Canada. Percentage Distribution of Earners By Earnings Groups and Gender — 1982.

Full-time Workers	% Males	% Females
Under - $1 000	1.0	1.2
$1 000 - 1 999	0.7	1.2
2 000 - 3 999	1.4	2.1
4 000 - 5 999	2.2	3.5
6 000 - 7 999	2.1	5.7
8 000 - 9 999	3.1	8.6
10 000 - 11 999	3.6	9.2
12 000 - 14 999	7.0	18.2
15 000 - 19 999	16.2	24.1
20 000 - 24 999	17.7	12.5
25 000 - 29 999	15.4	7.8
30 000 and over	29.7	6.0
Average earnings	$25 096	$16 056
Median earnings	$23 608	$15 075

What is the overall trend in these statistics? Why is this the case? In which income areas are a majority of women located?

Why might this be the case? Research the percentage distribution of earners for the past year by examining census records or other statistical information. How do the most recent figures compare to those in 1982?

4. Examine the following table compiled by Statistics Canada. In which area of the workforce have women increased the most? Why did the number of women entering the workforce increase so sharply between 1971 and 1981? Since more women entered the workforce, what social changes have occurred?

Leading Female Occupations, 1981*				
	Females	Males	Percentage of Males	% Increase in Female Employment, 1971-1981
Secretaries & Stenographers	368.	4.	1.1	53.5
Bookkeepers & Accounting clerks	332.3	73.5	18.1	143.6
Tellers & Cashiers	229.3	18.2	7.4	121.7
Servers/food & beverages	200.7	33.5	14.3	90.4
Graduate nurses	167.7	6.1	4.6	67.3
Elementary teachers	139.6	34.1	19.6	16.2
General office clerks	115.	27.8	19.5	45.
Typists & Clerk/typists	103.	2.2	2.1	21.3
Cleaners	96.7	138.2	58.8	76.5
Sewing machine operators	93.	5.1	5.2	62.2

* for occupations with 50 000 or more female workers

5. Cartoon Analysis: What points does this cartoon make?

"I tried doing laundry once. My voice got higher and I wanted to have babies."

APPLY

1. Divide the class into groups of four to six students. Examine the chart on page 149 which lists the factors that women cite for their lack of progress in the workforce. Discuss the reasons given by analyzing why these situations occur. For example, why is there "sex discrimination" in the work force? Develop an organizer which lists the conclusions reached by your group. In a separate column speculate how these restrictions on women could be reduced or eliminated.

COMMUNICATE

1. Research four of the following Canadian personalities and present your findings to the class. (1-2 pages)
Charlotte Whitton, Margaret Birch, Rosemary Brown, Iona Campagnola, Ellen Fairclough, Margaret Scrivener, Claire Kirkland-Casgrain, Judy Lamarsh, Flora MacDonald, Jeanne Sauve, Rejane Laberge-Colas, Karen Kain, Maureen Forrester, Anne Murray, Buffy Sainte Marie, Joni Mitchell, Carole Pope, Sylvia Tyson, Barbara Frum, Genevieve Bujold, Betty Kennedy, Barbara Ann Scott, Karen Magnussen, Petra Burke, Barbara Underhill, Gerry Sorenson, Laurie Graham, Lisa Savijarvi, Debbie Brill, Carling Bassett, Sylvie Bernier, Lori Fung, Linda Thom, Anne Ottenbrite, Marie Dressler, Gabrielle Roy, Margaret Laurence, Margaret Atwood, Gwendolyn MacEwan, Emily Carr, Maryon Kantaroff, Barbara Hamilton, Ethel Catherwood, Fanny Rosenthal.
Use the following questions to guide your research. Who are they? What did they do? When did they do it? What recognition have they received? How have they been role models for other Canadian women?
After all the presentations have been completed make a list of headings into which these women fit. For example: medical, entertainment, writing, law, politics, sports, art.

2. A position paper is a strong statement of your opinions about a topic based on selected facts to support your arguments. With a partner, pick one of the following statements. Once you have chosen a topic decide with your partner which side of the topic you will argue and which side your partner will argue. Once you have decid-

ed, write a position paper (1 page) which supports your view on the topic. For example, if you chose the second topic, "It is impossible to reduce or eliminate prejudice", you may show how and why this assertion is true, while your partner will show why it is not true and how prejudice can be reduced and eliminated. You will base your opinions on the materials in this chapter and on magazine and newspaper articles and books. Read your position paper to your partner, and then listen to your partner's paper. While the paper is being read, take notes and respond with challenges to your partner's arguments.

- Most people in Canada are not prejudiced toward minority groups.
- It is impossible to reduce or eliminate prejudice.
- The traditional socialization of women was unfair to them and to the societies in which they lived.
- Canadian women will achieve true equality with men in the near future.
- Male and female behaviour is learned and is not innate.
- Men and women have the same innate mental abilities and potentials.
- The present socialization process of people is determined by sex and has to be changed.
- The present socialization of Canadian women does not prepare them adequately for a career.
- Canadian society discriminates against working women and this must be changed and corrected.
- The feminist movement in Canada is helping to liberate both males and females from traditional sex role stereotypes.

Once the mini-presentations are completed, the class can be broken up into two sides and the topics can be further debated.

3. Write a formal essay of at least 1 000 words on the subject of discrimination against Canadian women in the workplace. Your essay should refer to specific personalities and data recorded in the text — outside sources can also be used to add to your ideas and facts. Remember to paraphrase or put the information down in your own words. You must not plagiarize and copy down someone else's work without recognizing through footnotes that it came from that source. The format or layout of your essay

should begin with an introduction which introduces the subject and gives an overview of the material you will present. The main part or body of the essay will contain the ideas and information that support your premise or hypothesis. Remember that separate paragraphs must be used when there is a change of direction in the essay, or when you go from one idea to another. For example, if you are referring to the *Abella Report* and you complete your point and move on to the *Bassett Report* to support another idea, the *Bassett Report* should be contained in a new paragraph. Finally, your conclusion should wrap up the general arguments and points of view in the essay to reinforce and summarize the focus or direction that your essay has taken.

CHAPTER

6

Aboriginal and Ethnic Groups

What is racism? prejudice? How have Canada's aboriginal peoples been treated? What does it mean to be part of a multicultural country?

INTRODUCTION   When you walk down a busy city street, do you notice the differences between people, or their similarities? Sometimes, in a crowd of strangers, you may feel quite separate and alone, and notice only the external differences between yourself and others. Perhaps you do not look like those around you, or perhaps you speak a different language. At other times, especially in smaller groups of people known to you, you may feel very close to the other people. Usually we feel closest to those with whom we share the bond of family ties or common culture.

Just as you may feel threatened and alone in the centre of an urban crowd, people raised in a different culture may feel threatened when surrounded by people unlike themselves. In the same way that you feel close to family or others who share cultural bonds with you, people from other cultures feel strong bonds with one another.

The differences among people should make life richer and more interesting. It is regrettable that, instead, these differences have often led to misunderstanding, prejudice, discrimination, and even violence. The strong bonds that make people feel they belong to a group, in some cases, have led to an exclusion and rejection of others who act or appear different.

In this chapter, we will examine racism, how it develops, and the forms it takes. We will look at how ethnic groups in some countries, like Canada, have had to struggle to be accepted. We will review the background to Canada's becoming a multicultural society and the responsibilities we share as members of that multicultural society.

6.1 *Racism and Prejudice*

THE MYTH OF RACE

You have learned how, over thousands of years, people have developed certain physical characteristics that have helped them to survive. These characteristics have been passed on to their offspring. Eventually this process, called natural selection, results in a population that has some common physical characteristics.

In the past, social scientists attempted to classify human groups. They created systems into which they could place different human physical characteristics in some sort of order. The most common types of physical characteristics that they used in their classification systems were skin colour, hair form and colour, eye shape and colour, head shape, and body height. The main groups they called races, and these were split into subraces. The number of races and subraces varied depending on whose classification system was being used. One system identified three main groups — the Negroids, the Mongoloids, and the Caucasoids. Other investigators added the Australian Aborigines and the American Indians. Still others favoured many more races and subraces.

Recent advances in genetics have shown that dividing the human species into races has no biological basis. The genetic traits that scientists thought distinguished one race from another are rarely found to be carried by one group of people exclusively. The genes responsible for so-called racial characteristics have combined and recombined over time, and it is apparent that there is a scale of differences throughout the world. For example, in the Old World, skin colour follows a trend of darkest, in areas of most intense ultraviolet radiation, to lightest in northern Europe (Scandinavia). Similarly, the frequencies of the different blood groups gradually change from Asia to Europe and into Africa.

On the basis of these observations, scientists began thinking in terms of overlapping and interbreeding populations of people. They recognize that humans are one species, that they all come from one common stock, and that they have, over the years, intermarried and migrated from area to area. Trying to separate them into races is therefore scientifically unfounded and useless.

THE REALITY OF RACISM

Attempts to classify humans into different races have led to the notion that certain races are superior to others. This notion of superiority/inferiority has given some groups of people the justification for discrimination, oppression, and extermination of other groups.

The tendency for some people to believe that their own particular group is superior and other groups are inferior is called **ethnocentrism**. People are likely to judge others by the standards of their own group. Often their judgments are negative if they do not understand the different beliefs, norms, and customs of the other group. The lack of understanding may result in distrust of the other group.

Racism is one example of ethnocentrism. It is based on the belief that the human species can be divided into major groups determined genetically rather than socially. It is the idea that one's own group (race) is superior to another group.

Another example of ethnocentrism is **ethnic prejudice** or having negative or unfounded feelings and opinions about members of other ethnic or cultural groups. Racism and ethnic prejudice are especially common in large cities where it is noticeable that immigration or shifts in population have changed the composition of the population.

During periods of economic uncertainty and decline, incidents of racial discrimination, hostility, and even violence tend to increase because some people single out certain groups as scapegoats for the economic problems.

There are also groups that are dedicated to promoting racist ideas and encouraging racial unrest. These extremist groups are especially active in urban areas where they may distribute hate literature, attempt to recruit members, and intimidate and harass members of target groups. One example of such an extremist group is the Ku Klux Klan.

Members of the KKK hold a cross burning ceremony in Alabama before President Carter's visit as part of his 1980 presidential campaign.

PROGRESS CHECK

1. How did social scientists in the past classify human groups? Why have they abandoned this practice?
2. Why is it stated that members of certain groups tend to feel superior to members of other groups? Do you agree?
3. Under what circumstances does discrimination against certain minority groups increase? Why does this happen?

CASE STUDY

Apartheid in South Africa

An extreme example of racism existing today is found in South Africa. The South African government, which represents five million white people, denies twenty-four million black people their basic human rights. The white people control the armed forces, the economy, and the government.

In South Africa, the blacks are kept separate from the whites, a form of social control called **segregation**. The particular type of segregatation officially practised in South Africa is called **apartheid**. The black and white populations are segregated from each other in housing, recreation, schools, health care, and jobs. There are neighbourhoods, schools, beaches, hotels, and restaurants that are reserved for white people only. Other facilities, generally inferior, are provided for blacks only.

The blacks, who make up 72% of South Africa's population, are not permitted to vote in elections and have no representation in the government. The white minority and their representatives rule the country and make laws which favour the white citizens of South Africa. Whites are generally middle class and occupy management positions. Blacks are usually unskilled, poorly paid and easily replaced in the workforce.

It is estimated that there is a shortage of 832 000 homes for South Africa's blacks and that 5 900 000 blacks live in 466 000 small housing units — an average of 13 people per house. There is an estimated surplus of 37 000 white homes.

More than half of the blacks are forced to live in ten "homelands" created for them by the white South African government. These homelands account for 13% of the total land area in South Africa, and they consist of desert-like lands poor in industries and resources. The low standards of living and health care in the homelands are reflected in the 20% mortality rate of black children under one year of age and the fact that 43% of black children in South Africa suffer from malnutrition.

Sores on the child are caused by malnutrition.

Black children attend different schools from white children. Classrooms in black schools are overcrowded and understaffed and this affects the quality of education black students receive. The South African government spends an annual average of $73.45 on education for each black child, while $1 106.00 is spent to educate each white child.

The South African government spends about $4 billion each year on its police forces to enforce the apartheid system. Those blacks who dare to criticize the unfair treatment they receive are arrested and imprisoned — usually without trial.

The African National Congress (ANC) is a political group organized in 1912 by four black African lawyers. The ANC was formed to unite the various African groups in their struggle against white domination. In 1960 the ANC was declared illegal by the South African government and the leaders were arrested, one of them being Nelson Mandela. He was freed the following year and went underground until he was arrested again in 1962. While still in jail he was convicted of treason in 1964 and sentenced to life in prison. At the trial, Mandela, a lawyer himself, conducted his own defence. He ended with the statement "During my lifetime I have dedicated myself to this struggle of the African people. I have fought against white domination. I have cherished the ideal of a democratic and free society in which all persons live together in harmony and with equal opportunities. It is an ideal which I hope to live for and to achieve. But if needs be, it is an ideal for which I am prepared to die."

Nelson Mandela is still behind bars and to his people he has become a symbol of the struggle of black South Africans. His wife, Winnie Mandela, has carried on his work in opposing apartheid in South Africa. In 1969 she was detained in prison without trial for more than a year, most of the time in solitary confinement. In 1970 and 1976 she was jailed for several months; in 1977 she was banished to another state of South Africa. The government hoped that this last measure would limit her ability to lead her people; however the international press sought her out and continued to publish her statements. In 1986 the banishment was lifted and she was allowed to return to her home in Soweto. She immediately resumed her work urging people to oppose the government.

Another South African leader, Bishop Desmond Tutu, has emerged as a strong opponent of racism in South Africa. Bishop Tutu won the Nobel Peace Prize in 1984 for his work in opposing apartheid. In recent years Bishop Tutu has called upon all countries in the world to impose economic sanctions on South Africa. He believes that if countries stop trading with South Africa and break diplomatic relations, the government may be forced to give the South African blacks equal rights.

Some countries like Canada have stopped trading certain goods with South Africa, and they have encouraged companies doing business in South Africa to withdraw from that country. Many believe that if economic pressure does not force South Africa to end apartheid, violence will break out and engulf the country in civil war. Already 2 400 people have died in riots since 1984 and 20 000 others have been arrested. Incidents involving violence are increasing in number.

Recently the South African government has introduced some reforms. Blacks are now permitted to own their own homes but they are still restricted to black neighbourhoods. It is no longer a crime for a white person to marry a black person — but once the marriage takes place, the white person must live in a black residential area. Some government segregation laws have been revoked but privately owned businesses and clubs carry on the apartheid policies with no interference from the government.

Bishop Desmond Tutu

In 1987 the South African government passed a new series of repressive laws against those who oppose apartheid in South Africa. These laws make it illegal for anyone to call for the release of prisoners being held without trial. Anyone breaking this law can be sentenced to ten years in jail and fined $13 000. These tough laws have further reduced the civil liberties of South Africans and many fear that it is just a matter of time before there is a massive explosion of violence. If this occurs there would be a blood bath, and there is no guarantee that the forces opposing apartheid would be victorious. To bring about the defeat of the unjust racist system in South Africa will require the combined efforts of people in countries all over the world.

"Unless the racist system in South Africa is ended soon, we're on the very verge of disaster. The situation may look calm on the surface but underneath it is extremely volatile... Those who support South Africa in any way at the present have to be asked: Would they actively support Nazism? Apartheid is just as vicious and evil as Nazism... There have been 3.5 million families uprooted and dumped like rubbish in the arid areas of Africa. There is violence on a monumental scale."

Bishop Desmond Tutu

"You find a terrible anti-black attitude among lower income groups. Part of the problem is that they see blacks as a threat to their jobs, and it's getting worse."

Johannesburg Mayor Monty Skaar

"It's a mystery to me how people in the homelands survive...The average person is lucky to get a bowl of mealie [corn meal] with or without some wild spinach or other vegetable the family can grow. Meals are almost entirely without protein, since eggs, milk, and meat and even bread are all luxuries."

**Ina Periman,
Director of Operation Hunger, South Africa**

1. What are some examples of discrimination against blacks in South Africa?
2. What have some countries done to oppose apartheid in South Africa? Discuss whether such actions help or hinder the case for equality in South Africa.
3. How do the three quotations add to your knowledge of the current situation in South Africa?

CAUSES OF PREJUDICE

We are not born prejudiced. Toddlers will play happily with other youngsters regardless of how different their culture or appearance. However, by the time they are in the primary grades, children have often acquired prejudices.

As children are socialized, they learn about the norms and behaviours of their own group. At the same time, they absorb certain negative attitudes about other groups from parents, teachers, other children, reading materials, television, and movies.

Most parents do not deliberately teach their children to be prejudiced, but youngsters listen to and copy the opinions of those around them. At first they use words and labels they do not understand; nevertheless these labels often help to support and reinforce prejudice that develops at a later age. Some of the negative labels they learn that have been applied to a wide variety of cultural or ethnic groups are "lazy", "drunk", "cheap", "stupid", "dirty", "untrustworthy", "untidy", and "greedy".

Prejudice is an attitude we have formed about a group of people that prevents us from seeing them as individuals, with strengths and weaknesses similar to members of our own group. Often people are prejudiced against a group about which they know nothing. They may not have met any members of the group, but they have a negative stereotyped image of them. They assume that all members of the group fit the stereotype.

Once a prejudice against a particular group has formed, it is highly resistant to change, since it is based on opinion, not fact. We can meet an individual from another group who does not fit the stereotyped image, but we will probably decide that this individual is an exception to the rule.

Stereotypes that are meant to be positive about members of certain groups can be as damaging as negative ones.

Members of certain groups are sometimes labelled as being good at sports, mathematics, music, or dancing. These stereotypes sometimes limit the abilities of individuals to certain areas and exclude them from being recognized and accepted in other activities. All stereotypes are damaging because they do not recognize individual abilities, needs, and potentials.

Most prejudice is an unconscious reaction toward others who are different. Sometimes it stems from the insecurity felt by members of one culture. Particularly in adolescence, it is reassuring to feel that you belong to an "in-group" that is superior. But this feeling of superiority creates feelings of inferiority, pain, and insecurity among those who are labelled as different.

Sometimes people do not wish to admit that they are prejudiced. They will explain that they are excluding someone from a particular group because that person would not feel comfortable, or that other members of the in-group would object if the person was included. They may not even recognize that they are prejudiced, or that they are acting in a way that might hurt the other person's feelings. In other cases, people may not be willing to risk the possible anger of the in-group by introducing the outsider to it.

The following is a summary of an article from the **Toronto Star,** *November 26, 1987 that discussed the results of a Gallup poll on discrimination. What is your opinion of the results?*

"About one in four Canadians feel they have been discriminated against in one way or another ... this is an increase of 5% from a similar poll taken in 1981. The increase is likely due to a broader awareness of human rights. Of those polled, 26% claimed sex as the basis of the discrimiantion; 22% stated that age was the basis, while 20% claimed racial origin was the origin of the discrimination. Religious discrimination was cited by 7%. Another 23% of those polled said they experienced discrimination because of their physical disabilities, body mass, appearance, language, or occupation."

Sometimes prejudice can be replaced by understanding if there is an opportunity for people from different cultures to be together for an extended period of time. The interaction of people may occur in a neighbourhood social group, a community action group, a high school yearbook committee, a professional association, a sports team, or a student government executive. What is required to overcome prejudice is the ability to recognize that your assumptions may be wrong, and a willingness to re-evaluate on the basis of your current experience. You must also make the effort to get to know the other person, and be generous enough to allow that person to have strengths and weaknesses like everyone else.

Progress Check

1. How is prejudice created in the minds of people?
2. Why is it difficult to eliminate prejudice?
3. Distinguish between "negative" and "positive" stereotypes. What can be harmful about "positive" stereotypes?
4. How do people sometimes avoid admitting that they are prejudiced?
5. How can prejudice be reduced?

6.2 *The Aboriginal Peoples of Canada*

When the great powers of the world colonized other countries, they did so partly out of self-interest, and partly in the belief that they were bringing civilization and Christianity to other parts of the world. Empire builders, such as Great Britain, France, Spain, Holland, or Portugal, conquered undeveloped countries and were able to use their natural resources. Taking land from the original inhabitants of a country was often justified by a feeling of racial superiority.

The **aboriginal peoples** in Canada were the first to inhabit and live in our country. Section 35 of the Constitution Act, 1982 defines the "aboriginal peoples of Canada" to include the Indians, the Inuit and the Métis. Indians are classified by the government into two groups — Status and Non-status Indians. Status Indians are recognized by the government and have certain privileges. At present they are entitled to live on reserves, to receive certain treaty rights from the government, and, in certain provinces, to have certain hunting, fishing, and trapping rights. They do not have to pay taxes for money earned on the reserves. Non-status Indians and the Métis do not live on reserves and do not have the other considerations received by the Status Indians.

The Collision of Several Cultures

When Europeans first arrived in Canada the native people were treated as equals. Each tribe was regarded as a separate nation and their chiefs were viewed with respect. The Europeans depended on the native people, and from them learned to survive in an often bleak and harsh environment.

European goods — iron pots, weapons, and tools — were sought after by the native people. For these goods they traded furs, and the trade brought the two groups closer together. Some Europeans intermarried with the native people. The children of these marriages came to be known as the Métis.

Unfortunately, the native people had no immunity to the diseases brought by the Europeans. Many died from illnesses such as smallpox, tuberculosis, typhus, and measles. These deaths greatly reduced the native populations and weakened them politically.

Some native people became dependent on the fur trade; other tribes were as sophisticated as the Europeans in trading; a few tribes concerned themselves primarily with trapping and delivering the furs to the Europeans. In doing so, they lost much of their old way of life; the customs of their culture were interrupted. As the European countries fought each other for control of the fur trade, they encouraged the various tribes to fight each other for control over their part in the trade. The Iroquois virtually exterminated the Hurons in the seventeenth century in the struggle over who would deliver the furs to the Europeans. Other tribes were skilled at "playing off" one European power against another. They would threaten to change sides in order to obtain better trade concessions.

The Europeans also encouraged the sale of alcohol to the native people. Not used to this drug, some Indians became addicted. Their culture was further disrupted by those social problems that can accompany alcoholism: arguments, fights, and beatings, forgetting to provide for one's family, loss of self-respect, and disregard for the structures of one's society. The majority of native groups, however, were relatively untouched by the impact of the Europeans or were able to blend satisfactorily aspects of the new culture with aspects of their traditional culture. They continued to practise their way of life.

Missionaries tried to convert the native people to European religions. They adopted a paternalistic or fatherly attitude toward the native people, and did not try to understand their beliefs. Instead, the native people were treated like children who could not make independent decisions. Most of the missionary attempts during this period failed to convert the Indians to Christianity.

Most Indians had a nomadic tradition, moving from one place to another to fish, hunt, and trap. They followed the seasons of nature and the migration paths of the animals they needed for food and clothing. The land and its resources were for everyone to share; the concept of owning land was not part of the native people's culture. Native people considered themselves to be the "stewards (guardians) of North America".

Nahanni woman is smoking a hide; hides become transparent after they are scraped and opaque during smoking. The task of smoking hides was done by young girls and old women.

Since the Europeans made few attempts to understand the culture of the aboriginal people, they simply tried to impose their culture believing it to be superior. Buffalo and beaver were hunted relentlessly to fill European market demands for furs. The very survival of some tribes was threatened. They depended on the buffalo, for instance, for food, shelter, clothing, and religious ceremonies. As the number of buffalo dwindled, the native people had to travel further to hunt — and then for fewer rewards.

As the Europeans decided to settle more and more land, they moved into the traditional territories of the native people. During the 19th and early 20th centuries, the governments of the settlers pressured the native people to leave their ancestral lands and live on smaller land areas, called reserves. This was usually done by treaty, by an agreement between the native people and the government. In exchange for giving up these aboriginal rights to the land, the native people received certain treaty rights, including reserves, a small amount of treaty money (usually $5.00 per year), some agricultural tools, and free education.

Prejudice and Discrimination

The Europeans had changed their views on the aboriginal peoples. As the fur trade developed, the Europeans used their own "middle men", the coureurs de bois, to transport furs. The native people were not as essential to the business of fur trading. As more missionaries tried to "civilize" the native people and met with failure in their terms, it was decided that the native people were "primitives", "savages", and "children". Such labels were reinforced by the Europeans' lack of understanding of native beliefs. As more natives were influenced by the whisky traders, they acquired labels such as "lazy", "drunk", and "unreliable". By the time settlement was reaching the interior of the country, some native people began to assert their rights. So the presence of the natives was considered to be threatening. New labels appeared — "bloodthirsty", "hostile", "obstacles to progress in the West".

The two views, native and European, on how the land and its resources should be used could not be reconciled. With the continuing weakening of the native way of life and the growing European strength, it became easier for Europeans to take over the lands. The stereotyped labels stayed with the native people; it was believed that they were inferior. Taking their lands or pressuring them onto reserves was considered appropriate behaviour.

The effect of isolating natives on reserves was to keep them apart from Canadian society as a dependent and powerless people. It also helped to provide them with a land base, and, because of the isolation, often protected the native identity. The majority of reserves are located in rural parts of the country. These reserves are still held in trust for the native people by the federal government.

The native people in Canada have been treated as inferiors, not able to handle their own affairs. Their culture has not been respected, and their unique way of life has often been ridiculed or ignored. Many native people became alienated and some turned to drugs, alcohol, and violence as a way to escape a society that refused to understand their differences or to recognize their rights.

The Indian Act of 1876 made it illegal for Indians on reserves to drink alcohol in public places (changed in 1970), to practise their religious ceremonies (changed in 1951), or to vote in federal elections (changed in 1960). Status Indian women and their children lost their privileges if they married non-Indians, although this law did not apply to Indian men who married non-Indians (changed in 1985).

THE NATIVE PEOPLE TODAY

Because of the poor living conditions on many reserves, life for the native people has tended to be harder than for most of the rest of Canadian society. Those who have chosen to leave the reserves have had to deal with discrimination in employment and housing.

Statistics show that only 20% of Indian students finish high school compared to 75% for other Canadians. There is a higher unemployment rate for native people. The average life-span of natives is 10 years less than that of other Canadians. Violent deaths among native people are three times the national average. The suicide rate in the 15 to 24 age group is six times the national average.

Many Canadians are aware that these statistics do not reflect on the native people themselves, but on the treatment they have received in the past. In 1987 a public opinion poll revealed that 60% of Canadians agreed that Canada has treated the native people in a shameful manner; 77% supported the idea of aboriginal self-government; and 66% agreed that the aboriginal peoples should have control over their own education, social services, hunting, fishing, and the running of their own reserves.

Aboriginal leaders are calling for land areas for Non-status Indians and Métis, control of the resources on their lands, and self-government. Such groups as the Assembly of First Nations, which represents Canada's Status Indians, the Native Council of Canada, which represents the Non-status Indians and the Métis, the Inuit Committee on National Issues, and the Métis National Council have gained some rights for Canada's aboriginal peoples.

These groups fought successfully to have the Constitution Act of 1982 recognize that aboriginal rights existed that were unique from the rights of other Canadians. It has not yet been finalized what these rights are. Native people see them as including ownership of the mineral rights of the land they own, compensation for lands that were taken from them, and self-government. For example, the Inuit, Dene (Indians of the Mackenzie District of the Northwest Territories), and Métis of the north want to divide the Northwest Territories into two territories over which they would have political control.

In 1987, a conference was held by the federal government, the provincial governments, and the leaders of the aboriginal peoples to discuss how the Constitution Act of 1982 could be amended to guarantee self-government for the native people. Some provincial premiers were unwilling to agree to such a constitutional amendment because they were afraid their own power would be weakened. The conference ended in failure and many leaders were bitterly disappointed.

As many Indian bands teach their children the history, language, and customs of their ancestors, the unique nature of Indian culture will, it is hoped, be renewed. Hundreds of Indian bands have set up their own schools on reserves and now control the quality and type of education given to children. As aboriginal students receive an education that is meaningful to them, there has been a dramatic decline in the drop-out rate. Some reserves have set up successful industries, such as wild rice, fishing, cranberry, and trucking operations. The Department of Indian and Northern Affairs is handing over some of its control to the Indian bands and is considering self-government proposals for 139 of the Indian bands.

Inuit schoolchildren at Povungnetuk, Quebec.

A Métis artisan makes nests for mallard ducks, Manitoba.

The native people are emerging as people with two cultural identities. Their sense of identity is firmly rooted in their ethnic background, but their skills and knowledge are enabling them to succeed in Canadian society as a whole. Certainly, Indians in Canada are the most disadvantaged ethnic group in Canada, but the past ten years have brought a dramatic improvement in their social and economic living conditions.

What do the following quotations add to our knowledge of the native people in Canada today?

"I started drinking at the age of seventeen, I guess mostly from frustration. I was caught up in two worlds. I didn't really fit into the white man's world, and I was losing my Indian ways".
Woodrow Goodstriker, from southern Alberta

"In a democratic age, it is incongruous to maintain any people in a state of dependency ... Old, distorted and paternalistic notions abhor the 'protection" of Indian people and nations and must be discarded. Ending dependency would stimulate self-confidence and social regeneration".
Report of Special Commons Committee, 1983

"We're at a historical crossroads ... A momentum has been built up among the Indian nations, and no power on earth will be able to stop it now. It's irresistible. We will never go back."

Joe Mathias, hereditary chief of British Columbia's 2 000 Squamish Indians

"For a time, much like winter when the land rests, the Indian people were sleeping. But now we have awakened."

Joe Miskokomon, President of the Union of Ontario Indians

PROGRESS CHECK

1. How were the aboriginal peoples treated by the Europeans when the first colonists arrived in Canada?
2. What evidence is there to support the idea that the white people were guilty of ethnocentrism in the treatment of the aboriginal peoples?
3. What differences exist between the white man's culture and the cultures of the native people? What were the consequences of these differences?
4. How did the Indian Act of 1876 discriminate against the native people?
5. What are some of the problems experienced by Canada's native people?
6. What rights are Canada's aboriginal people demanding from the governments? What success have they had?

6.3 *Canada, A Multicultural Society*

With the exception of the aboriginal peoples, Canada is a land of immigrants. The two original groups of immigrants, the British and the French, together make up the majority of the population. Immigrants from other European countries make up the next largest group. Most arrived from their native countries in "waves", often as a result of political, economic, or religious pressure. More recently, immigrants have come seeking a higher standard of living and greater political freedom than they expect in their home country.

Canadian Immigration Policies

Traditionally Canada has been a home and a refuge to many groups of immigrants. But over the years certain individuals and groups have been excluded because they were considered undesirable. They were prevented from entering Canada because they were criminals, poor, sick, or merely belonging to an ethnic or religious group that the Canadian government felt was undesirable. Many of these people who were prevented from entering this country were victims of prejudice and ethnocentrism.

Figure 6.1 gives an overview of Canada's immigration policies from 1869 to 1919.

Figure 6.1

1869 - Immigration Act	• said nothing about who should be admitted
	• did not even call for a medical examination for immigrants from the U.S.
1872 - Amendment	• prohibited the landing of criminals
1879 - Amendment	• prohibited people with no money from entering
1885	• Immigration Act
	• placed a $50 tax on each Chinese person entering Canada
1900 - Amendment	• tax on Chinese increased to $100
1902 - Amendment	• people with diseases prohibited
1903 - Amendment	• head tax on Chinese increased to $500
1906 - Amendments	• more control along U.S. and Canadian border
	• criminals and the sick still excluded
	• specified the amount of money immigrants should have on landing. (This was $500 for Chinese, $300 for other immigrants, and $25 for British immigrants in summer and $50 in winter.)
1907	• informal agreement between Canada and Japan in which Japanese government volunteered to restrict numbers immigrating to Canada to 400 a year
1910	• government stated that it intended to keep out of Canada those who were physically, mentally, or morally unfit; those belonging to nationalities that were unlikely to assimilate or fit into Canadian society; those who may cause unemployment or a lowering of living standards
1911	• border guards were to refuse entry to all U.S. Blacks on medical grounds
1914	• *Komagata Maru* with immigrants from India prevented from landing in Vancouver by Canadian military
1919	• Mennonites, Doukhobors, and Hutterites (religious groups from Europe) prevented from immigrating to Canada from Europe
	• alcoholics and illiterates banned

MULTICULTURALISM

Since the 1970's many Canadians have viewed themselves as a **multicultural society** in which people of different ethnic backgrounds are encouraged to preserve some of their customs, traditions, religions, and languages. You will have seen multiculturalism at work in the education system in Canada. Increasingly, our school system recognizes that schools can teach the language, history, and literature of ethnic groups, and that young people can celebrate their own cultures by means of festivals and international evenings.

Differences between ethnic groups are seen as strengthening the bonds of Canadian society. At the same time, immigrants are expected to obey Canadian laws, learn either one or both of Canada's official languages, obtain employment, and get along with other Canadians who might be members of other cultures.

Over the years, many immigrants and their descendants have lost their original identity and become Canadian. For instance, Toronto has the highest population of Danish people outside of Denmark, yet we are not usually aware of these people as an immigrant group since they have been assimilated into Canadian society. Other ethnic groups have expressed their dual identity by becoming hyphenated Canadians, like Greek-Canadians, Italo-Canadians, Polish-Canadians, and Portuguese-Canadians. Still others refer to their ancestry when explaining their ethnic heritage, and say that they are of British, Ukrainian, Chinese, German, or Dutch descent.

Some Canadians identify themselves by referring to their religious affiliation — Jewish, Moslem, Hindu, Christian, or Buddhist. Despite all these differing backgrounds, most Canadians share many of the same beliefs and values — they believe in democracy, freedom, peace, fair treatment, and negotiation.

PREJUDICE IN A MULTICULTURAL SOCIETY

Not all people in Canada live in harmony with their neighbours. As new immigrant groups settle in Canada, they are generally viewed as being different from those who arrived in the country earlier.

Some Canadians are prejudiced against newly arrived ethnic and racial groups because they feel threatened. They believe that these immigrants take away jobs from

Canadians and compete for positions and promotions that Canadians might want. Those who feel insecure about their jobs are the most likely to feel this way, especially in difficult economic times when jobs and opportunities become more scarce.

Canada's varied history reflects the strife that has occurred when a new ethnic group arrives on these shores. In fact, there is evidence that the Vikings came into conflict with the Inuit when the Vikings tried to establish settlements in northeastern Canada about 1000 A.D. Certainly there has been conflict between the two major groups of settlers in Canada, the French and the British. In general, they viewed one another with suspicion because of their different languages, religions, and customs.

More recently, when boatloads of Sikh refugees from India arrived on Canadian shores in 1987, many people reacted with hostility. The Sikhs were regarded as "queue-jumpers" who had not followed the proper immigration procedures. Some people thought that the Sikhs were a threat to the Canadian job market.

Progress Check

1. Why is Canada considered to be a multicultural society? Do you think that it is inevitable that ethnic groups will be assimilated into the larger Canadian culture?
2. What attitudes did the early immigrant groups have toward each other? Give examples.

Some Ethnic Minorities in Canada

The Irish Immigrants

A terrible potato famine in Ireland in the middle of the nineteenth century drove many thousands of Irish to British North America. They had little food, money, or possessions and many suffered from diseases such as cholera and typhus. Thousands died on the rough voyage across the Atlantic and were buried at sea. Others died of disease and malnutrition after reaching the colonies. Over 107 000 Irish people sailed for British North America in 1847 alone. Between 1841 and 1851 it is estimated that one million Irish died of starvation and disease and one million people emigrated to escape the intolerable conditions.

> *"The population were like walking skeletons, the men stamped with the ... mark of hunger, the children crying with pain, the women, in some of the cabins, too weak to stand ... All the sheep were gone, all the cows, all the poultry killed ... A mob of men and women, more like famished dogs ... whose figures, looks and cries all showed they were suffering the ravening agony of hunger."*
>
> **A contemporary account.**

The Irish came to the colonies because, like Ireland, the colonies were part of the British Empire. The Protestant Irish from Ulster (Northern Ireland) came from a sense of loyalty to the British Empire. The poor and the sick came because the local colonial governments could not deny them entry.

Since labour was generally scarce and expensive in British North America, some Irish laborers were welcomed. They found jobs building the roads, canals, and railways to earn money to buy a farm of their own. But, the Irish were blamed by some people for bringing the typhus epidemics to the colonies. Some people refused to hire Irish workers in the 1840's and 1850's for fear of getting the plague themselves. It was common to see signs in windows of stores and factories that read "No Irish Need Apply" or "No Irish or Dogs".

On the other hand, the Irish did receive considerable assistance from Canadians, both financial and medical. Doctors, nurses, nuns, priests, and ministers helped look after the sick at personal risk to themselves. French-Canadian families in Canada East (Quebec) opened their hearts and their homes to the hundreds of Irish children orphaned by the epidemic.

Despite the deaths, pain, and suffering, the survivors managed to start a new life in a new land where there was hope for a job, a home, a farm, and a future.

IRISH EMIGRANT DEATHS, 1847	
Numbers embarked	106 812
Died in passage	6 116
Died in quarantine	4 149
Died in hospital	7 180
Total deaths	17 445

The Ukrainians

Today the area of land called the Ukraine is located in the U.S.S.R., but in the 1890's part of the Ukraine was in the Austro-Hungarian Empire. Many Ukrainians were poor peasant farmers who worked on small plots of land to try to feed their families.

In 1895, Dr. Joseph Oleskow, a professor living in the Ukraine, heard about the free land offered in Canada and saw an opportunity to help the Ukrainians. He wrote and spoke to Ukrainians at meetings, encouraging them to emigrate to Canada. After a visit to Canada, Oleskow told Ukrainians about the climate, opportunities, and problems involved in moving to a new country, and advised them on how to emigrate.

Clifford Sifton was the politician in charge of bringing immigrants to Canada to settle the West in the late 1890's. He admired the farming abilities of the Ukrainians who were used to farming the dry, flat land in the Ukraine. He wanted them to come to Canada because he realized that they were tough, knowledgeable farmers who could develop and populate Canada's West. Thousands of poor and largely uneducated Ukrainians took up the Canadian government's offer of free land, and emigrated. Often entire villages emigrated and resettled together.

The Ukrainian settlers appeared very strange to the average Canadian of the time. They wore sheepskin coats, spoke little or no English, and were for the most part extremely poor. Canadians generally offered little help to these seemingly strange people.

Despite the harsh conditions, the Ukrainian settlers worked together and built their houses and barns and cultivated their fields. They tended to stick together for mutual support, and this isolated them even more from other Canadians. While the women and children took care of the farms, the Ukrainian men earned much needed cash by working at manual jobs in mining and in building railroads and highways. Many worked ten-hour days for .25 an hour. Often the jobs paid wages that were judged unacceptably low by other Canadians. Some Canadians resented the competition the Ukrainians presented in the workforce, and discriminated against them and made fun of them.

Despite the discrimination, the Ukrainians prospered and helped to develop the Canadian West. They built schools and churches, established their own newspapers, and developed a strong and healthy ethnic culture within the larger Canadian society. Today there are over 600 000 Canadians of Ukrainian descent who have managed to retain many of their traditional customs.

PROGRESS CHECK

1. Why did Irish immigrants come to Canada? What were some of their experiences here?
2. What encouraged the Ukrainians to come to Canada? What role did Dr. Joseph Oleskow play?
3. Why did Clifford Sifton want Ukrainians to settle in Canada?
4. How did other Canadians view the recently arrived Ukrainian settlers?
5. How have the Ukrainians preserved their subculture in Canada?

The Chinese

In the 1870's and 1880's, thousands of male Chinese workers came to Canada to work as unskilled laborers on the railroads and in the gold mines. Many died of disease, exposure, mine cave-ins or explosions. The average wage paid to a Chinese worker was $1.00 a day. From this he had to pay for tools, clothes, and living expenses.

At first the Chinese were viewed as good and conscientious workers, but later they were resented by Canadian workers for their willingness to work for lower wages. The Chinese were believed to be taking jobs away from Canadian workers.

The appearance, culture, and lifestyles of the Chinese were different, and were viewed with suspicion by many Canadians. Most of the Chinese workers settled in British Columbia where they soon became victims of racist feelings. Anti-Chinese riots broke out and demands were made to stop immigration from the Orient.

In 1885 (the year the transcontinental railroad was completed) the Canadian government placed a $50 "head tax" on every Chinese person entering Canada. Only a few Chinese immigrants could afford to pay this fee. In 1901 the head tax was raised to $100, and in 1904 to $500. These taxes were successful in reducing Chinese immigration to Canada, and some Chinese left Canada because of the prejudice and discrimination they experienced.

In 1923 the federal government passed a law which prohibited Chinese immigration into Canada. This law was effective in preventing the wives and children of Chinese workers already in Canada from joining their husbands and fathers. Between 1924 and 1947 only eight Chinese immigrants were permitted to enter this country. Between 1931 and 1941 the Chinese population in Canada dropped from 46 519 to 34 627.

Despite their poor treatment, more than 400 Chinese joined the Canadian armed forces in World War II to fight for their adopted country. In 1947 the Canadian government finally decided to permit Chinese wives and unmarried children to join their husbands and parents in Canada. It was only in 1949 that the Chinese were given the right to vote in British Columbia. It was not until 1967 that the federal government began to treat Chinese applicants who wanted to come to Canada in the same manner as other immigrants.

In 1979 the Canadian government did agree to permit 50 000 "boat people" escaping from Communist Vietnam to enter Canada. Many of these refugees were of Chinese descent. This merciful act was reinforced by the generosity of thousands of Canadians who contributed their time and money to help these desperate people adjust and become established in this country.

The Japanese

The Japanese immigrants, who began arriving in Canada in the late 1880's, experienced the same type of prejudice and discrimination as the Chinese. Their appearance and customs set them apart and they were often viewed with suspicion and mistrust by other Canadians.

Most of the Japanese settled in British Columbia and were employed in railway contruction, mining, logging, and later in farming. In 1901 a government report stated that many people in British Columbia resented the Japanese because they viewed them as a competitive threat to their jobs and businesses. In 1907 the Vancouver Anti-Asiatic League, with the support of some trade unions, took part in a demonstration in Vancouver protesting the immigration of Orientals. The demonstration soon was out of control and angry mobs damaged stores and smashed windows of homes owned by Chinese and Japanese.

In 1908 the Canadian and Japanese governments agreed to limit the number of Japanese immigrants to Canada. There were to be no restrictions on wives and families emigrating to Canada, however. So unlike the Chinese, the population of Japanese-Canadians grew rapidly, until by the end of the 30's, it exceeded the Chinese-Canadian population. But, discrimination against the Japanese-Canadians still ran high. Canadian citizens of Japanese descent were not permitted to vote in elections, become civil servants, lawyers, accountants, pharmacists, teachers, or policemen. Nor could they obtain liquor licenses. Restrictions placed on Japanese-Canadians in the fisheries industry severely limited their ability to make a living.

When Japan attacked Pearl Harbour in Hawaii on December 7, 1941 there were 23 224 people of Japanese ancestry living in Canada. Of these, 14 119 had been born in Canada. Since most Japanese-Canadians lived on the West Coast, many Canadians viewed them as potential collaborators with Japan. Hysteria broke out in Canada about this "threat" to Canadian security, despite the fact that most of the Japanese-Canadians were Canadian citizens and regarded themselves as Canadians. In that same year, the RCMP registered all persons of Japanese descent living in Canada, and they were required to carry identity cards at all times.

NOTICE

TO ALL PERSONS OF JAPANESE RACIAL ORIGIN

Having reference to the Protected Area of British Columbia as described in an Extra of the Canada Gazatte, No. 174 dated Ottawa, Monday, February 2, 1942:-

---

1. EVERY PERSON OF THE JAPANESE RACE, WHILE WITHIN THE PROTECTED AREA AFORESAID, SHALL HEREAFTER BE AT HIS USUAL PLACE OF RESIDENCE EACH DAY BEFORE SUNSET AND SHALL REMAIN THEREIN UNTIL SUNRISE OF THE FOLLOWING DAY. AND NO SUCH PERSON SHALL GO OUT OF HIS USUAL PLACE OF RESIDENCE AFORESAID UPON THE STREETS OR OTHERWISE DURING THE HOURS BETWEEN SUNSET AND SUNRISE:

2. NO PERSON OF THE JAPANESE RACE SHALL HAVE IN HIS POSSESSION OR USE IN SUCH PROTECTED AREA ANY MOTOR VEHICLE, CAMERA, RADIO TRANSMITTER, RADIO RECEIVING SET, FIREARM AMMUNITION OR EXPLOSIVE:

3. IT SHALL BE THE DUTY OF EVERY PERSON OF THE JAPANESE RACE HAVING IN HIS POSSESSION OR UPON HIS PREMISES ANY ARTICLE MENTIONED IN THE NEXT PRECEDING PARAGRAPH. FORTHWITH TO CAUSE SUCH ARTICLE TO BE DELIVERED UP TO ANY JUSTICE OF THE PEACE RESIDING IN OR NEAR THE LOCALITY OR TO AN OFFICER OR CONSTABLE OF THE ROYAL CANADIAN MOUNTED POLICE.

4. ANY JUSTICE OF THE PEACE OR OFFICER OR CONSTABLE RECEIVING ANY ARTICLE MENTIONED IN PARAGRAPH 2 OF THIS ORDER SHALL GIVE TO THE PERSON DELIVERING THE SAME A RECEIPT THEREFORE AND SHALL REPORT THE FACT TO THE COMMISSIONER OF THE ROYAL CANADIAN MOUNTED POLICE. AND SHALL RETAIN OR OTHERWISE DISPOSE OF ANY SUCH ARTICLES AS DIRECTED BY THE SAID COMMISSIONER.

5. ANY PEACE OFFICER OR ANY OFFICER OR CONSTABLE OF THE ROYAL CANADIAN MOUNTED POLICE HAVING POWER TO ACT AS SUCH PEACE OFFICER OR OFFICER OR CONSTABLE IN THE SAID PROTECTED AREA, IS AUTHORIZED TO SEARCH WITHOUT WARRANT THE PREMISES OR ANY PLACE OCCUPIED OR BELIEVED TO BE OCCUPIED BY ANY PERSON OF THE JAPANESE RACE REASONABLY SUSPECTED OF HAVING IN HIS POSSESSION OR UPON HIS PREMISES ANY ARTICLE MENTIONED IN PARAGRAPH 2 OF THIS ORDER, AND TO SEIZE ANY SUCH ARTICLE FOUND ON SUCH PREMISES:

6. EVERY PERSON OF THE JAPANESE RACE SHALL LEAVE THE PROTECTED AREA AFORESAID FORTHWITH:

7. NO PERSON OF THE JAPANESE RACE SHALL ENTER SUCH PROTECTED AREA EXCEPT UNDER PERMIT ISSUED BY THE ROYAL CANADIAN MOUNTED POLICE:

8. IN THIS ORDER, "PERSONS OF THE JAPANESE RACE" MEANS, AS WELL AS ANY PERSON WHOLLY OF THE JAPANESE RACE. A PERSON NOT WHOLLY OF THE JAPANESE RACE IF HIS FATHER OR MOTHER IS OF THE JAPANESE RACE AND IF THE COMMISSIONER OF THE ROYAL CANADIAN MOUNTED POLICE BY NOTICE IN WRITING HAS REQUIRED OR REQUIRES HIM TO REGISTER PURSUANT TO ORDER-IN-COUNCIL P.C. 9760 OF DECEMBER 16th, 1941.

DATE AT OTTAWA THIS 26th of FEBRUARY 1942.

Louis S. St. Laurent,
Minister of Justice

To be posted in a Conspicuous Place

In 1942, 21 000 Japanese Canadians were ordered by the federal government to be removed from the West Coast. Males between the ages of 18 and 45 were forced to labour on highways and farms in the interior, and women and children were placed in internment camps. Families were often separated and did not know the whereabouts of other family members. Their property and goods were seized and auctioned off for very low prices. Altogether 1200 fishing boats, 1500 cars and trucks, 770 farms, and all the homes and contents were sold by the government without the consent of those to whom they had belonged. All of this was done without evidence that Japanese-Canadians were a threat to Canadian security.

One of the Japanese-Canadian internment camps.

In 1944 Canada's Prime Minister, W.L.M. King, declared that "It is a fact that no person of Japanese race born in Canada has been charged with any act of sabotage or disloyalty during the years of war." The actions by the Canadian government were clearly racist and partly the result of pressure from anti-Asian groups in British Columbia. Similar actions were not taken against Canadians of German or Italian ancestry.

After the war ended in 1945, 4 000 Japanese-Canadians were "encouraged" by the federal government to leave

Canada and go to Japan. It was not until 1949 that the remaining Japanese-Canadians were permitted to return to the West Coast to try to rebuild their shattered lives. In that same year they were finally given the right to vote.

Today there are about 11 000 Japanese-Canadians alive in Canada who suffered through this humiliating experience. No public apology or compensation has been made to these people by the Canadian government, but pressure is growing to compensate those whose property was seized and sold without permission. It is hoped that if this is done, it will help to heal the wounds of those who suffered from these acts of racism in Canada.

THE QUEST FOR A JUST SOCIETY

In 1982 a law called the Canadian Charter of Rights and Freedoms was included in Canada's constitution, the Constitution Act, 1982. This law tried to ensure that the basic rights of all Canadians are protected and guaranteed. Equal and fair treatment regardless of sex, colour, or race are some of the rights that help to protect Canadians from prejudice and discrimination. This law also guarantees certain basic freedoms and rights to all Canadians that are considered vital in a free and democratic society. The federal Charter of Rights and Freedoms and similar provincial human rights codes are some of the laws that reflect changing values and beliefs that are now considered to be important in Canadian society.

The following are excerpts from the Canadian Charter of Rights and Freedoms, agreed to by the Canadian Parliament and nine of the provinces in 1982.

Fundamental Freedoms

Section 2
a. freedom of ...religion
b. freedom of thought, belief, opinion, and expression, including freedom of the press
c. freedom of peaceful assembly ...
d. freedom of association

Democratic Rights

Section 3.
... the right to vote

Mobility Rights

Section 6.(1)
... the right to enter, remain in, and leave Canada.

	Section 6.(2) Every citizen and every person who has the status of a permanent resident of Canada has the right a. to move to and take up residence in any province b. to pursue the gaining of a livelihood in any province
Legal Rights	*Section 7* ... the right to life, liberty, and security ... *Section 8* ... the right to be secure against unreasonable search ... *Section 9* ... the right not to be ...[without reason] ... detained or imprisoned *Section 10* Everyone has the right on arrest a. to be informed promptly of the reasons ... b. to obtain a lawyer ... without delay *Section 11* Any person charged ... has the right to b. to be tried within a reasonable time ... d. to be presumed innocent until proven guilty ... e. not to be denied reasonable bail without just cause h. if finally acquitted of the offence, not to be tried for it again and, if finally found guilty and punished ... not to be tried or punished for it again ... *Section 12* ... the right not to be subjected to any cruel and unusual treatment or punishment
Equality Rights	*Section 15. (1)* ... the right to equal protection ... of the law ... without discrimination ... based on race, national or ethnic origin, colour, religion, sex, age, mental or physical disability *Section 28* the rights and freedoms referred to [in the Charter] are guaranteed equally to male and female persons
Multicultural Heritage Rights	*Section 27* This Charter shall be interpreted in a manner consistent with the preservation and enhancement of the multicultural heritage of Canadians.
Aboriginal Rights	*Section 25* The guarantee in this Charter of certain rights and freedoms shall not be construed so as to abrogate or derogate from any aboriginal treaty or other rights or freedoms that pertain to the aboriginal peoples of Canada including a. any rights or freedoms that have been recognized by the Royal Proclamation of October 7, 1763 b. any rights or freedoms that may be acquired by the aboriginal peoples of Canada by way of land claims settlement

	Section 35.(1) ... aboriginal and treaty rights ... [are guaranteed]
Enforcement of Rights and Freedoms	*Section 24(1)* ... Anyone whose rights or freedoms, as guaranteed by this Charter, have been infringed or denied may apply to a court of competent jurisdiction to obtain such remedy as the court considers appropriate and just in the circumstances.

PROGRESS CHECK

1. List some examples of prejudice and discrimination that Chinese workers experienced in Canada.
2. Why did many Canadians pressure the Canadian government to pass laws against people of Japanese descent during World War II? In your opinion, were these laws justified?
3. Have any actions been taken by the Canadian government to compensate Japanese-Canadians for their losses in World War II? In your opinion, how should they be compensated?
4. Has Canada become a more tolerant society in its treatment of Japanese- and Chinese-Canadians? Explain.
5. Do you think that Canada has entered a new "age of tolerance" since 1980? Refer to the Charter of Rights and Freedoms in your answer.

OVERVIEW

Many people argue that prejudice and discrimination have always existed and that nothing can be done to end these negative attitudes and unequal treatment. They maintain that everyone is naturally prejudiced and it is useless to try to end it. But these negative attitudes are learned behaviours, acquired during the socialization process. There is nothing innate about these irrational behaviours. They are based on ignorance and fear. They are responsible for human suffering, destruction, and war all over the world. These negative feelings and actions can be reduced by making people aware that these attitudes are unfair and can cause enormous pain and suffering.

The period since 1980 in Canada has been referred to as the "age of tolerance" by many Canadians. Major breakthroughs are being made in making people aware of the unjust treatment of many minority groups. Laws have been passed to protect and help members of minorities, and Canada has gained a reputation as a tolerant society. At the same time, immigration of other ethnic groups has helped Canada's economic growth, social progress, and cultural sophistication.

It is the responsibility of all members of a multicultural society to become aware of prejudice and to try to understand and reduce it. It is easy to point to other people and cultures, labeling them as racist and prejudiced. It is far more difficult to point to oneself. Social changes begin with individual changes however.

KEY WORDS

Define the following terms, and use each in a sentence that shows its meaning.

ethnocentrism
racism
ethnic prejudice
segregation
apartheid
aboriginal peoples
multicultural society

KEY PERSONALITIES

Give at least one reason for learning more about the following.

Nelson and Winnie Mandela
Bishop Desmond Tutu
Dr. Joseph Oleskow
Clifford Sifton

DEVELOPING YOUR SKILLS

FOCUS AND ORGANIZE

Canada is described as a multicultural society where many different religious and ethnic groups live and work. Canada is a democratic and free society and is attractive to those who live in other countries where there are fewer freedoms, less space, and a lower standard of living. Many Canadians are actively working to achieve a society where everyone, regardless of skin colour, beliefs, and origins, is treated in a fair and equal manner. Hatred and discrimination do not occur only in other countries — they also exist in Canada. It is the purpose of this chapter to make students more aware of the background of the unfair treatment of some minority groups and to explore the attempts that are being made to create a more tolerant and just society for everyone.

(a) Create an organizer that lists examples of prejudice, stereotypes, and discrimination.
(b) What questions can be asked about these negative feelings, labels, and practices?

LOCATE AND RECORD

1. Class Project: Devote a week to searching for newspaper and magazine articles on current cases of prejudice, stereotypes, and discrimination. Have class members present a summary of their articles to the class and have them relate their information to the topics covered in this chapter.
2. Order the following films and find out more about these topics:
– *Bill Cosby on Prejudice*
   Produced by Pyramid Film
   Bill Cosby plays the role of a bigot who hates everyone.
– *Indian Speaks*
   Produced by the National Film Board of Canada
   Indian concerns about preserving and recovering aboriginal culture are examined

- *Enemy Alien*
  Produced by the National Film Board of Canada
  An examination of the history of the Japanese in Canada.
- *Prejudice: Causes, Consequences, Cures*
  Produced by CRM Productions
  Examines the nature of prejudice and its causes.

SYNTHESIZE AND CONCLUDE

Develop an organizer that compares the system of apartheid in South Africa to the experiences of Canada's aboriginal peoples. What similarities and differences exist? Assess whether one situation is more unjust and give reasons for your conclusions. Speculate about what would happen if the issues in both situations are not resolved. Share your conclusions with other class members.

EVALUATE AND ASSESS

1. Read over the background material in this chapter on the following groups of immigrants — the Irish, the Ukrainians, the Chinese, and the Japanese. Create a chart that records their experiences when they first arrived in Canada. Assess the reasons for their treatment. By listening to the radio, watching television, or reading the newspaper, assess whether or not recent immigrant groups receive a similar reception in Canada when compared to earlier immigrant groups.
2. Examine the Immigration Chart on page 194. Assess the attitudes of the Canadian government and society toward the groups listed. Discuss the reasons for those laws and attitudes.
3. Read *Obasan* by Joy Kogawa. Write a short book review that tries to evaluate and assess her story of Japanese-Canadians during World War II.

APPLY

1. Taking the Role of the Other: Divide the class into groups and have each select an ethnic group whose members have experienced discrimination. Research your ethnic group in the library and obtain information on its past experiences. Put yourself in the position of an ethnic group member.

What might your attitudes be toward members of your own group and toward the larger society? What experiences, opportunities, roles, values, interests, goals, and friends might you have as a member of this group? All research findings can be presented to the class as a whole.

2. The class can be divided into groups to discuss the causes of ethnocentrism and racism. Brainstorm why people who belong to specific groups discriminate against and feel superior to other groups. Speculate whether or not this treatment of others is learned or innate behaviour. Can anything be done to correct these negative attitudes and treatments? Try to determine whether prejudice and discrimination are declining. Are Canada and Canadians becoming more tolerant of ethnic groups? Record your conclusions and present them to the rest of the class.

3. Predictions: What would/should you do?
   - if you were with a group of friends and a joke was made about a member of an ethnic group?
   - if you saw your boss always throwing away the job applications of certain ethnic groups?
   - if you worked with ethnic group members and they were performing the same job as yourself, but you found out they were making less money?
   - if you saw a member of an ethnic group being harassed, called names, and threatened or beaten?

COMMUNICATE

1. Write an organizer or a short essay, or debate one of the following topics.

- Individuals will always be prejudiced and will always discriminate against others; this situation cannot be changed.
- Classifying people into races has no biological basis.
- War can be eliminated if ethnocentrism is destroyed in the minds of people. Ultimately all stereotypes are damaging.
- The native people of Canada are entitled to control their own affairs.
- Canada is a place where immigrants can find a place of refuge against persecution.
- The prejudice, stereotyping, and discrimination that exist in any society are a reflection of the values and attitudes of all the members of that society.

- Jokes that are made at the expense of any group reflect prejudice and should not be made.
- The Canadian government should take a more active role to preserve and support the identities of ethnic cultures in Canada.
- It is an illusion that Canada is a multicultural society. Canada is a bicultural country and all minorities will eventually be assimilated into one of the two dominant cultures.

2. Divide the class into groups and brainstorm the causes and the consequences of prejudice, hatred, and discrimination in the world today (eg. terrorist activities). Speculate about the kind of world there would be without these negative feelings and actions. Do you believe that prejudice, hatred, and/or discrimination will be reduced or eliminated in the future? Make sure that your thoughts and conclusions are recorded, and present them to the rest of the class for discussion and analysis.
3. What do you think the 21st century has in store for native people in terms of preserving their traditional ways of living? Based on information provided in this chapter and using additional references, write a brief essay in which you predict whether or not you feel traditional ways can and perhaps should be maintained.

UNIT

3

COMMUNICATION

CHAPTER

7

Human Communication

What is language and how do we acquire language skills?

INTRODUCTION Have you ever wondered what it would be like to live in a world without language? There would be no books, newspapers, telephones, or television; no conversations with friends, no family discussions.

We use language automatically, and depend on it so much in our daily lives that we take it for granted. Language is used to express our wishes, share our thoughts and feelings, and to give and take directions. What kind of interaction could we have with others if we could not use language? Imagine trying to think without using language. How much of who you are depends on your ability to think and to communicate with others?

In this chapter we will explore what language is, how we acquire it, and how language works for us.

7.1 *What is Language?*

Language is a method of expression, a means by which humans can interact. Humans can interact in two basic ways — physically or by using symbols. An example of physical interaction is a child grabbing a toy from another child. Another example is a child giving someone a toy. Using symbols is the usual way in which humans interact. Symbols can be used in spoken or written language (verbal communication) or through gestures, expressions, and appearances (nonverbal communication).

Phoenician	✛	◁	ꓘ	↳	⋈	ϟ	⋈
Hebrew	א	ד	כ	ל	מ	נ	ז
Arabic	ا	د	ك	ل	م	ن	ذ
Greek	A	Δ	K	Λ	M	N	Z
Russian	А	Д	К	Л	М	Н	З
Roman	A	D	K	L	M	N	Z

Phoenician, Hebrew, and Arabic are all Semitic languages and were in use by 1500 B.C. The Greek alphabet is derived from the Phoenician. The Russian and Roman alphabets are derived from the Greek.

By spoken language we mean the vocal sounds of human speech that symbolize objects, actions, and qualities. Written language consists of the written symbols that represent those vocal sounds. The combination of sounds and symbols produces a particular vocabulary for a language.

"I remember going on my first hunting trip. It was exciting to be alone with my father for a whole week and to come back to our family a successful hunter—a man."
An Arctic Childhood,
Norman Ekoomiak

LANGUAGE AND SOCIETY

Every human society uses language; it is the major form of communication between people. Societies could not function if members were unable to express themselves.

Each society develops its own set of symbols of expression. The meaning of these symbols will be shared by members of that society, but not by outsiders. Their language will not be understood by people from another society. Someone growing up in an English-speaking society cannot understand Japanese without first learning it. These two languages were developed in areas of the world far apart from each other and so were not influenced by each other. Each has different words to represent objects, actions, and qualities.

There are about 3000 different languages in the world today, 1000 of which are spoken by the native peoples of North, Central, and South America. Most of the languages of the aboriginal peoples are spoken by only a few thousand people. In addition, there can be many **dialects** spoken within a specific language. A dialect is a different way of speaking an established language, and it often exists in a particular region of a country.

These women in a small Moroccan village are speaking Maghribi, a local dialect that draws on both the Berber and Arabic languages.

Language and Values

Besides the literal meanings we give to words, there are also moral meanings we give to some words. The moral meanings reflect the values, the concepts of right and wrong, of the society in which we live. In most societies, the word "honesty" has a positive moral meaning. We encourage honesty because it is a quality that helps society to run smoothly and peacefully. The word "aggression" can have a positive moral meaning in a society that settles disputes physically, and a negative moral meaning in a society that values peaceful interaction.

Swearing and profanity are used by many people to indicate strong feelings. Profane language uses religious terms in an irreverent manner. Swearing uses profane language and negative references to body functions. For many people swearing and profanity are immoral, for others an unpleasant manner of communication, and for others a normal manner of communication. The frequent use of swearing and profanity tends to reduce the impact of the words both on the speaker and the listener.

By the time children become adults, they have learned most of the words they will use in their lifetime. The English language contains nearly 1 000 000 words, but it is estimated that no one person has a vocabulary that exceeds 200 000 words.

Language and Social Identification

Canada, being bicultural, has two official languages, French and English. Both the French and the English societies attach great importance to learning language. Without language, socialization cannot take place and people cannot fit into the rest of society. Language binds us more closely together and gives us a common sense of identity. By declaring Canada a bicultural country with two official languages, it is hoped that people will make the effort to learn both languages. In this way the two cultures can develop closer ties and identify to a greater extent with Canadian society rather than with French or English society.

Besides Canada's two main cultures, there are many ethnic groups in Canada that use their own language among group members. This is one important way in which members are drawn together and develop a common sense of identity and purpose. The Polish, Ukrainian, Jewish, Chinese, and Italian communities are just a few of the ethnic groups that have their own newspapers to reinforce and strengthen cultural links. Both spoken and written language are considered crucial to their survival as distinct groups in Canada's multicultural society.

Many of the words that make up the total vocabulary of a language are specialized terms, or **jargon**. They are used by professional and technical groups such as doctors, lawyers, sociologists, and computer programmers, or by special interest groups such as philatelists (stamp collectors) or baseball or hockey enthusiasts. The jargon helps to identify group members. People outside of these groups are not likely to understand the jargon.

Slang, the informal vocabulary that changes through time, can also provide identity. Slang is made up of words that are understood by most people, but that are considered inappropriate for use in business and school. Teenage peer groups frequently use slang words to identify themselves and to create a sense of belonging to the group. Sometimes swearing is used in the same way.

Counter cultures are groups that have values and norms that are different from the rest of society. Drug addicts and criminal groups are examples of counter cultures. They also have a language that is usually understood only by members of that group. Such a language prevents others from understanding the meanings of certain conversations, and provides a method of identifying members of that group.

LANGUAGE AND SOCIAL CLASS

People who are better able to **articulate**, or express themselves well, tend to be more successful in school and in obtaining jobs than those who have difficulty expressing themselves clearly. By using a richer vocabulary, longer sentences, and correct grammar, they make a better impression than those who do not have that ability.

A British language expert, Basil Bernstein, notes that teachers are mostly university educated and middle class. Their expectations tend to reflect middle class values and goals. As well, they often use long sentences and complex words in expressing themselves. He thinks this is true also of textbooks and classroom materials.

Bernstein points out that, in this kind of environment, children from a working class background are less likely to succeed and tend to become easily discouraged.

Bernstein feels that if the materials used were more meaningful to them, working class children would have a much better chance of obtaining an education that would develop their abilities and skills.

PROGRESS CHECK

1. Why is language important in human societies?
2. Why do cultures develop different languages?
3. How is language related to a sense of right and wrong?
4. Why do people swear? What reaction might people have to profanity?
5. How does language draw people closer together? Give examples.
6. Do you feel that Bernstein's observations are relevant to Canada? Explain your answer.

7.2 *Types of Symbols*

The human devised symbols which make up language can take many forms. They can be written or spoken words, body or facial movements, gestures, signs and visual images such as advertisements and illustrations. These various types of symbols, generally classified as verbal and nonverbal communication, make it easier for people to live and work together in society. Group members know the meanings of the symbols and can respond to them accordingly. The symbols permit us to solve problems, share our feelings and experiences with others, and develop and use new knowledge.

Often the situations in which we find ourselves determine how we use language. We choose the words, the tone of voice, the emphasis, and the body language according to the circumstances and the other people. A wedding or a funeral requires more formal language than would be used at a beach party with close friends. When we are with friends the language we use is often different from that used at home, at work, or in the classroom.

Body language too is affected by the situation. Would you wear very casual clothing, tap your fingers, and sprawl in the chair at a job interview? We often communicate unintentional messages to others through our body language. The way you sit, hold your head, walk, and look at others are types of body language that tell people how you feel about yourself and them. Your facial expression and the way you move your fingers, hands, and arms send messages to those around you.

Body language often interacts with verbal communication to increase the effect of the message that is sent. If we are expressing happiness, anger, or sadness, certain types of body language such as facial expressions, arm movements and body posture will interact with verbal language to produce a stronger message. Some experts now believe that body language can increase the level of the emotions we are experiencing and expressing.

Body language, just like verbal language, is learned. Often the similarities in body language are quite apparent in a family. At a very young age we imitate the body language of role models such as parents or siblings of the same sex. There are differences in the ways in which males and females walk, sit, and gesture, and these differences are recognized and imitated by the young child.

Body language is a powerful type of communication in most societies. Psychologists have found that in cases where body language and verbal language give opposing messages, the body language will be believed by the person receiving the messages.

Blissymbolics is a graphic, meaning-based communication system. Some of the symbols are pictographs, that is, they look like the things they represent. A word equivalent appears under each symbol to allow anyone unfamiliar with Blissymbolics to understand what the symbol user is saying.

Blissymbolics used herein derived from the symbols described in the work, Semantography, original copyright © C.K. Bliss 1949. Blissymbolics Communication International A Division of Easter Seal Communication Institute Exclusive licensee, 1982

This is the manual alphabet used in sign language.

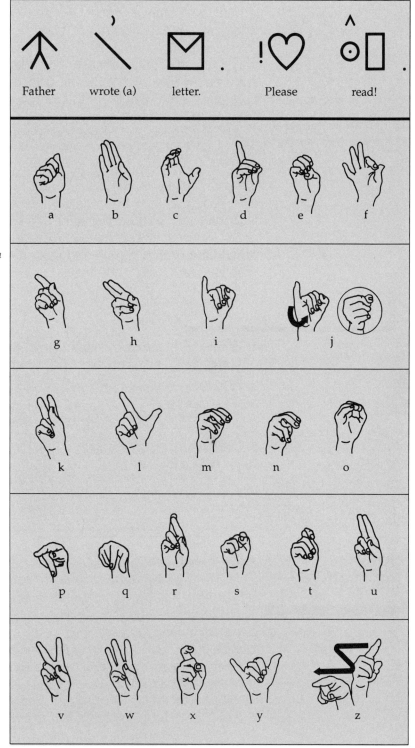

Sign language is a method of communicating through body movements. There are many forms of sign language in use today. American Sign Language is used by the deaf and mute to communicate. The Plains Indians of North America used a system of hand movements so that members of different tribes could communicate and understand each other.

Art and music are forms of communication. Artists try to convey certain images, feelings, and messages through their art. Painting, sculpture, architecture, drawing, and weaving are some of the media artists use to demonstrate their skill and imagination. Music can reflect culture, personal taste and interests. Written music, in the form of symbols called notes, is used to record musical sound and is an international language. It can be understood and played by anyone who has learned to read music.

Progress Check

1. What are the types of symbols that make up different languages? What functions do they perform?
2. How can different types of language reinforce each other? Give examples.
3. Is language learned or innate?
4. List examples of the types of language you use in three particular situations.

CASE STUDY

Koko the Gorilla

Koko, a female gorilla, was born in a zoo in California in 1971. She has made headlines all over the world because of her ability to use sign language and communicate with human beings. Her teacher is Dr. Francine Patterson, a behavioural scientist. Dr. Patterson claims that Koko can use up to 500 signs in American Sign Language to express her thoughts and wishes. American Sign Language is used by the deaf, and is the fourth most commonly used language in the U.S.A.

Koko is now able to use up to six signs together in short sentences. Her use of language has helped to refute the idea that animals are not sufficiently intelligent to learn language.

Other gorillas and chimpanzees have been taught to use sign language and are sometimes referred to as "the talking apes". These primates cannot speak because they do not have the vocal equipment to express themselves in words. But they can communicate with gestures and they can think. They are able to use sign language because they have hands with fingers and thumbs.

Koko made newspaper headlines in 1984 when her pet cat, which she had named "All Ball" in sign language, was killed by a car. She was deeply distressed when told the news, and kept using the words "frown sad" and "sleep cat" whenever she referred to her former pet. She repeatedly asked Dr. Patterson for another cat by using the words "tiger, please, tiger". In 1985 she was given another striped kitten which she refers to as her "baby". People all over the world were fascinated with Koko's ability to communicate her emotions in words. She and some other "talking apes" have learned to use language and to communicate and reason in a manner that is similar to human beings.

When Koko becomes angry she describes herself as "red rotten mad" in sign language and if she becomes annoyed with Dr. Patterson she calls her a "dirty toilet devil". She also has been known to tell lies. Once she sat on a sink which caused it to fall — she has a mass of 82 kg. When she was asked who broke the sink she blamed Dr. Patterson's assistant by gesturing "Kate there bad".

Koko and Dr. Patterson

In captivity gorillas live up to 50 years of age — in the wild they live to about 30. Koko still has a long life before her and perhaps her use of words and her knowledge will expand even further. Her example has raised many questions about the future use of apes. Could they be trained to perform simple tasks like cleaning and serving? Is it worthwhile to try to socialize large numbers of them and teach them to communicate with human beings?

One certainty has resulted from Koko's training and other recent studies — we now know that gorillas have the ability to think, reason, and communicate.

1. Why is Koko referred to as a "talking ape"?
2. What "human" qualities does Koko possess? Are these innate or learned?
3. How do you think apes should be treated or used in the future?

This three-year old chimpanzee, Lana, punches buttons to form a sentence. The language, called Yerkish, is composed of a complex system of geometric symbols. Lana now has a vocabulary of 75 words.

7.3 How Do We Express Ourselves?

Language permits us to exchange information, to be entertained, to argue, to socialize, and to earn a living. Without language we would not be able to identify with the rest of society.

As children learn language they also learn how to think. Having words for objects, actions, and qualities aids children's thought processes. But, not all thinking requires language. Some of the links in a chain of thought are unconscious and require no words. Many times we arrive at the answer to a problem in a flash of insight.

Language is used to pass on knowledge from one generation to the next. This has enabled various cultures to expand upon the skills and knowledge of their ancestors.

The development of written language has helped the transfer of knowledge from generation to generation. Written language also enables us to share knowledge with people in other parts of the world. It is more accurate than spoken language since it does not depend as much on memory. Written language cannot be misinterpreted as easily as spoken language.

Because language has helped in the accumulation of knowledge, the creation and development of complex technologies has been made possible. Language has, in turn, expanded because of the new vocabularies needed to deal with rapidly developing processes and products.

One of our basic needs as human beings is to have a feeling of belonging. Language allows us to establish friendships by communicating with others on a social level. We share our thoughts and feelings with friends. Being able to talk to a close friend, when we have a problem or are upset, can help us to feel better. When we confide in our friends, telling them things we do not wish others to know, we expect the information will not be passed on. If it is, we may feel angry and betrayed. How much of your day is spent talking about yourself or gossiping about others?

LISTENING

We not only talk to others about our opinions and feelings, we also spend a great deal of time listening to others expressing themselves. Listening is just as important to communication as speaking; without a listener, communication is incomplete.

One expert, Jesse S. Nirenberg, explains that people tend to listen at one of three levels: non-hearing, hearing, and thinking/analyzing. The level at which they will listen depends on who is speaking and the listener's interest in the subject being discussed. At the non-hearing level, people pretend to listen, commenting "yes" and "no" and shaking or nodding their heads, but they are not really hearing the conversation. At the hearing level, the conversation is listened to and remembered but it does not make any lasting impression on the listener. Third level listening involves thinking about and analyzing what is being said. According to Nirenberg, everyone takes part in all three levels of listening in the course of a day.

A person who is not interested in a conversation or is thinking about something else will "tune out" the speaker. This is a form of censorship with the listener controlling what is heard. Most people have been told on occasion "you only hear what you want to".

It is estimated that the average person's attention span is rarely more than 45 seconds. It is also estimated that each person spends about 50 to 80 percent of the day listening to others. It is said that any listener hears only half of what is said and understands only 25 percent of what is heard.

Another expert, Joseph A. DeVito, proposes three types of listening: listening for enjoyment and relaxation, listening to obtain information, and listening to friends who are seeking advice. The first type of listening could be listening to music. We use the second type of listening when we listen to our parents, teachers, and coaches to learn more about certain subjects and to improve our skills. When a friend has a problem, we listen to the circumstances and evaluate the information in order to offer advice or make suggestions. Again, the level at which we listen will depend on our interest in the subject and in the speaker (or musician).

Illiteracy

We learn language as small children and when we start school, we learn to read and write. **Literacy**, the ability to read and write, is important throughout our lives. Being able to read and write enables us to function in a literate society. We use these skills in many aspects of day-to-day living such as learning at school, reading newspapers, banking, and filling out forms for a driver's licence or an income tax remittance.

Illiteracy, the inability to read or write, is generally thought to be a problem of the underdeveloped countries of the world. Most Canadians do not think illiteracy is a problem in Canada. However, there are one million Canadians who cannot read or write at all. Of these, 70 percent were born in Canada. The difficulties illiterates face in the workplace are reflected in Canada's unemployment figures. Sixty percent of illiterate Canadians cannot find or keep a job and must rely on welfare and unemployment insurance. These welfare payments cost the government about $2 billion each year.

It is difficult to imagine how great a handicap illiteracy is in a literate society. Actions that are automatic for literate Canadians are impossible for the illiterate. They cannot read a newspaper, fill in an application form, write a letter or cheque, read directions on labels, or read the instructions for operating machinery and appliances.

In 1983 a federal government Commission reported that a Grade 8 education is the minimum level of education a person needs in order to function in Canadian and other western societies. Those who do not have this level of education are considered to be **functionally illiterate**; they cannot read or write well enough to function properly in society. According to the 1981 census results there are about 4 million Canadians with less than a Grade 9 education. It is also estimated that 40 percent of the 12 500 prisoners in federal prisons are illiterate or functionally illiterate.

What are the causes of illiteracy? Most experts believe that people born into poverty have a greater chance of being illiterate. Poor families are generally concerned with daily survival needs. They do not have the time or energy to encourage their children to remain in school. If the parents are illiterate, there are likely no reading materials at home to stimulate the children. As well, the parents do not have high expectations for their children.

Another cause of illiteracy is **dyslexia**, which is an extreme difficulty in learning to read. For a long time dyslexia went unrecognized and untreated. Now it is being detected in the schools and various treatments are conducted to try to correct it.

Sometimes, if children who are experiencing difficulty with reading and writing are not helped, they will fail to learn these skills. Professor Bayne Logan of the University of Ottawa reports that children can become anxious about learning new words if the meanings of the words are not properly explained to them. As a result they may doubt their ability to learn new words, and their doubts may lead to a learning disorder.

Another factor thought to contribute to illiteracy is television. Children who watch a great deal of television, and who are not encouraged to read for pleasure, often find it difficult to read.

What can be done to reduce illiteracy in Canada? Governments are now spending millions of dollars on programs to combat illiteracy. Teachers have been hired and volunteers recruited. Community and business groups are also concerned and are supporting volunteer programs. Many schools have special remedial programs geared to learning disabilities and illiteracy.

Progress Check

1. Why is it stated that not all thinking requires language?
2. What are the functions of language?
3. According to J. S. Nirenberg, what are the levels of listening? For each level, make a list of situations in which that level of listening might be used.
4. Do you agree that people tend to censor what they hear? Explain your answer.
5. What are the three types of listening J. A. DeVito described? Which type do you use most?
6. What problems do illiterate people experience?
7. What are the causes of illiteracy? How can illiteracy be reduced?

7.4 *Becoming Literate*

At birth, we know no language and we must communicate nonverbally using gestures and expressions of emotion such as cries and laughter. At first infants cry to draw attention to their needs. They make gurgling sounds which usually indicate contentment. Later they start to babble and make murmuring sounds.

Usually within the first year of life, babies begin to associate certain sounds with particular objects, people, and actions. Parents teach the baby by pointing to objects and naming them aloud. If the baby responds with a similar sound, the parents will praise the child and ask for a repetition. Through praise, encouragement, and attention the baby is motivated to learn more new words.

Each child learns to speak at a different rate. On average, children utter their first words at about eight months, usually "dada", "mama", or the name of a favourite toy. By the time children are about two years old they have a spoken vocabulary of approximately 300 words and are beginning to put words together to express their wishes, for example "mommy up" or "go outside".

Being able to express their wishes gives children a new sense of power. This provides additional incentive to learn more language. By the age of three most children are inquisitive about their environment and are constantly asking "why".

Between the ages of four and six, most children have a good command of language, and it is felt that they are ready to enter school. At six, the average child's vocabulary is about 2 000 words, as compared to an adult's of between 35 000 and 70 000 words.

Children who are encouraged by their parents to read, write, talk, ask questions, and explain what they are doing and observing are given an excellent chance to develop language skills. If their sense of language is well developed, they are likely to be confident about their abilities to follow directions, solve problems, and make sense of new situations. On the other hand, children raised in an environment which stresses that "children should speak when they are spoken to" and "children should be seen and not heard" have not been offered a proper chance to practise language, and their language skills will not be fully developed. They will be at a disadvantage in school.

Programs have been set up in many schools to help students with special needs. English classes for newly arrived immigrants and classes for those who need help to improve language skills are in place in many areas.

Experts agree that school makes enormous demands on students. They are required to speak and behave in a certain manner, and to observe the rules of the school. Those students who find it difficult to adjust to the school environment may develop behavioural problems. An inability to adjust and fit into the school community often results in feelings of dissatisfaction and inadequacy. The student may decide to simply give up and drop out of school.

Many people feel that if students who are experiencing difficulty at school are given encouragement and support both at home and at school they will be more likely to overcome their problems. Extra attention, sensitivity, and respect on the part of teachers and support staff may provide the necessary caring environment in which the student can develop.

"The word makes men free. Whoever cannot express himself is a slave. Speaking is an act of freedom ..."
Ludwig Feuerbach

"Language most shows a man; speak that I may see thee!"
Ben Jonson

PROGRESS CHECK

1. Outline the stages of language development in children.
2. What encourages children to learn more words and develop their language skills?
3. Why do some children find it difficult to succeed in school? What do you feel can be done to help these children?
4. Explain the meaning of one of the quotations shown above. How do you think the wording might be different if the statement were made today?

7.5 *The Development of the English Language*

LANGUAGE IS SHAPED BY CHANGE

The English language originated with two groups of German-speaking barbarians called the Saxons and the Angles who invaded England about 1500 years ago. (The familiar term "Anglo-Saxon" refers to descendants of the invaders.) The Saxons invaded England in 477 A.D. About 100 years later the Angles followed. The barbarians pushed the Celts, the people living in England at that time, north and west into Ireland, Wales, and Scotland. The Celts retained their language, Gaelic, and it is still spoken by descendants of the Celts in parts of

Britain. The language of the invaders became the language spoken in the conquered parts of England. Their vocabulary consisted of only about 2000 words.

Earlier, Britain had been part of the Roman Empire. Latin-derived words such as "sign", "pound", "street", "wall", "kettle", and "gem" were retained and added to the German vocabulary.

In 1066 A.D., the Normans, led by William the Conqueror of France, invaded and conquered England. Many French words were added to the language spoken in England.

All languages change as societies undergo change. New words are found to describe new products, ideas, situations, and methods. This has certainly been true of the English language.

It is interesting to read passages from the English poet Geoffrey Chaucer (1340 – 1400) and from William Shakespeare (1564 – 1616). The language used by the two writers is dramatically different although separated by only a few generations.

A Knight ther was, and that a worthy man
That fro the tyme that he first bigan
To ryden out, he loved chivalrye,
Trouthe and honour, fredom and curteisye.
Ful worthy was he in his lordes werre,
And therto hadde he riden (no man ferre)
As wel in Cristendom as hethenesse,
And ever honoured for his worthinesse.
The Canterbury Tales, Geoffrey Chaucer

When first this order was ordained, my lords,
Knights of the Garter were of noble birth,
Valiant and virtuous, full of haughty courage,
Such as were grown to credit by the wars;
Not fearing death, nor shrinking for distress,
But always resolute in most extremes.
He then that is not furnished in this sort
Doth but usurp the sacred name of knight,
King Henry VI, William Shakespeare

Change also occurs when two groups of people who speak the same language are isolated from one another over a long period of time. The English of British settlers arriving in Canada two hundred years ago changed with time. So did the English of those remaining in Britain. Today the accents and many of the words and expressions used by Canadians are remarkably different from those of the British who remained in Britain.

Some of the languages of earlier civilizations, such as Etruscan and Minoan, have died out whereas others, such as English, have developed and expanded. The success or failure of a language to survive depends on the fortunes of the group of people who use it.

CAREER PROFILE
HELEN KELLER

Helen Keller was born in 1880 in the state of Alabama. At the age of 18 months, she became permanently blind, deaf, and mute as the result of an illness. Being deaf and mute she was unable to learn language in order to communicate with others.

Despite the love her parents gave her, Helen was frustrated, unpredictable, and bad mannered. Lacking the ability to see, hear, or speak meant that she could neither understand right from wrong nor what was expected of her.

When Helen was seven years old, her parents hired a young woman by the name of Annie Sullivan to be her guardian. Annie had been nearly blind for the first 16 years of her life, but as she got older her sight had improved. She had experience working with blind people, although never anyone who was deaf and blind. At first Annie and Helen did not get along well because Annie refused to let Helen have her own way. She also insisted that Helen develop proper table manners. There were many physical fights between the two, but Annie was stronger and Helen soon learned to eat properly.

Whenever Helen wanted a particular item, she would simply grab for it. Annie started to teach Helen the word for the object by spelling it with her finger on the palm of Helen's hand. At first

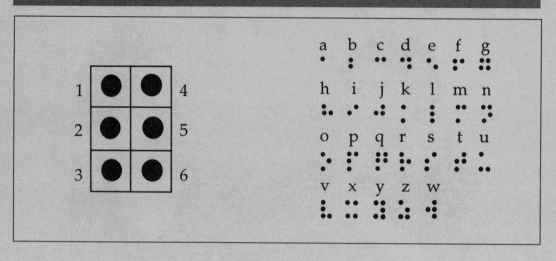

Six magic dots — the basis of the Braille language

Helen did not understand that the shapes Annie was tracing on her hand represented the objects she held. One day, after only a month of training, Helen was holding her hand under running water while Annie was spelling "water" on the other hand. Helen suddenly made the connection between the feeling of the word and what it represented, and became very excited. She had learned to communicate by finger language and the sense of touch.

Being very bright, Helen learned quickly to spell other words and to make her wishes known. Her temper tantrums ceased and her energy was directed to learning about the world around her.

Helen at this stage could only communicate in single words. Annie gave serious thought to the problem of teaching sentence structure. She knew that children learn to communicate in sentences by imitating their parents. They learn to understand the meanings of words other than those that represent objects. Then they begin to imitate the sentence structures and connecting words they hear. Annie decided to use sentences to communicate with Helen. At first Helen only understood the words that represented objects in the sentences Annie used. But after a while she learned to imitate the way Annie expressed herself and started to place words together in sentences and questions. It seemed that a miracle had occurred.

In addition to finger language, Helen soon learned **Braille**, the language used by the blind. It is a system of writing in which letters and numbers are represented by raised dots that can be identified by the sense of touch. Helen's desire to learn was amazing. She even learned to print letters on paper although she could not see what she was writing. Another amazing accomplishment was that she learned to speak, though unable to hear what she was saying. Her dedicated teacher Annie was always by her side to help her learn new things.

At the age of 16, she was admitted to Radcliffe College, graduating with honours in 1904. She began to write articles and books about her experiences as a blind, deaf, and mute person. Helen's energy seemed to have no limits, and she gave lectures all over the world. She also visited blind people and gave them encouragement, hope, and advice. After World War II she visited many hospitals and worked with veterans who had lost their sight in the war.

Helen lived a long and productive life. Her success was an example and an inspiration to others. Through determination, effort, and the help of those who loved her she overcame seemingly impossible handicaps and led a happy and fulfilled life. She died in 1962.

"That living word awakened my soul, gave it light, hope, joy, set it free .... Everything had a name, and each name gave birth to a new thought."
HELEN KELLER

"The two greatest characters in the 19th century are Napoleon and Helen Keller. Napoleon tried to conquer the world by physical force and failed. Helen tried to conquer the world by power of mind — and succeeded."
MARK TWAIN

Language and Culture

The language of a particular society often reflects its attitudes and beliefs and even its physical environment. In the Inuit culture, an accurate knowledge of snow conditions is important for survival. There are thus many words to describe the different types of snow; for example, hard crusty snow is called "sillik"; packed snow is "aniu"; and fresh powder snow is "nutagak".

The English language includes many terms that contain the word "man" — mankind, man-hours, manpower, manager, chairman, and policeman are a few. These terms reflect the traditional attitude of a society in which men had control of the marketplace and were the main wage earners. Only recently has there been progress toward replacing these words with non-sexist and non-biased terms to reflect our society's changing attitudes toward women.

Transportation is of great significance to Canada, the second largest country in the world. There are many words in our vocabulary that reflect the importance of transportation as an economic link in our society. A few examples of words developed to deal with motor travel are: "highway", "expressway", "throughway", "freeway", "underpass", "overpass", "cloverleaf", and "lane". Can you name some others?

Money, education, and communication are also topics of importance to Canadians. What are some of the words that come to your mind when you think of these subjects?

Often languages borrow words from other languages when the need arises for a new word, since borrowing a word is easier than coining one. The Australian aboriginal word "boomerang" and the Aztec words "chocolate" and "tomato" were adopted and incorporated into our language. Many English words are derived from terms used by the Amerindians, for example, "mocassin", "mackinaw", "skunk", "woodchuck", "jaguar", "kayak", and "toboggan". Countless other words have been adopted from other cultures, for instance, "spaghetti", "opera", "taboos", "tea", "coffee", "pajamas", and "samurai".

INTERNATIONAL LANGUAGES

As human knowledge expands, the world in which we live changes. The English language is one of several modern languages used to record the changes that are taking place. At international scientific and technological conferences, English is usually one of the official languages used by delegates.

Other languages like Russian, German, and French are also used because of the importance of science and technology in the U.S.S.R., Germany, and France. Chinese is probably the most widely spoken language in the world today because of China's large population (estimated to be more than one billion people).

PROGRESS CHECK

1. Why does language change over time?
2. How can the study of language reveal the beliefs and attitudes of the society that uses that language?
3. What are the main languages used at official international gatherings?

OVERVIEW

Language comes in many forms and is used for a multitude of purposes. Different types of language reinforce the messages that people receive and send. Language can reveal many things about people — their knowledge, attitudes, confidence, opinions, social class, education, interests, beliefs, feelings, and their ability to communicate. Language permits people to learn and to communicate with others. Without language, cooperation and group situations would be impossible. Human societies would not exist without language.

Industrial societies like Canada require people to be literate in order to read instructions and to communicate with others. Illiteracy in Canada is a serious problem, affecting millions of people.

Language changes as societies develop in new directions. New words are borrowed, combined, and invented to describe new situations, products, and processes. Language permits humankind to build upon existing information and to invent new knowledge. Language enables us to express our emotions and to explore new ways of dealing with both positive and negative experiences. Children gradually learn language and develop skills and abilities that permit them to function and fit into society. The English language is one of the most used languages in the world, and is constantly expanding and changing. A world without language is hard to imagine.

KEY WORDS

Define the following terms, and use each in a sentence that shows its meaning.

dialect	articulate	functionally illiterate
jargon	literacy	dyslexia
slang	illiteracy	Braille

KEY PERSONALITIES

Give at least one reason for learning more about each of the following.

Basil Bernstein
Koko
Francine Patterson
Jesse D. Nirenberg

Joseph A. DeVito
Bayne Logan
Helen Keller
Annie Sullivan

DEVELOPING YOUR SKILLS

FOCUS _____

The focus of this chapter is on the importance of language in human societies. Students and sociologists want to know about the types of language we use to express ourselves and about the functions of language. In order to investigate the importance of language in Canadian society it is essential to ask good questions. Divide the class into groups and brainstorm a list of possible questions that you might ask about the use of language and its importance. Each group should have someone write down all the suggested questions and ideas.

At the end of the brainstorming session select the best questions that were generated in your group about language. The best questions would be those that relate to the focus of the chapter. They should be clear, interesting, and answerable. One quesion might be "what would the world be like without language?" Another might be "what kind of person would I be if I had not acquired some language skills?" Both questions deal with the importance of language but they raise different points of view and observations — one question speculates about a group and the other about an individual situation. These separate but related questions may lead students to organize their speculations from different approaches. Each group in the class should compare the results of their investigations with the other groups.

ORGANIZE

1. Develop an organizer to compare the different types of language used in Canadian society and the situations in which they are used.
2. Design an organizer that lists the advantages of possessing a fully developed use of language and the disadvantages of limited language abilities.
3. Create an organizer that lists the influences and events that have changed the English language.
4. Create an organizer that compares situations in which language acquisition in children would be encouraged and discouraged.

LOCATE AND RECORD

1. Refer to the quotations, career profile, and case study in this chapter and identify the opinions and ideas that support the importance of acquiring a competent use of language.

2. Refer to newspapers, television, radio, library books, personal experiences, or conduct interviews to do the following assignment. Divide the class into groups and have each group select an occupation, sport, club, group, or activity. Have each group compile a list of jargon, actions, and ideas that are used in these particular choices. The meanings of the jargon and ideas should also be written down in an organizer.
3. Research how insects and mammals other than humans communicate. Present your findings to the class.
4. Divide the class into groups and research the contributions of other languages to the English language. Display boards can be used to illustrate the findings of each group.

EVALUATE AND ASSESS

1. Is silence a form of language? Compare the meanings of the following quotations on silence. Which authors are most correct in your opinion and why?

    Silence
    Don Fabun noted that "the world of silence may be a cold and bitter one; like the deep wastes of the Arctic regions, it is fit for neither man nor beast. Holding one's tongue may be prudent, but it is an act of rejection; silence builds walls — and walls are the symbols of failure." Thomas Mann in one of the most often quoted observations on silence said, "Speech is civilization itself. The word, even the most contradictory word, preserves contact; it is silence which isolates." The philosopher Karl Jaspers observed that "the ultimate in thinking as in communication is silence." Max Picard noted that "silence is nothing merely negative; it is not the mere absence of speech. It is a positive, a complete world in itself."

2. Colours are often associated with both negative and positive images and situations. Make a list of colours and the various meanings associated with them; for example, purple is sometimes associated with royalty. Compare your data with those of other students. What colours do you use to send messages when expressing yourself?

SYNTHESIZE AND CONCLUDE

1. Compare the different tones of voice that are used when communicating with others. What moods or messages are conveyed through tones of voice?
2. Carefully observe the ways people communicate with each other both inside and outside the classroom. From these observations what conclusions can be reached about their levels of education, maturity, peer group, roles, and social class?

APPLY

1. Write a poem or short essay on the importance of language to human beings and their various activities.
2. Canada is sometimes described as a multicultural society which encourages the use and preservation of different ethnic cultures, languages, and traditions. But some sociologists believe that assimilation into Canada's bicultural society is inevitable as the media, the schools, and peer groups promote the use of English and French. They argue that as soon as children of ethnic groups learn and use Canada's two official languages, the process of assimilation has begun. They believe that the use of certain words changes and influences our values, beliefs, and attitudes. Discuss this idea with your classmates and present evidence to either support or reject it. Is it inevitable that all minority groups will be assimilated eventually, and lose their original ethnic language?
3. Arrange with your teacher to have a blind person speak to the class about living and working in your community.

COMMUNICATE

1. The class can be divided into groups that will each research one of the following statements. Research information can be found in this chapter and through library sources.

- Age and social class affect the learning and use of language.
- Language is essential to represent oneself to others.

- Language is a vital tool in the socialization process.
- Body movements and gestures do not merely reinforce other types of language — they are a language in themselves.
- A person's manner of speaking will depend on the particular situation. Language facility determines people's confidence and abilities.
- Words create images in the minds of people.

Each group will summarize its findings and conclusions, and make an oral presentation to the class.

2. On the basis of these statistics ask yourself whether you think that Canada is (a) unilingual, (b) bilingual, or (c) multilingual. Offer reasons to support your point of view.

1986 Population Census for Canada – A Report on Languages Spoken in Canada

Total Population 25 400 000	Number	Percentage of Population
Number of Canadians who said English was their first language	15.3 million	61%
Number of Canadians who said French was their first language	6.2 million	24%
Number of Canadians who spoke more than one first language (English and French)	1 million	4%
Number of Canadians who speak a first language other than French or English (2.1 million of these spoke a European language, 634 000 spoke a language from Asia or the Middle East, 138 000 spoke an aboriginal language)	2.9 million	11%

3. Predict how human communication will likely change in the future, based on your knowledge of today's technology, and proposed future technology. Include in your brief essay how you think increased communication through advanced technology will change our attitudes about other societies.

CHAPTER 8

The Influence of the Mass Media

How are we affected by mass media? Are we being manipulated? How does mass media reflect our Canadian society?

INTRODUCTION   Your life would be affected by the **mass media**, even if you never watched television, never went to the movies, or never read a newspaper. That is because so many people do these things, that a mass audience for the media is formed. The noun "medium", from the Latin, stands for a way or means to do something. The plural form of the word is "media". The mass media, therefore, are the several ways of communicating with the greatest number (the mass) of the people. They bring us the comics in the daily paper and weather reports by satellite.

Television and radio networks, the film industry, and the newspaper and magazine publishing businesses provide the modern means of communication. They also provide a strong influence in our lives, because people are affected by the images presented to them. Our society is reflected by these images. Is the reflection an accurate one? Are we **manipulated** or managed skilfully by the media to see ourselves in certain ways? Do we have some control over the images presented by the media? To answer such questions we must examine the various media and how they interact with us.

8.1 Radio

It seemed a miracle to be able to hear news, sports, music "over the air waves". There are people who can recall still Foster Hewitt's first hockey broadcast in Canada on March 22, 1923. Hockey as a national sport could be said to have been created by radio as thousands of Canadians listened to "Hockey Night in Canada".

On December 23, 1900 a Canadian named Aubrey Fassenden successfully made the first radio broadcast from an experimental laboratory in the U.S.A. At first radios were powered by crystals and batteries and people used earphones to listen to radio broadcasts. Speakers soon replaced the earphones, and electricity became the new source of power for radios. By 1919 regular radio programs began in Canada, and by 1922 radio broadcasting was in operation throughout the country. By 1929 there were 300 000 radios in Canada.

Technicians with radio equipment around 1920.

But there was a wider choice of programs on American radio stations that attracted Canadian listeners. In 1932 the federal government established the Canadian Broadcasting Corporation, the CBC, to support and encourage Canadian performers — a frontal attack on the increasing popularity of American shows.

Radio, in general, and the CBC, in particular, did bring Canadians closer together. They listened to the same newscasts and shows, coast to coast. The intensity of this linking of Canadians was likely no more apparent than during World War II. The wartime speeches of government leaders, Churchill, Mackenzie King, and F. D. Roosevelt, were heard across the nation. Canadians felt united by the broadcasts they shared.

The popularity of television in the 1950's forced radio stations to change. Radio was no longer the only electronic medium in the country. Radio stations began to specialize in the programs they offered — rock music, symphonic music, news broadcasts, political commentaries.

Despite the in-roads made by television on the time people spend listening to electronic media, radio remains popular. Radio is particularly suited to the individual listener. You can use your headsets, your "ghetto blasters", or your clock radio. In each case, you have a personal communication system. As Peter Gzowski of the CBC's *Morningside* says: "when people listen to the radio they listen alone, in a car or while they're doing the housework, in the background at the office, or on earphones on the tractor."

The intensity with which whole families listened to a radio broadcast in the 1940's may have changed. Radio is no longer the only means of communication. Yet, the personal quality of radio as you listen to your favourite station, your favourite disc jockey, your favourite program, is still strong.

"God gave man two ears, but only one mouth, that he might hear twice as much as he speaks."
Epictetus, the Stoic, an early Greek philosopher

"If the issue of Canadian identity is less an obsession than it was twenty or thirty years ago, it's because the CBC has been providing us with an accurate mirror in which to see and know ourselves. No other Canadian institution comes close to doing as good a job."
Morris Wolfe in *Radio Guide*, November 1986

8.2 Films

One hundred million individual tickets are sold at Canadian movie theatres each year — 70% of them to people under the age of twenty-five. Many of the films made today are directed at young people. The way actors speak, dress, and behave, and their attitudes towards each other and the rest of society, contain powerful messages to young people.

By trying to assess current attitudes, movie producers have made films that reflect the values and concerns of society. Consider the many recent films on Viet Nam. Together they represent the still unresolved feelings of society to this conflict. Think about the making of *On Golden Pond* at a time when society is beginning to concern itself with an aging population. What kinds of statements on our society have been made by movies like *Moonstruck*, *Three Men and a Baby*, or *Dirty Dancing*?

8.3 Television

Television was invented in 1920, but the research and development of this new electronic medium was slowed down by World War II. However, by the early 1950's television was available to most Canadians.

Today television is one of the most popular, and powerful, forms of media. It is estimated that by the time of high school graduation, a student will have watched about 15 000 hours of television. This compares to about 11 000 hours spent in a classroom. There are television sets in 98.5% of Canadian households, half of which have more than one set.

TV's ability to present information almost instantaneously with an event gives viewers a sense of being in touch with the world. An individual living room becomes a window on one's community and nation, as well as on nations kilometres away. But, much of television programming is entertainment — plays, music, situation comedies.

As with films, television programs project images of certain careers, certain personality types, certain decision-making situations. There have been many debates about whether people and jobs are stereotyped on television, about whether TV images reflect accurately the values and mores of our society. The debates will continue, but one point of agreement seems to be universal. Viewers must maintain their ability to differentiate between what is simply entertaining and what is realistic. For example, an hour-long program with a theme of alienation or loneliness, or family fights may be presented in an entertaining fashion. Yet, because the program took an hour to resolve an issue does not mean that the same social issue could be resolved within an hour of real-life time.

> "Television has turned out to be one of the most phenomenal of all human inventions. It touches lives more intimately and pervasively than anything that predates it, with the possible exception of the discovery of fire."
>
> Ellen de Franco in *TV On/Off*

ROCK VIDEOS

Rock videos are short, three- or four-minute television programs that promote rock singers, bands, and their albums. The music is combined with images that display rock stars as young, fast-moving, and living in luxury. Videos are designed to sell records and make singers and bands more popular.

Enormous sums of money can be spent on a single video: for example, Michael Jackson's *Thriller* cost $14 million to produce. In North America alone, record companies spend over $100 million a year producing new music videos which, in turn, encourage the sale of more than $5.5 billion in records.

> *"All you really have that really matters are feelings. That's music to me ... That's what this music is all about. It's about feeling. It's about wanting. It's about needing and cramming yourself full of it."*
>
> Janis Joplin

8.4 Printed Materials

Johann Gutenberg invented the first printing press with movable type in 1450 in Germany. This invention encouraged the production of several forms of printed material which, in turn, expanded human knowledge. That process of learning continues for us as we read and use newspapers, magazines, and books on a daily basis.

NEWSPAPERS

Seventy percent of the Canadian population over 50 years of age (and 55% of those between 18 and 24) read a newspaper every day. It is apparent how influential papers are considered to be, when you look at the amount of advertising placed in them. More advertising is sold by newspapers than by any other medium — 25.1% of all advertising in this country (worth $1.5 billion a year) compared to only 17.1% for television.

Because they are so widely read, but also because they can be selective in their news coverage and arrangement, newspapers are worth careful analysis. This is important, because most of us tend to trust and accept what we read more than what we view on television.

Newspapers compete energetically for readers. Healthy competition means trying to get the best coverage of the news and hiring reporters and columnists who will inform or entertain their readers. But some newspapers are also known for competing for the lowest level of reader interest, a taste for sensationalism.

MAGAZINES

There is a magazine for every interest or collection of interests. Like newspapers, magazines can provide both information and entertainment. There is often a point of view that is taken by the writers of a magazine. Readers must be able to recognize that point of view and evaluate what is really being said. As Patrick Watson, a Canadian journalist, has said, "many journalists agree wtih deep conviction that their job is to report in a value-free and non-judgmental way ... but no journalism is ever free of value judgments or bias."

BOOKS

Books were the first medium of communication to reach a wide audience, after the invention of the printing press. But, today they are not considered a part of the mass media. A book is read privately or in a classroom situation. Although many books have attracted a large readership, their impact on society is not considered to be as strong as television, radio, films, or newspapers and magazines. Thriving bookshops and public libraries indicate the impact that a book can have on individuals, however.

Progress Check

1. How has radio programming changed over the years?
2. How many hours in a week do you listen to radio? Is it usually the same station? If yes, why? If no, why?
3. Why do you think many films are directed toward the teenage market?
4. Review three films you have seen in the last six months. Did you agree with the film's portrayal of society? Explain your answer.
5. Which videos have made you want to buy the record?
6. Why do you think fewer young people read newspapers than older people?
7. Why are books not considered part of the mass media? Do you agree?

8.5 *Media Literacy*

Courses in media literacy are now being taught in high school. These courses teach students to analyze television programs, film scenes, and newspaper/magazine writing. Critical thinking skills are needed to reduce and/or prevent manipulation by the media.

CAREER PROFILE
MARSHALL MCLUHAN (1911-1980)

Marshall McLuhan was born in Edmonton, Alberta and graduated from the University of Manitoba in 1934. He received his Ph.D. in 1942 from Cambridge, England. In 1946 he became a professor of English at the University of Toronto, where he taught until 1979.

In the 1960's Marshall McLuhan became a world celebrity for his ideas on communication and the effect of mass media upon people. He believed that if the media was used properly and its effects understood, a better world could be created. He thought that the media could unite people and create positive attitudes and goals. McLuhan believed that the media would also make people more independent of each other, more individualistic.

He wrote several books about the effects of the media on culture and individual behaviour. He observed that the media was changing peoples' values, attitudes, and lifestyles. He thought that human beings have a balance among their five senses — sight, hearing, smell, taste, and touch. The balance between the senses changes according to the environment. For example, if the visual sense is eliminated, as in the case of a blind person, the senses of hearing and touch are increased.

McLuhan believed that when print was invented by Johann Gutenberg, peoples' sense of sight increased at the expense of the other senses. Through the medium of print, technology, industrialism, nationalism, and individualism were made possible. The invention and use of radio and television in the twentieth century helped to correct the imbalance of the senses. In McLuhan's opinion radio increased the sense of listening and television increased the senses of hearing and touch, and decreased the sense of sight.

McLuhan referred to television as a "cool medium". He explained that television forces people to fill in the gaps in the messages received and stimulates the imagination. This creates more participation and involvement than in "hot mediums" that include radio and film. "Hot mediums" provide everything in their messages and permit little participation because there are fewer gaps. McLuhan argued that newspapers were used as "wrap arounds" that allowed people to escape from each other at public gatherings and on public transportation.

McLuhan believed that people could only obtain control of the media if they came to understand its effects. Until this understanding was achieved, the media would continue to control and change peoples' lives. He thought that television could be a positive force, but he also stated that children must be able to read and write fluently before they were exposed to television. Otherwise, they might lose their identity and become mere reflections of the television.

Marshall McLuhan provided the world with new ideas, and made more people think about the effects of the media upon individuals and societies. Robert Fulford, the editor of *Saturday Night Magazine* referred to McLuhan as the "most important thinker in communication". During his lifetime he received many honours: the Governor General's Award in 1962 and a Companion of the Order of Canada in 1970. He enjoyed a full career as teacher and writer.

Television can be viewed as entertainment placed within the context of our society. Many of its messages are accurate; some messages are subject to the length of the program, the needs of the advertisers, or the network's perception of the audience.

Special effects and stunts in television and in films can be studied. An understanding of how something is created can help in the understanding of what the particular effect is supposed to convey to the viewer.

Newspaper and magazine articles are often constructed to create certain impressions. The deliberate use of headlines, photographs, and size of type are meant to send a particular message to the reader. Student awareness of these techniques can develop the skills necessary to understand or question the motive behind any story. Are writers only trying to make us curious or appeal to our emotions with certain devices or techniques?

Mass media has created what Marshall McLuhan called a "global village". We are instantly aware of events in our community and throughout the world. It is essential that we develop the tools to understand what is being said in our village.

8.6 *Advertising in the Media*

Commercial radio stations, television networks, newspapers, and magazines depend on the revenues obtained from paid advertisements to remain in business. The goal of advertising is to create a connection in people's minds between their desires and the product that is being advertised. If this is accomplished, more people will buy the product and the profits of that particular company will increase. Advertisements are designed to appeal to people's emotional needs — the desires to be popular, attractive, happy, loved, and comfortable.

It is estimated that Canadians are exposed to an average of 500 ads a day from one source or another. Eventually, most of these commercials are ignored or screened out by the reading or viewing public, but the potential power to influence consumers through advertising in the media is enormous.

CHILDREN AND ADVERTISING
Children under the age of ten are easily influenced by advertisements directed at them. Young children find it difficult to separate the program they are watching from the commercials. They believe that the characters in the commercials are an extension of the program. Children cannot understand easily that companies make money through the sale of their products. They also find it difficult to understand or accept their parents' remarks that they cannot afford to buy particular toys or games. Some of them think that their parents receive their money from the stores as change, when they buy something. Children who watch three to four hours of television each day come into contact with about 20 000 ads each year.

In Canada, the government does not directly control the content of television commercials. Instead, television and radio stations and advertisers follow guidelines in *The Broadcast Code for Advertising to Children.* Under this code no ads can urge children directly to buy or to ask their parents to buy particular products.

There is a growing awareness among parents that certain programs originally designed to entertain children have turned into lengthy advertisements to sell toys based on the program's characters. Pressure is mounting on broadcasting stations and government to bring in regulations to control this combination of "show and sell".

DISCRIMINATION IN ADVERTISING

Commercials showing women only in non-decision making roles or as sex objects are also no longer acceptable. Advertisers who present these stereotyped images and who do not show women as decision-makers, professionals, or responsible business persons are being criticized by concerned citizens and groups. Some commercials have begun to show men caring for children, making meals, and cleaning the house. The goal is to have commercials that portray people, regardless of sex or ethnic background.

In 1987, leaders of ethnic groups in Canada decided to establish a national institute to improve the media's treatment of minorities in programs and commercials. When members of minority groups are not properly represented, this exclusion reinforces the attitude that they are not an important part of Canadian society.

People who are offended by particular ads are encouraged to file a complaint with the Advertising Advisory Board. Media Watch, a women's group that monitors images of women in the media, is also interested in comments from the viewing and reading public.

ADVERTISING AND THE QUALITY OF LIFE

Advertising is under scrutiny from several directions. A campaign of the federal Ministry of Health and Welfare is aimed at the overdrinking they believe is fostered by beer ads on television. The ministry recognizes that these ads are targeted to the younger end of the market scale, the 18 to 29 year olds.

The Ministry is concerned with health, but also it is disturbed by the quality of life portrayed in the beer ads. The ads seem to say that nothing is worth doing by itself; every activity is better if accompanied by beer. Current beer advertising promotes the idea that if you walk, swim, crawl, or dance more than ten metres, you deserve a beer in celebration.

Some people have stated that television is committed neither to inform nor to entertain. They state that instead, television is concerned only with selling an audience to a commercial sponsor. The following is an excerpt from the

January 1986 issue of *Harper's*. Do you think that television networks are this concerned with the purchasing power of any particular audience? "How much an advertiser pays for a 30-second spot depends not so much on how many viewers tune in , but on the quality of those viewers, affluent young adults being preferred. Women between 18 and 49, for example, watching during prime time on Fridays, sell for $16.50 per thousand; older women and teenagers, who buy less, sell for less."

8.7 *Violence in the Media*

There are great differences of opinion about the effects on audiences of violence shown in the media. Some people believe that violence portrayed in television programs or movies has negative and damaging effects, while others believe it has little negative effect.

YOUNG CHILDREN

Pre-school and elementary school children watch an average of 24 to 27 hours of television a week. At least six violent acts an hour can be witnessed in television viewing. The issue is whether these violent acts encourage children to be aggressive and violent, or whether children understand the difference between make-believe situations and real life.

Media experts have found that it is difficult for children under the age of ten to distinguish the artificial world of television from the real world. Gregory Fouts, a psychologist in Calgary, states that children under ten have a tendency to imitate violent behaviour seen on the screen. They might watch someone being hit with a baseball bat on television and later imitate that behaviour without realizing the harm they are doing.

Jay Bishop, a psychology professor at the University of Alberta, has concluded that television sets should be placed in family living rooms so that parents can control and regulate the programs viewed. Also, he thinks parents should ask their children about their reactions to the scenes on the screen. They can explain that violent programs are entertainment performed by actors, and are not real. Children who receive this attention and guidance from their parents are more likely to grow up secure and confident, and will not be likely to exhibit violent and aggressive behaviour.

Certain psychologists, like Professor Bishop, believe that cartoons such as "Road Runner" are potentially dangerous to young children. The characters in these shows are constantly battered and destroyed but somehow miraculously recover. Some children might be led to think that battering or hurting someone in real life has no long-term effects. But other experts disagree; they think cartoons like these show how the character being chased can outwit its pursuers and reach safety. These messages can teach children that solutions can be found to situations where they are under attack or facing a major problem.

Because parents often do not have the time to sit with their children to select and discuss programs, some critics want the government to censor shows to prevent any damage that might be done. Others disagree, and say that television teaches children that violence does exist in the real world and prepares them for it.

The following are comments on television viewing, particularly by children. What are your opinions of the comments?

"The TV set has altered the organization of daily habits more than any other discovery in recent times. Because of it children spend less time on family chores and at play."
Professor Jay Bishop, University of Alberta

"When parents sit down and actively participate in the viewing, children are much less likely to imitate violence and other negative things they see."
Professor Gregory Fouts, University of Calgary

"TV has to be recognized as a force for both good and bad. People have to learn how to use TV rather than be used by it ... Personally, I feel some people have to do without TV for a while, if they find it has taken over their lives."
Janis Nostbakken, former TV Ontario supervisor and producer of *Polka Dot Door*

> *"Most young children have been watching TV all their lives, literally since they were infants, and their lives have been guided and molded by the medium. TV has told them what to eat, what to wear, how to behave, how to talk, and even how to think. TV has been their principal teacher of values."*
>
> Claudine Groller, elementary school teacher, Scarborough, Ontario

TEENAGERS

Parents and teachers are also concerned about the effects on high school students of violence in the media. They think it may make young people less sensitive to real-life violence and keeps them from identifying with people who are experiencing terror and pain. It can also make them more suspicious about the real world.

The combination of violence and sexual images portrayed by rock videos has many people concerned. Some rock videos promote and reinforce negative stereotypes for both males and females. Males sometimes display extreme aggression and violence, while females are shown as decorative bystanders.

The violence of some videos has roused worries that some viewers will be stimulated to copy violent scenes they witness on the screen. Constant exposure to such scenes can make people less sensitive to brutality in real-life situations. At the least, there is concern that the apparently glamorous and easy lifestyles rock videos portray can confuse young people about traditional values surrounding marriage, family life, and the need to work. Media Watch and other concerned groups want to make the public aware of the potentially damaging influence of rock videos. They want parents to discuss and question the images used to entertain their children. They also want teenagers to learn to think about and analyze the content of rock videos and compare it to real-life situations.

Some people point to the positive effect of rock videos used to raise great sums of money for charity. In 1985, Canadian performers made the record and video "Tears Are Not Enough" and raised millions of dollars for the victims of famine in Africa. Both the critics and supporters of rock videos believe that rock stars and recording companies have an obligation to the viewing public and must assume much of the responsibility for the content and the messages of their videos.

A Gallup Poll taken in 1984 showed that two-thirds of Canadians polled believed that exposure to television violence can turn children into angry adults. How would you have voted in such a poll as Figure 8.2? In 1988, CBC polled viewers about sex and violence. Assess your votes if you have been polled according to the questions in Figure 8.3.

GALLUP POLL

"Do you think children under 10 years of age who are exposed to a considerable amount of violence on television are more likely to become overly aggressive adults than are those not so exposed?"

If yes: "Do you think it is the responsibility of the parents to prevent or control such exposure, or do you think the government should act to control it?"

MORE LIKELY TO BECOME AGGRESSIVE

	National 1984	1975	With children under 10	No children under 10
Parents responsible for control	33%	17%	34%	33%
Government responsible for control	9	20	6	11
Both parents & government responsible	21	23	18	21
Don't know	3	1	3	3
TOTAL MORE LIKELY	66%	61%	61%	68%
NOT MORE LIKELY TO BECOME AGGRESSIVE	29	30	35	26
DON'T KNOW	5	9	4	6

Figure 8.2

CBC-TV SURVEY ON SEX AND VIOLENCE

Percentage thinking scenes including sex, violence, or swearing should not be shown before various times of night.

	Should never be shown	Before 11p.m.	Before 10p.m.	Before 9p.m.	Before 8p.m.	Before 7p.m.	Should be allowed anytime
SEX AND NUDITY							
Long embraces and kisses	6%	13%	26%	50%	64%	67%	33%
Scenes in which sex is talked about	7	20	39	65	79	83	17
Scenes in which homosexuality is talked about	19	31	47	70	80	82	18
Scenes with nudity	26	50	68	86	91	92	7
Scenes of couples having sex (no nudity)	29	50	67	85	91	93	7
Scenes of couples having sex with nudity	37	63	79	92	96	96	4
VIOLENCE							
Cars being wrecked in car chases	11	16	27	50	64	67	33
Shoot outs or gun fights	13	24	38	62	76	83	18
Fist fights	14	24	38	61	76	80	20
Violent scenes of war	18	30	47	75	85	91	9
Scenes of people being killed or injured violently	30	47	65	88	94	1	4
Swearing	33	47	61	80	89	92	7

Figure 8.3

CASE STUDY

The Story of Terry Fox

Terry Fox was born on July 28, 1958 in British Columbia. He was a very determined and stubborn child. In high school he was an average student but through determination and hard training he became a good basketball, soccer, and rugby player. Terry enjoyed sports, and planned to become a high school physical education teacher. His plans for the future were suddenly interupted. While jogging one day, he experienced a sudden, terrible pain in his right knee. He was diagnosed as having a rare form of bone cancer called osteogenic sarcoma. On March 9, 1977 his right leg was amputated and replaced, in time, with an artificial limb made of fibreglass and steel. This limb was attached to the stump of his leg and permitted him to walk and eventually to run.

Determined to fight his disease, he was resolved to run again. He thought that he could increase awareness of cancer if he ran across Canada raising money for cancer research. On April 12, 1980 he began just such a run across Canada in St. John's Newfoundland. He called his fundraising run for cancer research the "Marathon of Hope". At first Terry hoped to raise one million dollars for cancer research, but as the run progressed and donations increased, he revised the figure to $1.00 for every Canadian or a total of $22 million. His quest to run 8530 km across Canada to raise this huge sum of money seemed to be an impossible task. But Terry was fiercely determined to accomplish his goals — regardless of the price. He planned to reach his home town of Port Coquitlam, B.C. in six months by running an average of 48 km a day.

At first his run was plagued by snow, ice, and cold winds but he continued to run. His stump often bled, but he ignored the pain. As he ran over the roads and through the towns of Newfoundland, Nova Scotia, New Brunswick, Quebec, and Ontario the newspapers, magazines, and television cameras followed his journey. The media captured Terry's courage and determination to help other people beat cancer. Here was a young man who had risen above his own predicament and was raising money to fight a disease that takes the lives of one in every four Canadians. Media reports showed him being greeted by cheering Canadians. Radio programs followed his progress. Terry Fox's picture began to be a familiar sight to Canadians and many began to regard him as a national hero. His dream of a cross-country run and his hope for research money began to excite the imaginations of other Canadians.

By August 31, 1980 he had run 5330 km and was just outside Thunder Bay, Ontario. The next day he began his usual run, but after 29 km he began to experience sharp chest pains. Despite the pain he managed another 5 km, cheered on by on-lookers who had no idea that he was in serious trouble. Even before he was rushed to the hospital he had the premonition that the bone cancer had spread to his lungs. His guess was proven correct. He had wanted so badly to complete the remaining 3000 plus km of his journey but this would never happen. By September 2, 1980 $1 100 000 had been raised — far short of his target.

Terry was given chemotherapy and suffered painful side effects. But Terry's "Marathon of Hope" was just beginning. When Canadians heard that their hero was dying in hospital their hearts went out to him. The CTV network paid a five-hour tribute to Terry in a national fundraising program for cancer research — a total of $10 500 000 was raised. All over Canada, Canadians began to collect more money for cancer research through fundraising events. Altogether $23 400 000 was raised in 1980. Terry's target had been met, and surpassed.

Terry had become a national hero in a country that has few heros. He became a symbol of Canadian unity and strength. He was a role model to everyone — not just for those who are physically handicapped. Terry received honours and tributes because of his efforts and accomplishments; he was awarded the Lou Marsh Trophy for outstanding athletic achievement and his portrait was placed permanently in the Canadian Sports Hall of Fame. He was also made a Companion of the Order of Canada, a presentation made by the Governor General.

Terry Fox died on June 28, 1981, but his death was not in vain. The inspiration he provided and the pride that he instilled in Canadians continue to be remembered. Each year, millions of dollars are raised for cancer research in his memory. Terry was a great Canadian who refused to quit — his memory and his courage will always be remembered.

1. How did the media have a positive effect on Canadians during Fox's run? How could the media have had a negative effect on the "Marathon of Hope"?
2. What do you most admire about Terry Fox? Do you believe he deserves to be called a "national hero"?

8.8 Related Questions

CENSORSHIP

Two things combine to produce censorship: public concern about moral standards and governmental decisions to protect the public of all ages from harmful material. Censorship is an issue that divides people. You have probably already debated this subject in class with a topic like: "Censorship is an answer to violence in the media" or "Censorship is a denial of the public's right to know."

A censorship bill was introduced into Canada's Parliament in 1987; called Bill C 54, it may have been drafted to respond to the fears of people who link the violence in the mass media with violent crimes and abuse. The strongest resistance to the bill came, however, from professionals in public institutions, like libraries and art galleries, and from writers, poets, publishers, and editors. They expressed concern that a new censorship law would not remove pornography from magazine displays or degrading videos from the shops. Instead, it would silence and punish expression in literature and art. The question remained: should the public judge for itself, or should the government act as its protector?

CULTURAL SOVEREIGNTY

Any medium read, watched, or listened to should have been made for a particular audience. Books, for example, reflect and at the same time create, the culture and community out of which they come.

Cultural sovereignty means having mass media that grows out of your own country's heritage and culture. Canada is not the only country trying to keep its cultural sovereignty in the face of the spread and influence of American films and television. But, because we are the next-door neighbours of the United States and share the same language, it may be hardest for us to do. It costs large sums of money to make movies and television shows. Because our population is small, considering the size of the country, we cannot always fully support a cultural industry.

This is the reason why the subject of culture was raised during the recent free trade debate. When a country's doors are fully open to trade, questions arise. Is there enough protection for Canada's cultural industries? Should we be concerned with keeping a cultural identity separate from that of the U.S.? Mass media influences each one of us. Do we want a collection of Canadian media that reflects our society's goals and mores, and that we can control in the best interests of our cultural identity? There are no simple answers to such questions.

Progress Check

1. What might be the benefits of a media literacy course?
2. What are the purposes of advertising?
3. Look back to your television viewing as a child. Do you think that (a) the amount of viewing, or (b) the content, was harmful to you? Explain your answer.
4. Marshall McLuhan said that "Ads are really doing their work when you don't notice them". Was he right? Explain your answer.

5. Choose an ad that is guilty of sexual or racial stereotyping. Could the company change the stereotype and still sell its product? Explain.
6. Why are violence and sexual images often combined in rock videos?

Alouette I, launched in September 1962, is part of the satellite communication system that allows news to be broadcast around the world in a matter of minutes.

OVERVIEW

The mass media have become steadily more powerful as disseminators or messengers of information, advertising, and entertainment. Television, radio, computers, films, and satellite communications have become a vital part of Canadian society. They are making people more aware and better informed than ever before, and are now considered essential to our way of life. They also draw an audience for sports and entertainment in a way that was previously impossible. Millions of people can be watching the same events at any given time and experiencing similar thoughts and feelings about what they are watching and hearing.

In spite of the many benefits derived from the media, many people are worried about the manipulation and control that the various types of media can have over our lives. The media have the power to influence our thoughts, attitudes, and opinions. Because television and film can interact with people so directly, young people can be influenced and possibly damaged by them. A growing number of Canadians are concerned about the negative stereotypes in the media. In the short run, these images can be thought amusing, but in the long term they can have destructive consequences.

Concerned citizens continue to put pressure on governments, advertisers, and those who control the media to restrict messages that are inappropriate for a free and equal society. The development of media literacy courses and the critical judgment of all people who use the media are essential if individual thinking and actions are to be preserved. Linked to this critical sense is concern for what is Canadian in the media. Over time, and despite influences to the contrary, cultural bonding within the country must continue to take place.

KEY WORDS

Define the following terms, and use each in a sentence that shows its meaning.

mass media
manipulate
cultural sovereignty

KEY PERSONALITIES

Give at least one reason for learning more about each of the following.

Foster Hewitt
Aubrey Fassenden
Johann Gutenberg
Marshall McLuhan
Terry Fox

DEVELOPING YOUR SKILLS

FOCUS _____

The focus of this chapter is the media. Canadians spend much of their leisure time reading, listening, and viewing various forms of the media. Many Canadians are concerned about the quality and content of the materials in the media and are seeking controls to improve their messages.

Many Canadians are worried about the high level of American content and its influence upon Canadian culture. Some believe that governments should play a more active role in controlling the media.

Divide the class into groups and brainstorm a list of questions that can be asked about the media. Each question will bring out different points of view. Answering these questions through discussion, reading, research, and writing will lead students to a better understanding of the media and its effects upon people.

ORGANIZE

1. In an organizer describe the possible effects of advertising and violence on children and young people. In another column list the controls that are placed upon the media to reduce or eliminate these effects. In a third column give your opinion on these existing controls.
2. Create an organizer that summarizes the views of Jay Bishop and Gregory Fouts. Give your opinions of their views.

LOCATE AND RECORD

1. Examine the different newspapers in your community. How are they similar and different? Examine their layouts — are they designed to cater to people's emotions and curiosity? Explain.
2. Interview people who grew up without television. Ask them what they remember about a television set first appearing in their homes. How did their way of life change? Find out what early television shows were like. Do these people have a different attitude to television today from when they first started watching TV? Was their childhood without television different from yours? If so, in what ways?
3. Divide the class into groups and have each group write the following agencies and organizations requesting information on their activities and the literature they have available. Have each group research the information they receive and present a seminar to the class on the goals, purposes, and functions of these agencies.

Advertising Standards Council
350 Bloor St. East
Suite 402
Toronto, Ontario   M5W 1H5

Alliance of Canadian Cinema,
  Television and Radio
  Artists (ACTRA)
2239 Yonge Street
Toronto, Ontario   M4S 2B5

Association for Media Literacy
40 McArthur Street
Weston, Ontario   M9P 3M7

A.C. Nielson Ltd.
160 McNabb Street
Markham, Ontario   L3R 5V7

BBM Bureau of Measurement
1500 Don Mills Road
Suite 300
Don Mills, Ontario   M3B 3L7

Canadian Broadcasting Corporation
  (CBC)
1500 Bronson Avenue
P.O. Box 8478
Ottawa, Ontario   L8N 1H6

Canadian Daily Newspaper
  Publishers Association
890 Yonge Street
Suite 1100
Toronto, Ontario   M4W 3P4

Canadian Radio-Television
  and Telecommunications
  Commission (CRTC)
Central Building
Promenade du Portage
Hull, Quebec   K1A 0N2

Canadian Television Network
  (CTV)
42 Charles Street East
Toronto, Ontario   M4Y 1T5

Canadians Concerned about
  Violence in Entertainment
  (CCAVE)
1 Duke Street
Suite 206
Hamilton, Ontario   L8P 1W9

Children's Broadcast Institute
234 Eglinton Avenue East
Suite 405
Toronto, Ontario   M4P 1K5

Department of Communications
Journal Tower North Building
300 Slater Street
Ottawa, Ontario   K1A 0C3

Department of Consumer
  and Cultural Affairs
Place du Portage
50 Victoria Street
Hull, Quebec   J8X 3X1

EVALUATE AND ASSESS

"We are on the point of winning a gift that history has seldom yielded and never on such a scale. For centuries men have dreamed of a universal language to bridge the linguistic gap between nations ... Man will find his true universal language in television, which combines the incomparable eloquence of the moving image, instantly transmitted, with the flexibility of ready adaptation to all tongues. It speaks to all nations and, in a world where millions are still illiterate or semi-literate, it speaks clearly to all people. Through this eloquent and pervasive universal language, let us strive to see, in the words inscribed over the portals of the BBC, that "Nation Shall Speak Peace Unto Nation"." *Problems and Controversies in TV and Radio*, Skoria and Kitson, editors

1. Do you feel that television is a form of communication that speaks clearly to all people? Support your answer.
2. What changes would you make to television programming to achieve the goal stated by the BBC "Nation Shall Speak Peace Unto Nation"?
3. If you were sent from another planet to evaluate Earth as a place to peacefully colonize, would television give you an honest impression of human society? What report would you send back to your own planet?

SYNTHESIZE AND CONCLUDE

1. Some TV "soaps" are considered to be more realistic than others. A realistic "soap" tries to present us with believable characters in typical situations so that we can identify with them and their problems. Characters in realistic "soaps" have the same types of occupations, concerns, problems, attitudes, and values as we do. Name a realistic "soap" and describe the ways in which it is realistic. Some of the questions that might be asked are:
   – Do you think that having more realism makes the "soap" more or less attractive than unrealistic "soaps"?
   – What would draw some viewers to a realistic "soap"?
   – What would draw other viewers to one that emphasizes more fantasy elements?
2. You need to be able to distinguish real-life experiences from those shown on television.
   (a) Divide into groups and list the individuals who are currently featured on television, for example, the police, teachers, doctors, nurses, teenagers, widows, single parents.
   (b) Use four columns in an organizer to compare these individuals by their television images and their real-life counterparts. If there are differences, explain why. Use an organizer such as the following:

Subject	TV Image	Real Life	Explanation
the police	pursue criminals in fast car chases	routine tasks, such as speeding tickets	we prefer to see the exciting to the commonplace

(c) Discuss the reasons for the differences between the TV image and the real-life characterizations.

APPLY

Discuss the meanings and accuracy of the following statements. At the end of your discussions develop an organizer that summarizes the points of view expressed. In a separate column comment on the accuracy of the article and in a third column speculate on what should be done to deal with these concerns. In a final column predict what would happen if your recommendations were put into effect.

> A recent concern for many parents is a growing awareness that certain cartoon programs serve as advertisements to purchase a likeness of the cartoon characters in doll or puppet form. Many people think that programs originally designed to entertain children have turned into lengthy advertisements to sell toys. Pressure is mounting on broadcasting stations and governments to bring in regulations to control these types of programs.

> People who are offended by certain commercials are being encouraged to file their complaints with the Advertising Advisory Board. This Board is actively involved in reducing and eliminating stereotypes from advertisements. But despite the efforts of agencies and broadcasting stations to eliminate stereotypes in commercials, they continue to exist. A study was done between 1982 and 1983 by the Advertising Advisory Board on voiceovers in commercials on radio and television. Voiceovers are offscreen voices on television commercials and all voices heard on radio commercials. It was found that voiceovers, in 1 457 television commercials had 88.3% male voices while only 11.7% were female. In 7 919 radio advertisements 78.1% were male voiceovers and 21.9% female. Advertisers and businesses obviously still believe that male voices carry more authority and credibility with listening audiences.

> Censorship does limit the freedoms of Canadians to write, read, and watch certain materials. Obviously young and sensitive minds have to be protected from damaging materials. But do adults require cen-

sorship of the media? Many believe that a small minority of adults will be encouraged to commit harmful acts if they are exposed to violent and pornographic materials. But should these materials be censored or banned for the majority of people when only a small minority will be influenced by what they read or view?

Some Canadians want stricter controls on newspaper reporting. They believe that the press is sometimes irresponsible. They want to forbid the publication of the names of persons accused of committing criminal offences until they are convicted by a court of law. Those who support tighter controls also want the press to stop publicizing private lives of prominent people, such as politicians. They argue that their right to privacy is being denied. Scandalous stories might sell newspapers but they can also ruin careers. Such stories do not serve the interests of the public — they merely entertain at a very high cost.

Many of those who support censorship of the media support the efforts of the Canadian Radio-Television and Telecommunications Commission (CRTC) to control radio and television broadcasting in Canada. One of its purposes is to ensure that Canadians listen to and view Canadian programs on Canadian television and radio shows. It requires that 30% of Canadian AM radio programs and 50% of television shows be Canadian. It has no control over American channels that are brought into Canada by aerial or cable. Canada has the highest rate of cable transmissions in the world (61%). American programs continue to be the most popular for Canadians. Canadians watched an average of 24.2 hours of television a week in 1986 and 64.1% of these hours were spent watching foreign shows (Statistics Canada).

COMMUNICATE

1. Each class member can select and read a book on the media. Below is a list of suggested texts — additional titles can be suggested by the school librarian. As you read your selected text, summarize the information and ideas presented by the author. When you finish your reading, put a summary of what you have read on a ditto that can be copied for each member of the class. These copies should be handed out at the beginning of the class so that everyone has time to read the summary. Each student is responsible for giving an oral presentation. Before presenting the seminar the student should write down key points on index cards. During the presentation the presenter is encouraged to address the class, referring only oc-

casionally to the index cards. Class members can ask questions at the end of the presentation.

McLuhan, M.	*Counter blast*
McLuhan, M.	*Understanding media*
Watson, P.	*Conspirators in silence*
Lumsden, I., ed.	*Close the 49th parallel*
Meier, R.L.	*A communications theory of urban growth*
Haggart, R.	*The uses of literacy*
Finnigan, B., comp.	*Making it: the Canadian dream*
Mann, W.E.	*Canada, a sociological profile*
Irving, J., ed.	*Mass media in Canada*
Winn, M.	*The Plug-in-Drug*
Wolfe, M.	*Jolts*

2. Write a short essay and/or debate one of the following topics.

- Censorship is part of everyone's daily life — choices, attitudes, opinions, interests, selection of friends are all forms of censorship.
- Advertisers use sex images, sex stereotyping, and violence to sell their products.
- Advertising affects the types of people we become.
- The government should strictly censor all programs and advertisements that have negative and damaging images.
- The government should not permit American television programs to be viewed in Canada.
- People are merely reflections of the images they see on television and in the movies.
- Cartoon programs that merely advertise the sales of particular toys should be banned.

UNIT 4

CULTURE

CHAPTER

9

What is Culture?

How can we define culture? How do anthropologists study various cultures?

INTRODUCTION Culture is one of the most important aspects of being human. Without it we would not be so successful as a species, for through culture we are able to adapt to a wide range of conditions. Through culture we distinguish ourselves from other species — only humans have developed culture.

All social scientists are interested in the study of certain aspects of culture. Anthropologists explore culture through time and through space, to discover cultural differences and similarities.

In the first part of this chapter we will examine the meaning, origin, and development of culture. In the second part, we will explore how anthropologists study culture.

9.1 *Defining Culture*

One of the most important principles anthropologists apply when they study a society is **cultural-relativism**. This is the assumption that each culture is important and deserving of respect. Anthropologists avoid passing judgment on cultures. They believe that no culture is better or worse than another one, and that each culture, regardless of its state of development, contributes to our understanding of what it means to be human.

When people assume that their culture is better than someone else's, they engage in what is called ethnocentrism. Ethnocentrism has no place in the study of cultural anthropology.

Most of us are so caught up in our own culture we are often unaware of how it affects our lives. Culture controls, directs, and protects us. It may also make us quite different in attitudes, beliefs, and behaviour from people who have grown up in another culture.

Culture can be defined in three ways: as a text for living, a program for survival, and a way of life.

A TEXT FOR LIVING

Every day of our lives we take part in hundreds of routine activities and interactions at home, at school, at work, and in the community. For each activity and interaction we have been given a text, an outline of what to do. This text is called culture. Culture tells us what to wear and when to wear it, what to eat, how to communicate, what to think, and how to behave in many different situations. We also interpret the communication and behaviour of others in terms of our culture.

Since we are not born with culture, it must be learned. Sociologists call the process of learning culture socialization. Anthropologists call it **enculturation**. But we do internalize our culture; it becomes a part of us, guiding our actions and thoughts. Enculturation is necessary for social and physical survival. Since we all live in societies, we must be taught how to live with other people. Culture gives us a predictable and orderly structure, a text to follow, within which we can interact with others. It also provides a pattern to follow for survival in a hostile environment.

A PROGRAM FOR SURVIVAL

In early human societies survival often depended on how well people cooperated with each other and how well they could communicate. Banding together in groups or tribes made it easier to protect themselves and to hunt. They developed weapons, tools, and hunting techniques for protection and for securing food. Animal skins were used to protect them from the weather. They learned how to harness fire and used it for keeping animals away, for light, for heat, and for cooking. Their skills and techniques were taught to their children and so, were passed from generation to generation.

Eventually fire was used to soften, melt, and shape metal. Better weapons, tools, and utensils could be made and these made life somewhat easier. Metal ornaments were also fashioned, and they increased the pleasure people found in life.

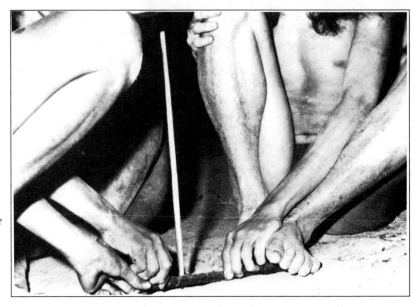

A twig is rubbed on a piece of wood until it produces smoke. Dry moss is put in the hot place and blown until it smoulders and then burns.

Today, although it is not as apparent, culture still helps us to survive and to meet our basic needs. The foods we consume and how we go about obtaining them are part of our culture. Clothing, even though it may be decorative, is still a survival tool because it protects us from the weather. The work people do to make a living today may not seem similar to the basic activities of early humans. However, the work of today provides people with a means of satisfying most of their basic needs.

Aspects of our culture such as sports, religion, music, and art do not necessarily help us survive physically. They do, however, help bring people closer together.

A WAY OF LIFE
Eventually culture developed into more than a program for survival. It became a way of life. It became ways of speaking, of doing things, of acting, and of thinking that were specific to various groups of people. Humans began to show differences one from another, according to their culture. They had different values, customs, and ways of doing things.

As we grow up, we develop two personalities. One is our individual and distinctive way of acting and relating to the people around us. This individual personality makes us different from our friends and siblings. The second personality is one that we share with all the other people in our society.

Each human group has a shared personality. The food, clothing, language, and customs of a French person will differ from those of a Scottish Highlander. The Indian sari, the Japanese kimono, the Inuit parka, the Hawaiian muumuu, and the Middle Eastern aba all function to cover and protect the body. Yet, they are very different in design. Each of these garments functions most efficiently in the particular culture for which it was designed.

Except for variations in skin and hair colour, facial features and size, humans are physically alike. We walk on two feet, have similar brains, and similar drives. But there are variations in the way human groups relate to their particular environments. Different religious beliefs, childrearing practices, family structures, food, and dress are a few of the variables that combine to create different ways of life, or cultures.

Food

In the example of food, the practices involved with eating and the choice of foods differ from one culture to another. For example, in many Canadian homes eating hamburger is an appropriate way to satisfy hunger. But Hindus in India

do not eat hamburger because the animal from which it is prepared, the cow, is considered sacred and cannot be slaughtered. How food is eaten varies from one culture to another. It may be eaten with the fingers, with various types of cutlery, or with chopsticks, and it may be eaten from communal dishes or from individual dishes.

Culture can determine when it is appropriate to eat as well. In Moslem and Christian cultures members of the faith are required to fast at certain times of the year. In the Hebrew culture, kosher foods, those prepared according to Jewish rules and rituals, must be eaten. Eating habits are also modified in some cases to achieve social goals. Since being fat is not socially acceptable in some cultures, people will ignore hunger pangs to stay slim.

Reproduction
Like the drive for food, reproduction is a natural human drive. Without it our species would become extinct. Each human society, however, has different practices for the related aspects of the reproductive drive, for example, courting, marriage, and childrearing. In Saudi Arabia, a man may have many wives, while in Canada, monogamy (one husband-one wife) is the acceptable practice. The average age for marrying also varies from one culture to another. In Latin America it is acceptable for females as young as 13 to marry, whereas in Canada 16 is usually the youngest acceptable age. Some cultures have clear rules that dictate which sexual activities are acceptable during courting, when to mate, which individuals are suitable as mates, the ceremony that accompanies a marriage, and the number of children that should result from the marriage.

PROGRESS CHECK

1. What is the goal of cultural anthropologists?
2. How is culture a text for living, a program for survival, and a way of life? Summarize each of these areas.

9.2 *The Development of Culture*

To understand how and why culture developed, we must review the evolution of early hominids.

THE HUMAN BODY

Among the primates, humans are the only ones that have developed an erect posture and a bipedal movement (walking on two feet). This development meant that our arms and hands could be used for purposes other than walking and running. The evolution of bipedal creatures with flexible hands hastened the arrival of culture.

The human hand is a highly specialized organ for functions such as grasping, manipulating, or constructing. At first glance your hands may not appear very remarkable to you. In fact they are one of the most versatile tools in all of nature. Our hands have thumbs located opposite our fingers (called an opposable thumb) for grasping objects. In addition, the fingers are very well supplied with nerve endings. This makes the hand an ideal instrument for grasping or handling even the smallest objects with care. Hands also have a great deal of flexibility, since they rotate at the end of our wrists. The combined rotational abilities of the wrist, elbow, and shoulder, make complex movement of the hand possible.

What does the structure of the hand have to do with culture? With this remarkable device we are able to create, change, or shape objects. We can start a fire by making a spark, or we can make a wheel which will help us transport things. We can fashion pottery from clay, paint pictures, and write letters. With our hands we can signal other people, throw a football, or cook a meal.

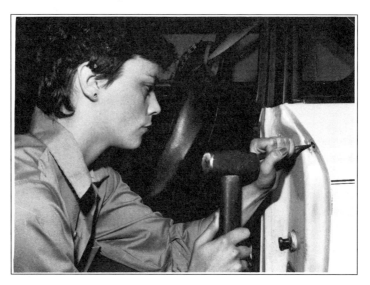

THE HUMAN BRAIN

Hands cannot think — they are merely tools. A crane or a bulldozer would not move were it not for someone controlling the hydraulic levers that activate the bucket or the shovel. In humans the brain serves a similar function. It controls and directs the hands, enabling them to perform tasks.

Humans have a more complex brain than any other creature. Our brain can interpret nerve impulses from other parts of the body, remember, think abstractly, and use symbols. Within the brain are various centres that enable humans to perform such functions as speaking and reasoning.

LANGUAGE

A complex brain, hands capable of fine manipulation, and bipedal walking and running have been important factors in the development of culture. Language is another important factor, especially in teaching all individuals in a society their culture and transmitting it from one generation to another.

To function in a society, members must learn the culture. Language is essential to this learning process — it is a person's cultural admission ticket. People who are deaf or blind learn language using hand signals, lip reading, or Braille. Language, whether it is speech, written words, hand signals, or Braille, is based on the use of symbols to stand for objects and concepts. Using language we are able to communicate thoughts about past events and possible future events.

The ability to use language, combined with the human desire to know and understand, has produced works of literature and achievements in mathematics and science. Language helps us to preserve our cultural traditions and our history, and to expand our knowledge to speculate about the future. Forms of symbolic communication are important aspects of all cultures. The symbols used by one culture will not necessarily correspond to those used by another culture however.

The Earliest Uses of Language

Just when and how did early humans first begin to use language to communicate ideas? There is no actual evidence, but social scientists have offered some interesting hypotheses.

It may have been that the first language of humans was a system of hand signals. Or, the first language could have consisted of a variety of calls similar to those made by other primates. Whichever was the case, as human experience and abilities expanded, the system would have become overloaded eventually. That is, the number of gestures or calls required to express objects and ideas would have multiplied to the point where some would become indistinguishable from each other. Such overloading would result in several possible meanings. When misunderstandings happened frequently enough, people would become aware that the old system of communication had to be expanded and changed.

For language to develop, there had to be nerve pathways between areas in the brain concerned with visual images and sound patterns. This was necessary in order that a given sound could be associated with a mental image. As the human brain evolved, it increased in complexity, and more intricate forms of communication were possible. The brain of the later Homo erectus was almost twice the size of Australopithecus, and the increase may be related to the need for a larger memory for vocabulary.

Social scientists have speculated about the pressures that might have contributed to the development of language. Some feel that children at play, using actions and objects to symbolize the real world experiences of their parents, might have created a more elaborate system of calls.

Our early ancestors were hunters. When they moved out of the forests and became hunters of large game animals, precise communication would have been important to their success. Language would have made the sharing of hunting skills and the planning of strategy much more efficient.

As language became more elaborate, people could also begin to share ideas and beliefs, and to develop traditions. Culture and language expanded, and with them the ability of humans to function more efficiently together.

Hominid reconstructions

The Far Side, copyright 1988 Universal Press Syndicate. Reprinted with permission. All rights reserved.

PROGRESS CHECK

1. How did certain biological changes lead to the development of culture in humans?
2. Why was the use of symbolic language important in the development of culture?
3. How did the early hominid hunters help to develop culture?

9.3 How Anthropologists Study Culture

In order to investigate a given culture, cultural anthropologists engage in research at a site or sites where people of that culture live. This is called **field work**. Two anthropologists, James P. Spradley and David W. McCurdy, describe field work in this way: "(It) has come to have an almost sacred significance to anthropologists, for many it is like a rite of passage into the discipline because it is the single most important factor for acquiring the anthropological perspective."

Anthropological information gathered in the field is used to test hypotheses about behaviour and culture. It is also used to increase the existing knowledge about various cultures. As new information is collected, it enables anthropologists to describe a culture more accurately, and to analyze it. New information about a culture may also be compared to similar information about other cultures.

ETHNOGRAPHY

When cultural anthropologists go into the field to study a living culture, they must learn the language and take part in the daily life of the people. By living with the people and participating in community life, the anthropologist can observe the people closely and develop a feeling for the culture.

The description of a culture is called an **ethnography**. Field anthropologists prepare an ethnography by collecting first hand a variety of data and observations on the culture they are researching. They take part in events and get to know most of the individuals in the village or community. They might, for instance, map out the physical arrangement of a village, take a count of the inhabitants by age and gender, measure the agricultural crop yields, describe the activities of individuals in a household, make observations on the religious practices of the community, or gather any other data that relate to the particular aspect of the culture they wish to understand.

Anthropologists in the field use various strategies to study the way of life of a group of people. Following is a partial list of their methods.

1. Genealogical Method: Anthropologists are very interested in the identities of individuals and the relationships of

these individuals to each other through marriage and birth. This data makes it possible to trace the members of a society over a long period of time and to create a "family tree" for all families.
2. Life History Method: By recording the life history of one or a few individuals, the anthropologist is able to reconstruct some of the major cultural influences in their lives.
3. Projective Testing: Originally developed by psychologists, such tests as the Rorschach ink blot test give anthropologists insights into individual personalities.
5. Visual and Audio Method: Taking photographs or recording behaviour on film or videotape are technological methods of data collection that have become valuable field tools because they can offer a more reliable survey of people and events than verbal recollections.
6. Questionnaires: Anthropologists try to establish links with individuals in the particular society who regularly provide them with reliable information about the culture being studied. Sometimes a questionnaire is used for this purpose.

ETHNOLOGY

Once the ethnographic study, the description of a culture, is complete, an anthropologist may use it to compare that culture to other cultures, and to make generalizations about culture. This line of study is called **ethnology**. It is through ethnology that anthropologists are able to explain unique aspects of human behaviour as well as more general behaviour patterns.

Ethnology has been compared to cryptography (deciphering codes) because both disciplines try to decipher hidden meanings. Just like a cryptographer, the ethnographer seeks to find the meaning behind a message. For the cryptographer the message may have to be translated from Morse Code into English. For the ethnologist, the significance of behaviour, clothing, and beliefs of a culture has to be deciphered and explored.

CLASSIFYING CULTURE

When we analyze culture we discover that it consists of many different elements. To simplify things, anthropologists usually classify culture into two categories. The first category is material culture while the second is called non-material culture.

Material Culture

Material culture is the visible part of culture. It consists of all the objects and products manufactured by humans. These may range from a simple abacus to a sophisticated computer. Material culture includes art, tools, weapons, clothing, ornaments, houses and anything else produced by a society.

Non-Material Culture

Non-material culture is more abstract. It deals with how we view the world. It includes precise knowledge about the natural world around us, for example, knowledge of biology, chemistry, physics, and mathematics. Beliefs, such as those involving the creation of the universe or life after death, are also part of non-material culture.

Non-material culture is also concerned with the behaviour of a society. A society holds certain values that determine what is considered good, right, moral, beautiful, or ethical. For example, in North America many people value activity and work, humanitarianism, neatness, efficiency, practicality, freedom, and individualism. These values are held by many Canadians and Americans, but they are not necessarily held by other cultures.

Norms are rules shared by a society which regulate behaviour. For example, norms tell us whether we should shake hands, eat with knives and forks, or stand during the national anthem. Folkways are the less important norms that regulate everyday behaviour, for example, conventions of dress or table manners. Folkways are customs that are learned by each new generation, and often are changed by succeeding generations. People who break folkways are often gossiped about, ridiculed, and ostracized.

Some behaviour is considered extremely harmful to society. Examples are theft, vandalism, rape, and murder. The rules governing this type of behaviour are called mores. People who violate mores will be punished, in many cases severely. The most important mores in our culture become laws. In this way there can be no mistaking the strong feeling about violating them. Laws are formal codes of behaviour. They distinguish between right and wrong behaviour and prescribe a punishment judged appropriate to the seriousness of the act.

PROGRESS CHECK

1. What is the purpose of field work done by cultural anthropologists? Briefly outline the various field work methods discussed.
2. How are ethnography and ethnology related?
3. Differentiate between material and non-material culture.
4. Give one example of the following: values, norms, folkways, mores.

9.4 Analyzing Culture

Generally, humans do not live alone, but in groups. They are social animals who depend on others and seek the company of others. Living in groups requires a structure for our relationships and interactions with others. Each person will have a role to play within that structure. Anthropologists study the ways in which different cultures are socially structured and organized to understand better those cultures.

Anthropologists must also understand the environment of the culture they are studying, as well as the economics and the religious customs to be able to understand the particular culture. All parts of the society come together for the anthropologist to study.

SOCIAL STRUCTURE

Social structure is the manner in which the individuals and the groups within a society are organized. Some form of organization is necessary in a society, for without it, members would have a difficult time surviving in the world.

Ethnographers who study cultural groups have found that there are different systems of organization possible. When they conduct field studies, one of their first tasks must be the identification of the different social divisions within a population. Once the various groups have been identified, the ethnographer must work out how individuals relate to the groups and how the groups relate to each other.

Kinship

The most important type of group in a preliterate society is called a **kinship** group. This is a group composed of members who are linked to each other by blood or by marriage.

Those who share a common ancestry are called **consanguineal**, or blood kin; and those who are related through marriage are called **affinal** kin.

One kind of kinship group is the **nuclear family**, consisting of parents and children only. The **extended family** is a kinship group that might consist of the parents, their children, the children's spouses, and the children's children, or any other close relatives.

Another kinship group, called by anthropologists a **descent group**, consists of all those individuals who are descended from a common ancestor. A descent group is a permanent group, and members remain part of the group for their lifetime. Membership changes as new members are born and old members die, but the descent group remains. In contrast, the nuclear family breaks up with the death of the parents, or when the children move away to create nuclear families of their own.

There are several other types of kinship groups that may contribute to social structure. **Kindred** is a kinship group which includes an individual's relatives from both the father's and the mother's side. Another type of group, important in some cultures, is the **clan**. Members of a clan trace their lineage back to a common ancestor, or sometimes to a mythical being. Clan members may live in one community or spread out in several communities. A **tribe** usually refers to a larger group composed of a number of clans or descent groups.

Associations

There are several other ways besides kinship which can contribute to the structure of a society. Non-kin groups, sometimes called **associations**, can be composed of people linked by a common occupation, interest or goal; by gender; by age; by political affiliation; by religious belief; or by citizenship in a country.

At some universities and colleges, male students may join a fraternity and female students a sorority, to form new friendships and gain a sense of belonging. Similarly, in parts of Africa, there are secret societies composed exclusively of men or of women.

In our society children are frequently grouped with others of a similar age. In school they may be associated with the same classmates over a period of several years. Members of a particular graduating year may remain good friends after they leave school.

One of the most frequent associations in which groups are organized is the sharing of common goals, interests, or occupations. Examples are a chess club, a school board, a sports team, a political party, a union, and a religious group.

SOCIAL INTERACTION
Social structure influences the way in which individuals interact. People learn to behave in social interactions according to their particular relationship to those which whom they are interacting. When talking to the principal of your school you would normally address this individual by an appropriate title — Mr., Mrs. Ms., or Miss — followed by a surname, you would not engage in physical contact, and you would avoid using profanity or slang. However, when addressing a fellow player on the basketball team, the social interaction would be different. You would use the player's first name, and you might slap the player on the back, and use slang expressions. The principal's social identity and role called for one type of interaction. Your friend's social identity and role called for a different type.

Progress Check

1. What is the difference between consanguineal and affinal kin?
2. Other than kinship how can social groups be organized?
3. Give two examples of social interaction other than the ones mentioned.

THE ENVIRONMENT OF A CULTURE
When anthropologists study culture they try to absorb every aspect of that culture. They are interested in discovering how populations interact with their physical environment. They want to learn what influence the environment has on the culture, and whether the culture has changed the environment.

Ecology deals with the interactions of animals and plants with each other and with their environment. Anthropologists who take an ecological approach to the subject of culture concentrate on the relationship between a population and its environment. How has the population adapted its way of life to fit the environment? Have other populations living in similar environments adapted in similar ways?

A culture that develops in a harsh environment, such as a desert or the arctic, will be heavily influenced by the people's struggle to survive and their need to adapt. Populations living in arctic and subarctic areas, where the ground is snow-covered for many months of the year, must develop a deep understanding of environmental conditions and must learn the necessary strategies for coping. They must know the various forms ice and snow can take.

Most people from a warm climate think of ice as being either thin or thick, rough or smooth. However ice comes in many forms. For example, the Slave Indians of the Northwest Territories, who live about 200 km south of the Arctic Circle, have many categories into which they classify ice. During hunting and trapping expeditions in the fall and early winter, their ability to discriminate between the various types of ice is vital to their safety. Decisions about what rivers can be crossed, or which lakes are safe and for how long, are life and death matters.

THE ECONOMIC STRUCTURE OF A CULTURE

Cultural anthropologists also study the economics of a population: the way in which goods and services are produced, distributed, and consumed. They investigate how the economic system is organized to provide the necessary materials for survival. They are also interested in knowing how each person in the society contributes toward the production, distribution, and exchange of goods and services. The economy may involve a complex production-distribution network of people and activities as is the case in our industrial society. Or it may be based on one or two activities such as hunting, fishing, the gathering of vegetable material, or herding.

Food Supply

One of the basic resources on which an economy depends is food. For a population that makes its living from hunting and gathering the economic structure is simple. The population lives in small mobile groups, sharing food, owning no property, and having few personal possessions.

Permanent settlements are likely to be located in areas where food is plentiful and the supply is stable. For instance, seacoasts offer good opportunities for fishing and for gathering shellfish year-round. If individuals in a society specialize in how they make their living, then surplus food might be traded within the population for other goods or services. Surplus food might also be traded to outside populations for different types of food or goods.

Once humans learned how to domesticate certain plants and animals, they had a fairly stable food supply. However, some agricultural practices can lead to problems for some cultures. Over-grazing by herd animals can lead to soil erosion and the growth of less desirable plants. Continuously growing crops on poor soil can result in sterile soils incapable of supporting edible crops. A population that is dependent on only a few species of plants and animals is vulnerable to famine if those species fail for some reason. A culture can be changed and even destroyed by its management of the basic economic structure that deals with food supply.

Disease and extreme weather conditions that affect livestock and crops can also change the various aspects of a culture.

Land Ownership

Once populations live in permanent settlements, they usually devise a system of land ownership, and the land becomes part of the economy. In tribal societies, land is often owned by a descent group, and individuals within the group are given access to it. The Hopi of Arizona cultivate land located on an arid plateau. Their agricultural techniques are specialized for the short growing season and the dry conditions of the plateau. The area they farm is owned by clans, and each clan gives land to individual families.

Systems of Economic Exchange

The goods and services of an economy must be distributed to those who require them by some system of economic exchange. The systems of exchange vary depending on the culture. Many anthropologists use the following three categories to describe and compare the different exchange relationships: **market exchange**, **redistribution**, and **reciprocal exchange**.

Market exchange is the system on which industrial economies of the West are based. The value of an item is based on the law of supply and demand. If an item is scarce and buyers want it, the price will be high. If an item is plentiful the price will be lower to increase buyer demand.

With redistribution, goods move from the individual level to a central collection point, and then are redistributed back down the system. This type of exchange is common in societies having a central authority such as a chief. The goods flow from the lowest ranking people to those of highest rank, then they are redistributed to all ranks.

Reciprocal exchange is a common form of exchange used by tribal societies, and involves an exchange of goods between social equals. The goods or services given in return, are not necessarily offered immediately. This is especially true if the exchange is between closely related individuals or within a tribe in which sharing is the norm.

Market in Aswan, Egypt

Progress Check

1. Why do anthropologists study how a society interacts with its environment?
2. What can anthropologists learn about a culture by studying its economic structure?
3. Give an example of each of the three systems of economic exchange. Can more than one of these exchange systems operate at the same time in a culture? Give examples.

CAREER PROFILE
BIRUTÉ GALDIKAS

Biruté Galdikas is an anthropologist of Lithuanian parentage, who was raised in Canada. Her area of specialty is primatology, specifically the orangutan. All anthropological work done on non-human primates is of interest to those who study human societies and cultures. Cultural anthropologists are interested in non-human primate behaviour because it provides them with an idea of how early hominids may have behaved before they developed language. Physical anthropologists are interested in the anatomical similarities and differences between the apes and humans.

Like Jane Goodall who studies chimpanzees and the late Dian Fossey who studied gorillas, Biruté Galdikas was chosen by Louis Leakey to undertake longterm studies of the great apes in their natural environment. He was impressed by her interest in the great apes, her enthusiasm, and her academic record. For the past 15 years she has spent much of her time in the rain forests of Borneo studying the way of life of the solitary orangutan.

Galdikas left for Borneo at the age of 22 with her new husband, Rod Brindamour (also a Canadian), and a newly earned Master's degree from UCLA (University of California, Los Angeles). The two of them set up camp in an abandoned loggers' hut in a remote forest area of southern Borneo. Galdikas searched for orangutans in the swamps while Brindamour made trails, managed the camp, mapped the area, and took photographs. (At that time the area was a forest reserve. Now, thanks to Galdikas' work, the reserve is a national park, protected from the logging industry.)

Two years after she arrived in Borneo, additional support from the National Geographic Society allowed Galdikas to expand her camp and hire additional workers from the Dayak tribe. With the added help more data could be collected, and Galdikas could stay in the field for a few days at a time.

Finally, after more than seven years in Borneo, Brindamour decided to return to Canada and to his interests in being a helicopter pilot and computer expert. Binti, their young son, followed him a year later. Since then, Galdikas has married a Dayak tribesman, Pak Bohap. The camp has further expanded to include a dormitory and eating hall, twenty-five Dayak assistants, and several American helpers. Every year Galdikas returns to Canada for four months, during which time she lectures at Simon Fraser University, spends time with her son Binti, and visits orangutans in zoos.

Orangutans leave their nests in the trees at dawn, so Galdikas rises at 3:30 a.m. to be ready to follow them. Once an orangutan is spotted Galdikas follows it from one food source to the next. The animals spend most of their active hours eating or looking for food, and travel up to 2 km a day doing so. Each day is spent tracking adult orangutans through the forest and waist-deep swamps. It is often difficult trying to keep up with the animal who is swinging through trees about 30 m overhead. She records what the individual is doing at one-minute intervals. Activities might include foraging, moving, resting, vocalizing, mating, grooming, and caring for the young. She also makes note of other conditions, such as, where the animal is, how high off the ground, and whether there are other animals in the vicinity. Her studies concentrate on reproduction of the orangutan, their feeding habits, and the impact they have on the forest through their feeding activities.

Galdikas is regularly sent orangutans that have been captured for the illegal pet trade. Many that came to her sick have died before they can be helped to re-adapt to the wild. Usually the ex-captives that inhabit the camp are safe to be around, but some can cause problems. Like wild orangutans, they love to wrestle, and the young ones do not know their own strength. Once when Galdikas was knocked down while wrestling with a young ex-captive orangutan, another orangutan rushed in and took a bite out of her arm.

Orangutan males can weigh up to 90kg. If the wild ones were allowed into camp, they could be very dangerous. Galdikas has been threatened by orangutans in the wild. A female who did not like to be followed tried to topple a dead tree onto her. Another time an adult male, named Ralph, when pursuing an ex-captive female into the camp, met up with Galdikas on a bridge. Had Galdikas made the slightest movement, he would have attacked her; however, she managed to stare him down and he retreated. Male orangutans frequently have staring bouts to test each other's will.

Life in Borneo has not disappointed Galdikas. She says that she is much happier, less rushed, and more patient. Her life is devoted to studying the orangutan. As she says, she "feels a special kinship with orangutans ... They've gone off on a path divergent from ours and developed this incredible strength of character denied us because we are a gregarious species. I always prided myself on my ability to withstand solitude. I enjoyed it, craved it. But I've followed orangutans day in, day out for a month, and they'd never meet another of their kind, and if they did, they'd run away to avoid contact." When asked if she has turned her back on human society, she responds "I've never been disillusioned with humankind. Working with orangutans has made me much more aware of humankind's compassion. The higher in my estimation orangutans rise, the higher we do."

Galdikas has studied one population of orangutans for longer than any other researcher, yet she has still not been in Borneo long enough to follow an individual orangutan from birth to death. (Orangutans live for fifty to sixty years.) That could take her a lifetime, and she is content to do so.

SOCIAL CONTROL

In any society there must be standards and rules that regulate behaviour in order that people can function together productively. The regulation of behaviour, or social control, operates at three different levels: the individual, the political, and the legal. Anthropologists can learn a great deal about a culture by studying its social controls.

Social control at the individual level is exercised by the conscience. Individuals are taught the values, norms, folkways, and mores of their culture and will be guided by their conscience to observe them.

At the political level, social control takes the form of public policy. Decisions must be made about all areas of life, from how food is to be produced and distributed to what festivals and holidays are to be recognized. All cultures develop a political system to address the various needs of their populations, and to maintain social order. Nomadic hunters and gatherers who live in small groups do not usually have a political leader. Small agricultural communities consisting of a few kin groups may also not have a central political authority. A political system that involves all members would exist in this situation. In larger societies kinship ties are weaker. Individuals tend to look to a centralized political organization to maintain social order.

Social control at the legal level involves enacting and enforcing laws to maintain social order. In some cultures there is no separation of the political and legal levels of control. The head of state may be the supreme legal and political, and sometimes religious, authority. In other cultures, such as ours, there is a separation of the political and legal levels.

In our culture the legal system is far more complex than in pre-industrial cultures where the law was often not written down. We have developed a legal system based on elected law makers, appointed judges and juries, lawyers, courts, and prisons. As sophisticated and complex as the system may be, its principal purpose is the same as in tribal societies. A decision must be made on whether people are innocent or guilty, right or wrong. Once this decision is made, a sentence or judgment is passed.

THE RELIGIOUS CUSTOMS OF A CULTURE

Every culture has a body of knowledge and set of activities called religion. Religion is concerned with defining, explaining, and influencing supernatural beings and powers, and helping to cope with the fundamental problems of existence.

Anthropologists study the religious life and customs developed by different cultures. They examine how the religious beliefs of a culture address issues such as: the meaning of life and death, values and group goals, tragedy and destruction, and evil. They learn about the particular supernatural beings and powers of a religion; they investigate religious behaviour in ceremonies, rituals, and prayer. The myths, legends, and writings are also studied.

In 1985, a Gallup poll showed that 87% of Canadians believed in God and in heaven. Twenty-nine percent believed in reincarnation, that is, returning to this world as another living being after death. If you were a cultural anthropologist, what would you conclude from this survey of Canadian society as shown in Figure 9.1?

GALLUP POLL RESULTS

The question was: "Which, if any, of the following do you believe in?"

		AGE		
	National	18-29	30-49	50+
Believe in God				
Yes	87%	81%	88%	91%
No	10	17	8	6
Don't know	3	3	4	4
Believe in Heaven				
Yes	71	66	69	77
No	24	31	23	19
Don't know	5	3	8	5
Believe in Hell				
Yes	39	36	38	42
No	56	61	56	52
Don't know	5	4	7	6
Believe in Devil				
Yes	33	32	33	34
No	62	65	61	60
Don't know	5	3	7	7
Believe in Reincarnation				
Yes	29	31	31	25
No	59	58	58	62
Don't know	12	12	12	13

Figure 9.1

PROGRESS CHECK

1. When anthropologists study political organization what aspects of culture are they examining?
2. What aspects of religion do anthropologists study?

OVERVIEW

Culture can be defined as a text for living, a program for survival, and a way of life for a group of people. It is a kind of collective personality of a cultural group. Human populations developed cultures because they were biologically equipped to do so. As humans developed larger brains and the ability to walk on two feet, such distinct human skills as toolmaking and abstract language followed in time.

Cultural anthropologists study various cultures in a systematic way in order to learn as much as possible about a particular culture and its relationship with other cultures. They study social structures and the social interactions within those structures. They investigate how a culture interacts with its environment, and what its economic structure is. Learning about a culture's social controls and religious practices adds more information to a cultural anthropologist's understanding and definition of a culture. What can we learn from the work of cultural anthropologists?

KEY WORDS

Define the following terms, and use each in a sentence that shows its meaning.

cultural relativism
enculturation
field work
ethnography
ethnology
material culture
non-material culture
kinship
consanguineal
affinal
nuclear family

extended family
descent group
kindred
clan
tribe
associations
ecology
market exchange
redistribution
reciprocal exchange

DEVELOPING YOUR SKILLS

FOCUS

What questions can be asked about the origins of culture, its development, the effects of the environment on culture, the functions of culture, the influences of culture, and the methods that are used to study culture?

ORGANIZE

Create an organizer which traces the development and growth of the use of language. In a separate column list the advantages that language usage gives human beings. In a final column speculate about how language helped to develop the growth and spread of culture.

LOCATE AND RECORD

1. Create a genealogical chart of your family and ancestors. Indicate whether they are related by consanguineal or affinal ties. Identify their ethnic and cultural identities — are they similar or different from each other?
2. In groups, list examples of the folkways and mores that regulate and control Canadians. In a separate column list the punishments for breaking these types of norms.
3. A group of volunteers can refer to the Criminal Code of Canada and list twenty crimes on subjects that the students find interesting. They can write out the definitions of these offences in their own words and describe the minimum and maximum penalties that persons receive if convicted of these crimes. The group can write down its research on a ditto and give each member of the class a copy. The class can then discuss whether the punishments fit the crimes listed. Are the punishments too harsh or not harsh enough?
4. Individual members of the class can conduct separate ethnographic research projects on such primitive cultures and societies as the Masai and the Tasadays. Information for these anthropological studies can be found in books and magazines in the library. Once the research is completed, a summary of the findings and conclusions can be shared with the rest of the class.
5. List all the activities that you perform during a day that require the use of your hands. In a separate column list

those activities that you would be unable to perform without hands. In another column comment on the advantages that hands give human beings.
6. In groups, conduct a brainstorming session about the types of controls, directions, and protections that Canadian culture places on individuals living in Canada. Record your conclusions under separate headings. Using a separate heading, comment on whether your examples are really required to keep our society operating.
7. How do the different religions of the world regulate and control the behaviours, life-styles, food consumption, and beliefs of people living in different cultures? The class can divide itself into groups and each can do research on a particular religion. Prepare a written report and communicate the results of your research to the rest of the class.
8. Divide the class into groups and have each group choose a type of food(s) that is consumed by a particular ethnic group. Each group can list the various types of food (associated with the ethnic group), the methods of food preparation, and the manner in which the food is consumed. Group members can find information in the library or from restaurants, or they can record actual eating experiences. The group then can record the effects of the environment on the types of food used by different peoples. The various groups might even cook and bring in samples of the food they are describing. Are the types of food, their preparation, and consumption an important part of culture?

EVALUATE AND ASSESS

1. Examine the Canadian Charter of Rights and Freedoms. Make a list of the ten most important rights guaranteed in this document. What does the Charter of Rights and Freedoms reveal to us about Canada and its culture?
2. Examine the following survey on favourite spectator sports in Canada. On a piece of paper record which sports are the most popular and least popular. What are some of the similarities and differences in answers given by men and women, regions, and income? In a separate column give an explanation that might account for the differences between the various groups. Does physical environment (region), sex, or income affect

people's preferences for certain sports in Canada? Discuss your conclusions in class.

GLOBE-ENVIRONICS POLL
"What is your favorite spectator sport?"

	Total %	Men %	Women %	Atlantic %	Quebec %	Ontario %	West %	Toronto %	Montreal %	Alberta %	Less than $15,000 income %	More than $35,000 income %
Hockey	36	44	28	31	44	30	36	27	40	46	29	35
Baseball	17	16	17	20	12	21	15	25	14	11	19	17
Football	7	9	5	3	4	6	13	6	5	13	4	9
Wrestling	4	4	3	7	3	4	3	2	3	1	9	2
Skating (figure/speed)	3	—	6	3	5	3	2	4	5	3	4	3
Soccer	3	4	3	1	2	4	4	3	3	6	2	4
Tennis	3	2	3	1	3	3	2	4	5	3	3	3
All others	16	12	17	18	16	17	12	11	3	11	11	12
No answer	12	7	17	15	11	12	12	16	11	7	17	12

3. Examine the following and summarize the legal rights that citizens are entitled to in Canada. What do these rights tell us about society?

---

1. NOTICE UPON ARREST
   I am arresting you for ....
   (briefly describe reasons for arrest)

2. RIGHT TO COUNSEL
   It is my duty to inform you that you have the right to retain and instruct counsel without delay.
   Do you understand?

CAUTION TO CHARGED PERSON
You (are charged, will be charged) with ....
Do you wish to say anything in answer to the Charge? You are not obliged to say anything unless you wish to do so, but whatever you say may be given in evidence.

SUMMARY CAUTION TO CHARGED PERSON
If you have spoken to any police officer or to anyone with authority or if any such person has spoken to you in connection with this case, I want it clearly understood that I do not want it to influence you in making any statement.

SYNTHESIZE AND CONCLUDE

1. Think about the types of conversations, programs, and activities that you are involved with in any given week. What types of cultural activities (sports, music, school, etc.) are most important? Are these learned behaviours and interests? Do they bring you closer together to others, and if so, in what ways? How would your life change without these cultural activities? Answer these questions in writing and share your conclusions with your classmates.
2. Identify the meanings and importance of the following actions, symbols, and objects in Canadian culture. What common images do they create in the minds of those who see them? How do they bring people closer together? Make a list of any other objects, symbols, and actions that create images that bring Canadians closer together.

the Canadian flag	a wedding band	a smile
a Christmas tree	a handshake	a cross
a police uniform	a menorah	

APPLY

1. Age grading is used by the schools to group people of the same age together in the same classroom. Speculate about the advantages of age grading. Do students of the same age have a separate youth culture which can be said to be different from the rest of Canadian culture?
2. Canadian society has developed a progressive income tax which requires people to pay higher taxes as their incomes rise and increase. These taxes are partly used to create a "safety net" for those who find themselves unable to care for themselves or to provide essential services. Discuss whether or not the present tax system is fair. List examples of Canada's "safety net" and discuss whether it should be strengthened or weakened. Comment on whether this "safety net" is an important part of Canadian culture.
3. Speculate about the functions and purposes of religion in primitive societies. Compare your conclusions with the modern religions of today. How are they similar and/or different from the religions of more primitive societies?
4. Divide the class into groups and speculate about the ways in which culture makes it easier for human beings to interact with each other and makes them more predictable.

Remember, culture is made up of values, customs, and traditions. List your conclusions and share them with the rest of the class.

COMMUNICATE

1. Write an organizer or short essay, or debate one of the following topics.
   - no culture is better or worse than any other
   - only human beings have cultures
   - without culture human beings would not be able to live with each other in group situations
   - if human beings were not able to stand erect there would be no culture
   - humans have a more complex brain than any other creature
   - religion plays an important role in Canadian society
   - the death penalty is supported by a majority of Canadians and should be brought back by Canadian lawmakers
   - individuals are merely a reflection of the culture in which they are raised
   - there is no Canadian culture — Canada is a country of regions with no common culture or identity
   - it is unfair for women to be addressed as "Miss" or "Mrs."
   - Canadians are too materialistic and should share more of their wealth with poorer people living throughout the world

2. Through a class discussion, begin to create an ethnology of Canadian culture. What values, customs, activities, and goals do Canadians think are important? How does the Canadian environment help to determine our occupations, preferences, and lifestyles. The class can then create posters and charts to display the results of their investigations, studies, and conclusions.

3. Draw or make a collage of different clothing styles worn in various areas of the world. Identify the types of clothes and explain how the environment has helped to affect the design and material of the clothing.

4. Write a brief essay in which you predict changes in Canadian culture by the year 2050. What kinds of changes will have occurred? For example, how will people make use of leisure time, how will our "world of work" have changed, how different will the living conditions of most Canadians be, and what kinds of cultural events will attract large groups of people?

CHAPTER

10

How Anthropologists View Cultures

What can anthropologists learn about various cultures? What can their work teach us about human society?

INTRODUCTION   Many people have little or no knowledge about cultures different from their own. Through the study of other cultures we can come to appreciate the different ways in which people have approached life and have learned to deal with problems that face them. Some of these problems are common to all humans and others are problems specific to a particular environment.

The study of other cultures can also show us how adaptable humans are as a species. We learn how humans in all cultures have used their intelligence and skills to solve problems, and their creative powers and imagination to produce beauty and quality in their lives.

Anthropology seeks to reconstruct human culture and experience throughout time. For instance in North America, anthropologists trace human culture back many thousands of years. The introduction of European culture to North America a few hundred years ago is a very recent event in relation to the length of time humans have been living in North America.

In this chapter, we will examine the way of life of the Tasaday tribe of the Phillipines. Although this society has existed for many hundreds of years it was only discovered by outsiders in 1971.

Another culture we will discuss is the Masai tribe of East-Central Africa. These people are self-sufficient cattle herders whose way of life has existed largely undisturbed for centuries.

10.1 *The Tasaday Tribe*

In mid 1971 the world learned of a Stone Age culture existing in a remote valley of the island of Mindanao in the Phillipines. These people, called the Tasaday, are a small tribe, only 24 in number at the time of discovery. Prior to a visit by an Ubo tribe hunter, Dafal, in the 1960's, the tribe had had no contact with the outside world.

They live in caves in the side of a cliff. Two fires which are used for cooking during the day and for heat at night, are kept going in the largest cave. This cave has an entrance about 4.5 m high and 9 m wide, and it extends about 15 m back into the cliff.

The Tasaday travel through the almost impenetrable jungle of their valley by swinging on vines. Since they have no enemies, they have no weapons, only stone tools.

SUBSISTENCE PATTERNS

Food Gathering

Their simple economy has been based on what they could gather in the forests and streams. They collect wild bananas, ginger, rattan fruits, palm cabbage, and a variety of other fruits, nuts, berries, and leaves. Some of the flowers and fruits are eaten raw at the time they are picked, other items are cooked back at the cave. Wild yams, their staple food, are dug up using a digging stick. To supplement their diet of fruits and vegetables, they collect and roast small fish, frogs, tadpoles, and crabs from the streams, grubs from fallen logs, lizards, and snakes.

Their only form of agriculture is the replanting of the tops of yams to allow the plant to regrow after most of the root has been eaten.

Food is shared among the group, but if only a small amount is found, children eat first. Usually the Tasaday, being such a small group, can satisfy their food needs in a short period of time, so they have lots of leisure time.

Clothing

Clothing could not be simpler. Men and women wear a pouch made of orchid palm leaves over the genitals, and over this the women wear a skirt of leaves. The men pierce their earlobes to insert wooden plugs, but the women did not wear any ornaments until Dafal's visit. He gave them bronze earrings and necklaces of beads at that time.

Technology

The technology of the Tasaday is as simple as their economy. For knives they use split bamboo; for scrapers, stones; for axes, large split stones rubbed to a rough edge, inserted in a cleft stick, and held in place with a rattan strip. The stone axes are useful for a number of operations — splitting firewood, opening soft logs to find grubs, breaking open fruit shells and nuts, cracking bamboo, sharpening digging sticks, and shredding bark for fires.

There is no sense of private property among the Tasaday. All the stone axes and other tools are used jointly. Nor do they have any need to use the tools as weapons. The Tasaday have no enemies, and no word in their vocabulary for war.

Fires are made by the Tasaday by rotating a stick in a notched log until dry material, such as bark, can be lit from

the heat of friction. They carry and store water from streams in sections of bamboo. Some of their drinking water comes directly from living bamboo stems.

Their technology has already changed since 1966 as a result of their first visitor. Although their environment supplied them with plenty of food, Dafal taught the Tasaday to trap and kill deer, pigs, and monkeys, and to preserve food by smoking it. Before learning to hunt and trap, the animals had been their friends; now the animals are wary of the Tasaday.

SOCIAL STRUCTURE

Balayem, one of the members of the tribe, describes its structure. "Our men are (he listed 10). Of these (he listed 5) have wives. Most of our children are male. I, Balayem, have no wife, and some other men have no wives. But I also have no father or mother or brother or sister. All others have someone. Only I am alone." When asked whether the men shared women because there were fewer females than males, he said: "No. A man and a women stay together until their hair is all white."

Apparently there is no ceremony when a couple marries, even though it is for life, and there is no sex before marriage. There is a taboo against incest, and several of the women appear to have come from one of two other groups that the Tasaday claim exist. These tribes have not been seen by the Tasaday for some time however, and they may have died out.

The Tasaday have no chief or leader, probably because there are few decisions to be made, and only a few adults to make them. Decision making is done communally. Responsibilities are shared equally, with no divisions of labour unless certain individuals happen to be more skilled or more interested in some activities than in others. Both parents help with child rearing and food gathering.

RELIGIOUS CUSTOMS AND BELIEFS

When a Tasaday tribe member dies the body is covered with leaves and left in the forest. There is no one place, but various places where the body might be left. When asked whether they had souls, the response was, "We do not know. All we know outside of our daily life comes from the dreams we have. But we do not know what dreams are."

There is a particular species of bird that influences them. If this bird sings when the people are preparing to go somewhere, they will remain in the camp for a few hours.

Their fears appear to be of a practical nature rather than linked to spiritual beliefs. They prefer to stay in their cave at night — there are poisonous snakes and thorns they would not see; and they do not venture out when they hear thunder — there is the danger of falling trees.

The Tasaday believe in a creation myth. "In the beginning the sun and the moon were husband and wife and had many children who were gods. These gods came down from the high places in the sky to make the mountains and the forests. From clay they made a man and two women who came to life when the greatest of the gods — Moogood Mani breathed upon them. All people were children of the same father — some were lifted by the magic of Moogood Mani to the high mountains and others floated down the rivers where they stayed and became different from one another."

They believe firmly that they should always remain where they are. An ancestor had a dream which predicted that the people would be safe from disease if they did not leave their home, and that a great one would come to visit them and would bring good fortune.

THE FUTURE OF THE TASADAY

Many anthropologists speculate that the Tasaday tribe mirrors our earliest human ancestors. One observer of the Tasaday culture sees in these people a lost aspect of human innocence. He believes that: "Maybe we ought to look back to primitive peoples to find out where the world went wrong. There seems to be a growing sense that it has gone wrong. Maybe we can learn from the Tasaday." Certainly the peaceful and harmonious way in which these people interact with each other and with their environment is to be envied.

Already outside pressures in the form of new ideas and new technology have brought about changes to their culture. As a result of one visit from the outside world they moved to a larger cave and built log platforms for each family — something that had not occurred to them to do before. Also, they have added hunting and trapping to their subsistence activities, and they are learning to use bows and arrows.

Unfortunately, as more and more people interact with the Tasaday and bring the outside world to them, their unique culture may soon disappear.

Progress Check

1. What kinds of technology did the Tasaday use?
2. What type of marriage relationship do they have?
3. Briefly recount their religious beliefs and customs.
4. What fears do outside observers have regarding the Tasaday culture?

10.2 *The Masai*

The Masai are a group of clans who migrated from the Upper Nile to the Great Rift Valley of Kenya and Tanzania in East Africa three or four hundred years ago. The clans who live on the produce of domestic mammals, the pastoral Masai, are considered to be the Masai proper. The Masai scorn the other clans who have a different way of life — the clans who are craftspeople, who cultivate crops, or who consume game and fowl. The Masai do not cultivate crops or hunt for food; they only hunt for sport or to protect their cattle.

SUBSISTENCE PATTERNS

The domestic animals kept by the Masai include cattle, sheep, goats, and sometimes donkeys. Cattle are by far the most valued of the stock, however. Each morning the animals are taken off to graze by the children of the village, and at night they are brought home. The village huts are arranged in a circle with a thorn fence around the circumference to protect the stock from predators at night. Every three or four years the villages are moved because of the build-up of mud in the cattle enclosure.

Milk from the cows is the basis of the Masai economy. The women do the milking. Occasionally butter is eaten, and blood from oxen if someone is in a weakened physical state. A couple of pints of blood are removed from the neck of the ox by shooting an arrow into the jugular vein, and collecting the blood with a gourd. Dung is used to coagulate the blood once enough has been taken, and then the ox is released.

Although cow's milk constitutes the basic Masai diet, some other foods, such as grains, tea, and sugar, are bought. Meat is eaten under certain circumstances: at ceremonies, in cases of illness to strengthen the invalid, and sometimes to welcome visitors.

The warriors, called the moran, must observe certain traditions with regard to food and drink. They can eat beef, but not together with milk. They are not permitted alcohol, but before a battle or a cattle raiding they chew the bark of the mimosa tree which is a stimulant. If too much is chewed, they fall into a coma. To reinforce close ties among themselves, the moran must always drink and eat with each other, never alone.

Clothing and Ornaments

Masai clothing is simple, consisting of homespun tunics and cloaks in shades of yellow, orange, red, and brown. Sandals are worn on the feet. For ceremonies cloaks are made from the hides of sheep and goats, from vulture feathers, or from black and white colobus monkey fur. Headdresses used in ceremonies are made from lions' manes or ostrich feathers.

Adornments are many and varied, for the Masai spend hours decorating their bodies. Girls and women wear heavy arm and leg decorations made from iron, copper, or brass wire, and a variety of necklaces, earrings, and headbands. Their heads are shaven.

The moran spend hours plaiting each others' hair, interweaving it with strands of ochered string. They too wear necklaces, bracelets, and earrings, although not the heavy metal bracelets and anklets worn by the women.

Hunting

Among the Masai hunting is not necessary for survival, since the meat from game animals is not eaten. However hunting to demonstrate valour is an important part of the culture. Various honours are bestowed on those displaying individual feats of bravery. During a lion hunt, for example, whoever spears the lion is entitled to wear the mane. Warriors wear a lion's mane headress with pride, and are the envy of others during a dance. Due to a decline in game, these lion's mane headresses are becoming scarce.

MASAI CUSTOMS AND SOCIAL STRUCTURE

Most Masai ceremonies are related to the stages of the life cycle through which an individual or a group passes. At the age of 16, boys are circumcised and join the age set known as the moran. At night they sleep in a rough camp rather than in the comfort of the village, and they are expected to keep themselves fit for fighting. In times of war, they are the army and they must defend their country. Sometimes they engage in tournaments with other clans, since there is no longer anyone to dispute their hold on the grasslands.

When new pasture must be found during dry spells, the moran help out. Often when the herds are far away from the villages in the dry season, the moran are responsible for them.

For the period in which the young men are moran, not only do they prove their manhood, but they also learn all that is necessary for an adult Masai to know in order to fully participate in clan life.

Between 14 and 20 years later, all the moran are promoted to the status of junior elders. As junior elders they do not have much political influence, but they are now able to marry and settle down. The age sets above the junior elders are the senior elders and the retired elders, the most influencial age set. There are a few individuals who reach the age set above the retired elders, and these men are consulted on religious matters and details of ceremonial behaviour.

Legal and political decisions are made democratically by councils, at which every man can have his say. Each age set elects a representative who will act as chairman at his age set councils.

Today many Masai will refuse to work for whites. A man who does so will never be able to take part in the council of elders.

Marriage

A man, once he has passed through the moran age set, may have several wives, but they must come from a different clan. The number will depend on his wealth in terms of cattle, for he must be able to provide enough milk for his wives and children from his herd of cattle.

A female will be betrothed at birth, or even before, and will marry at puberty, leaving her home to go and live with her new husband. With the help of other women she will build a house made of dried cattle dung for herself. Her husband will give her milking rights for a few of his cows. There is no divorce.

Labour is usually delegated according to gender. Women provide supplies of milk, honey, and beer, and do most of the household chores. The men tend the cattle and sometimes hunt.

It is not uncommon for the women and donkeys to walk 16 km to obtain water.

RELIGION

The Masai are **monotheistic**, believing in one god. The supernatural power they believe in is called Ngai, a word that means sky and air. They pray to Ngai for health and fertility, and for rain during periods of drought. Rain and grass, both held sacred by the Masai, are provided by Ngai. As a powerful moral force, Ngai punishes those who commit sins such as wanton killing of defenceless creatures.

For people other than a few great ones, there is no afterlife. The dead are not referred to, and deaths are ignored as much as possible. Both corpses and the hopelessly ill are unceremoniously left outside for the hyenas. Great men, when they die, are sometimes buried with grass, a stick, and new sandals. The bodies of laibon, hereditary seers, are buried under a stone cairn when they die. People passing by will toss a pebble on the cairn out of respect.

The role of the laibon is to act as an intermediary between the Masai and Ngai, and as a judge in major disputes among the Masai. He may suggest the size of fine, often in terms of head of cattle, that should be paid to a wronged party. Other duties include foretelling the future, authorizing ceremonies, ensuring fertility, sanctioning war, bringing rain, and advising the moran.

THE MASAI AND THE MODERN WORLD

Kenya and Tanzania are trying to bring the Masai into their national economies. They are offering schools and hospitals, livestock innoculation programs, disease control through cattle dips, and training in herd management. However, the value of the traditional Masai techniques of herd management is now being recognized.

Today, most Masai boys attend school, many going to secondary school as well. Some have gone on to government posts, business careers, and teaching. The majority of Masai, however, continue to live as herders according to their age old culture.

Progress Check

1. On what is the Masai economy based?
2. What place does hunting have in their culture?
3. Who are the moran? What do they do?
4. What are the Masai's beliefs about death?
5. What type of marriage practices do the Masai have?

CAREER PROFILE
FRANZ BOAS

Franz Boas made substantial contributions to the budding science of anthropology through his research on the Indians of British Columbia, and his training of a generation of anthropologists at Columbia University in New York City. The influence of his philosophy on the direction of anthropology is still felt more than forty years after his death. For these reasons he is referred to as the father of American anthropology.

Born in Germany in 1858, Franz Boas came to North America in 1883 as a geographer. His goal was to map the unknown west coast of Baffin Island. This goal was not accomplished. Instead, he lived with the Inuit on Baffin Island for twelve months, collecting ethnographical material on their way of life which he later published. During his stay he developed a deep respect for their way of life.

Disease brought by Europeans had swept through the local Inuit populations, reducing their numbers from thousands to hundreds. Boas sensed the importance of recording Inuit customs and legends. He returned to Germany convinced that anthropological studies were far more urgent than geographical work.

Back in Germany Boas catalogued a collection of Indian artifacts from British Columbia. He was impressed by the "flight of imagination" they displayed, and the "wealth of thought" represented in the masks. At this time nine Bella Coola Indians from the West Coast were visiting Germany, and Boas spent a great deal of time with them learning their language and discussing their culture.

When Boas next returned to North America in 1886, he came as an ethnologist to study the West Coast Indian societies. With the completion of the Canadian Pacific Railway and the influx of settlers, the cultures of the West Coast Indians were being threatened. Boas felt that "within a few years everything would be obliterated". (How wrong he was, however!) During the next three months he met with members of five native societies, and studied the culture of the Kwakiutl of northern Vancouver Island in some detail.

When he left Vancouver Island, armed with a collection of more than one hundred legends and extensive notes, Boas stopped in New York, and decided to stay. He married and began his long academic career which included forty years at Columbia University, where he trained a number of respected anthropologists, such as Ruth Benedict, Alfred Kroeber, Margaret Mead, and Edward Sapir.

Boas emphasized field work and first-hand observation of a culture as a whole, rather than the collection of selected types of information. He tried to gain an inside view of a culture by living with the people he was studying, and by collecting the beliefs and experiences of people in their own language. Thousands of pages of West Coast Indian myths, family histories, customs, dreams, and religious ceremonies were recorded in the language in which they were spoken, and translated by Boas with the help of an Indian assistant. This approach was named the "culture history school" of anthropology, and it dominated American cultural anthropology for most of the twentieth century.

Another area in which Boas made a significant contribution was the study of linguistics. He produced a *Handbook of American Indian Languages*, and a *Kwakiutl Grammar* (published after his death). In the introduction to the *Handbook* he pointed out the relationship between language and a society's way of categorizing experience and of viewing the world.

Boas was outspoken in his belief that all the human populations of the world showed equally the ability to develop cultural forms, and that differences in cultures were not due to genetic factors. His book, *The Mind of Primitive Man*, in which he voiced this enlightened view of race and culture. was burned by the Nazis in Germany in the 1930's. The book also influenced the civil rights movement in the United States in the 1950's.

One of the most important concepts taught by Boas was cultural relativism which stressed the importance of studying a culture in terms of that culture's values, not in terms of one's own cultural values. By doing so, Boas felt that people would be able to view their own culture more objectively as well.

A Kwakiutl Sun-Spirit mask

A Salish mask

OVERVIEW

In this chapter we have talked about two different cultures that have been studied and documented by anthropologists. As well, we took a brief look at the work of Franz Boas

The Tasaday tribe of the Phillipines, when they were discovered in 1971, were thought to be the last "stone-age" people. Their simple economy based on gathering food from the forest and streams required nothing more complicated than a stone axe. They lived in caves, clothed themselves with a few leaves, and travelled by foot or by swinging from vines. What perhaps has been the most enlightening aspect of studying their culture is the discovery that they are a happy, gentle, non-aggressive society. This discovery runs contrary to the popular notion that stone-age societies were by definition rough and aggressive.

Like the Tasaday, the Masai are a non-technological society. Once known as fierce warriors, they used to plunder the surrounding countryside to add to their stock of cattle, upon which their economy is based. With no written language their customs and traditions are passed on by word of mouth. Their lives are strongly influenced by the spiritual. Although some members of the tribe have embraced modern technology, most retain their traditional way of life.

Cultures have been coming into contact with each other throughout human history. The process of **acculturation** that takes place when two cultures come in contact always produces change. When a technological culture and a non-technological culture meet the results can be dramatic. Whether it is seen as beneficial to the non-technological society is a question many people have asked.

In the case of Canada's native people, the meeting of two cultures may result in people with bicultural identities. The native people can have an identity that is firmly anchored in their culture while possessing the skills and knowledge necessary to succeed in the larger Canadian society.

KEY WORDS

Define the following terms and use each in a sentence that shows its meaning.

monotheistic   acculturation

DEVELOPING YOUR SKILLS

FOCUS AND ORGANIZE

1. What questions should we ask about the cultures and lifestyles of the Tasadays and Masai?
2. Create an organizer that lists the major characteristics of each culture.

LOCATE AND RECORD

Divide the class into groups and do library research on one North American native society, for instance, the Six Nations, the Dene, the Inuit, the Plains Indians, the Haida, the Kwakiutl. Each group should find out about the beliefs, customs, food, and lifestyles of the society they have chosen to research. The effects of the environment on the culture of the society should be determined. Be sure to take notes so that you will have a record of your findings. Share your research with other members of the class. After all the groups have made their presentations, discuss the similarities and differences that exist among the different societies.

Give examples of the continuation of native cultures within the larger Canadian society.

EVALUATE AND ASSESS

Identify the points of view expressed in the following quotations on customs in other cultures and assess whether similar situations exist in Canadian culture.

"Anthropologist Victor Uchendu tells us that the Igbo children of southeast Nigeria share two worlds. The Igbo children participate in the world of the children and in the world of their parents. They take active part in their parents' social and economic activities. One could find them almost anywhere: at the market, at the village or family tribunal, at funerals, at feasts, working on the farm, and at religious ceremonies. They share with their parents the responsibilities of entertaining their parents' guests. There are no children's parties from which the parents are excluded, nor are there any parents' parties from which the children are excluded. The children do not even have separate sleeping quarters. Igbo children take an active and important part in the work of their village."
A Social View of Man, Alan J. C. King and Walter W. Coulthard

"The whole way of life of the Hopi is cooperative and peaceful; this, indeed is the spirit of their religion. A boastful, aggressive Hopi is unknown. In the first place, his culture provides him with little or no opportunity to learn such characteristics and should such learning occur, members of his society put considerable pressure on him to "unlearn" them."
The Study of Society, Blaine E. Mercer and Jules J. Wanderer

"Among the eastern highlanders of New Guinea, character ideals are such that the child most liked and admired by adults ... is the one who commands attention by tantrums, by a dominating approach to his fellows, by bullying and swaggering, by carrying tales to his elders. These actions epitomize the characteristics so desirable in the "strong" man and woman ... They are the mark .. of the fighter and warrior..."
Excess and Restraint: Social Control among a New Guinea Mountain People, Ronald M. Berndt

"In Bali, on occasions ... when the strictest etiquette must be observed the children are not much in evidence, so quiet are they ... The hierarchy of age recognized in any family group .. [is such that] every child knows that he is allowed to "speak down to", scold, and order about his younger brothers and cousins, just as he himself is spoken down to and ordered about by brothers and all relatives older than he."
The Individual and Culture, Mary Ellen Goodman: "The Balinese Temper, Character and Personality" by Jane Belo

SYNTHESIZE AND CONCLUDE

Divide the class into groups and compare the cultural similarities and differences in the Tasaday and Masai cultures to those of the Canadian culture. List your conclusions and rank their importance. Number one would be the most important. Determine whether there are more similarities than differences.

APPLY

The following article reflects on the values, customs, and rules that exist in Saudi Arabia. Canadian and Saudi Arabian societies exist far apart from each other, and in different environments. Either individually or in groups create an organizer by dividing a page into three columns. The first column should record the values, customs, rules, and legal system of Saudi Arabia; the second column should do the same for Canada. In a third column speculate about the changes that would take place in Canadian society if the values and practices of Saudi Arabia were put into effect in Canada.

"Saudi Arabia is a conservative society and has a very strict sense of right and wrong. Its laws and punishments are based on Moslem law. If people are convicted of a crime, the punishments they receive are harsh by the standards of Canadian society. Public beheadings, floggings, and amputations are the punishments given to those convicted of committing crimes. Some murderers and adulterers are beheaded in public squares. Some thieves have their right hands cut off, and those who commit lesser crimes are whipped in public or are jailed. There is also an element of mercy which is absent in some legal systems. If the family of the victim of a murderer forgives the murderer, the death sentence will not be carried out.

The Moslem religion strictly forbids the consumption of alcohol, and anyone found consuming it will be flogged or imprisoned. The eating of pork is also forbidden because it is considered unclean and unfit for human consumption. Women are expected to dress modestly and must keep their arms and legs covered when walking in public places. Religious police with wooden sticks patrol the streets and will use their sticks on women who have violated the expected dress code. Men who make unwanted advances toward women can expect to be whipped.

The Saudi Arabian legal system is also very different from Canada's. Suspects are arrested and tried by a judge; there are no jury trials. The accused does not have a lawyer during the trial, and is kept in prison the entire length of the trial. There is no release on bail.

Punishments in Saudi Arabia may seem harsh by Canadian standards, but crime rates are also lower. Saudi Arabia has one of the lowest crime rates in the world. In 1982 there were only 11 220 major and minor crimes committed by a population of 7 000 000. Statistics Canada reported that, in 1983, 2 143 236 crimes were committed by a population of 24 000 000.

Saudi Arabia is a religious society and its punishments are based on the Koran, the sacred book of the Moslem religion — "an eye for an eye, a nose for a nose, an ear for an ear, a tooth for a tooth". To people living in Saudi Arabia these punishments are just and fair."

COMMUNICATE

1. Write a one-page position paper on one of the following assertions. In your presentation either agree or disagree with the statement. Make sure you support your views with evidence.

- "We have much to learn from the simpler, less complicated lifestyles of pre-literate societies."

- "Canada's native people met the technologically more advanced Europeans and, as almost always happens in these meetings, the European way became the accepted way of life. But no simple "westernization" of the natives occurred. A way of life built up over generations in response to conditions of the land was not immediately dropped in return for new ways."

- "An understanding of human diversity is urgent ... Whatever wisdom cultural anthropology has learned in its sweeping study of human ways is wisdom about diversity — its extent, its nature, its roots."

- "For better or worse, Western standards of living are rapidly becoming universal aspirations ... The impact of the West has broken down economic self-sufficiency, sent the young away from village settings to cities. Kinship groupings larger than the family have disintegrated or lost their functions. Old religions have been abandoned ... The old order ... is gone ..."

- "Most individuals are so influenced by the norms with which they are familiar that any other mode of behaviour is unthinkable. When informed that certain Tibetan tribesmen exist all winter without a bath, that the Caribs relish the eating of certain tree worms, that the Inuit eat decayed birds, feathers, flesh, and all, the average North

American is inclined to be disgusted or incredulous. Yet one of the writers has seen aboriginal peoples in the jungles of South America who were nauseated by the taste of Grade A canned peaches, who laughed in derision at the practice of tooth-brushing, and considered the white man's firm refusal to pluck out his eyebrows an example of rank exhibitionism."

- "The government policy of placing native peoples on isolated reserves contributed to the maintenance of a distinct identity by reinforcing the boundaries between the native peoples and the larger society. This allowed some degree of protection for their communal way of life."

2. In the future, anthropologists of the time will be analyzing our cultures of today. Using the ways anthropologists have described the two cultures analyzed in this chapter, put yourself in the position of an anthropologist working in the year 3000. Describe Canadian culture from the perspective of the anthropologist. (You may wish to work in groups, giving each member of your anthropological team special areas to study. For example, Professor Familial might study only family relationships, whereas Professor Religiosum concentrates her analysis on the religions in the culture.)

UNIT

5

SOCIAL INSTITUTIONS

CHAPTER

11

The Family

Why is the family unit a social institution? How is the family changing?

INTRODUCTION Think back to your early years. What are your first memories? Perhaps you recollect an outing or a celebration. Most likely members of your family are an important part of these early memories.

The family is the basic institution found in all societies, filling the needs of children and overseeing their socialization. Since most people spend a good part of their lives interacting within a family, it is important to study the family if we want to learn about how people live. As an institution, the family influences and is influenced by other institutions, such as the economy, education, and recreation. It follows, then, that studying the family will also help us to understand the structure of society.

In this chapter we will examine the origins of the family as an institution, and the functions that the family performs. We will also look at family structure in Canada, noting the ways in which it is changing to meet new needs and to cope with new stresses.

11.1 *Defining a Family*

In 1979, Statistics Canada used the following definition of a family in preparing for a census of Canadians: "a married couple with or without never-married children living in the same dwelling, or a lone parent living with one or more never-married children." Today, the need to clarify the definition of a family is more urgent as increasing numbers of people choose to be married but childless, or avoid a legal, formal union with or without children, or keep separate dwellings, or divorce one another.

TYPES OF FAMILIES

The nuclear family is normally established through marriage. It consists of husband, wife, and children living together in the same home. The extended family is made up of three generations — the nuclear family plus the parents of either the husband or wife. This is the strict definition of the extended family, however, households made up of a nuclear family plus an assortment of other relatives are also referred to as extended families.

The **reconstituted family** is composed of a nuclear family in which one or both of the parents have been married previously and have brought their children from the previous marriage into the new family.

The **single-parent family** may consist of a mother or father and the children from a relationship that is no longer functional due to separation, divorce, death, or desertion. In some cases, where the parents do not live together but wish to share in the upbringing of their children, two households may be set up to allow for joint parenting. The children's time will be divided between the two homes in some way, for example, part of the week with one parent and part of the week with the other one.

The **polygamous family** contains several husbands or several wives. If there are several wives and one husband, it is a **polygynous** (meaning several women) family. If there are several husbands and one wife, it is called a **polyandrous** (meaning several men) family. There are many cultures in which polygynous families are common, but very few cultures in which polyandrous families occur. In many societies that practise polygyny, the more wives a man has, the greater his status, since the number of wives reflects his wealth. The Todas of southern India practise fraternal polyandry, whereby women are shared by two or more brothers. Usually the eldest brother will be the acknowledged father of any offspring.

THE ELEMENTS OF A FAMILY

Sociologist Frederick Elkin, in his book *The Family in Canada*, claims that Canada's geography and history make it impossible to describe a typical Canadian family. In a multicultural society based on two founding cultures, there will naturally be several kinds of family.

Although there may not be a single typical Canadian family, all families have some common elements as their basis. First, the family is a social institution and the most important building block in society. Second, the family originates in marriage or a similar agreement by which people live together as a unit. Third, the family usually has some combination of the following parts: a mother, a father, and children. Fourth, the members of the family are usually joined together by a number of strong bonds, such as legal and emotional. The legal bond is two fold. Marriage in most societies is governed by law, as is the parents' responsibility for their children. The emotional bond within a family is the affection and caring support family members give each other.

11.2 *Functions of the Family*

An American anthropologist, G. P. Murdock, studied the structure of 250 societies. He found that each one of them had as its basis small social units composed of wife, husband, and children. This basic family unit appears to be universal (existing everywhere). Murdock attributes the universality of the family to the four important social functions it fulfils. These functions are sexual, reproductive, economic, and educational. We shall examine each of these functions in turn.

THE SEXUAL FUNCTION

The family begins with a marriage or an agreement on the part of the couple to live together. The couple can thereby establish a fairly stable sexual relationship and interpersonal relationship. Sexual relationships outside of marriage may or may not be approved by society, depending on the society in question. However, it is more common for societies to frown on having children outside of marriage than on having sexual relationships outside of marriage.

In this century there has been a marked change in North American attitudes toward premarital sex. In a 1965 study of college women, 70% thought premarital sex was immoral. In 1970 the figure was 34%, and in 1975 the figure was down to 20%.

There is a universal taboo against incest. Marriage or sexual relations are not allowed between family members, but some societies allow marriage between first cousins. The incest taboo is of practical value both biologically and socially. Genetic studies of other animals confirm that inbreeding for several generations (between brothers and sisters or between parents and offspring) results in weaker and less fertile offspring. The incest taboo, because it forces people to marry outside of their immediate group, also promotes the forming of alliances between different social groups. These alliances might create friendly feelings between groups, and make hostility and warfare less likely.

THE REPRODUCTIVE FUNCTION

All societies need children if they are to survive. The family is the accepted framework within which children are produced. The family provides a home and family members who take the responsibility of caring for the children. In fact, most societies require that parents be responsible for their children. Historically, there has been discrimination, both legal and social, against children born outside of marriage. Only those children born to parents who were legally married were considered legitimate. Some of the legal and most of the social discrimination has eased in Canada. However, since illegitimacy is held in lower esteem, reproduction is encouraged within the legal family unit in Canadian society.

THE ECONOMIC FUNCTION

People require food and shelter, and the family shares responsibility for obtaining the necessities of life and for performing household duties. When one spouse works outside the home and the other within the home, there is likely to be a sharp division of labour. Children frequently have responsibilities within the family, and perform tasks appropriate for their age. Older children living at home may be expected to contribute financially to the running of the family.

In North America it is now common for both husband and wife to contribute financially to the running of the family. One income is often insufficient to maintain what is accepted as a comfortable living standard. There has been a steady increase in the number of two-income families from the early 1900's until the present day.

TABLE 11.1 **Canadian Workforce, 1986**

Number of women in workforce	5 608 685
Number of men in workforce	7 441 170

THE EDUCATIONAL/SOCIALIZATION FUNCTION

Children need regular contact with caring adults if they are to learn how to become fully functioning adults in society. The caring support and guidance provided by the family best performs this function.

In our society, the educational function of families has been mostly taken over by the state once a child reaches school age. But early training, such as learning language, is usually carried out within the family. Other aspects of the socialization of children, such as teaching basic skills, attitudes, and values, are also considered the responsibility of the family.

THE FAMILY 327

PROGRESS CHECK

1. What are some types of families?
2. List the four common elements for any type of family.
3. Name the four key social functions that families perform, and give some explanation of each.
4. Account for the incest taboo.

11.3 *The Changing Nature of the Family in Canada*

In the nineteenth century Canadian families were quite different from those we are used to today. The extended family was more common then, especially among the wealthy in rural areas. The extended family in rural areas typically included the nuclear family plus grandparents, but also other relatives, hired hands, and servants. It made economic sense to have more people involved in food production, in the running of the household, and in caring for the children and the elderly. In the cities, married couples often lived with one set of parents until they could afford to set up a separate household.

In this century, with increasing industrialization and urbanization, there has been a shift from the extended family to the nuclear family. Many young people left the farms and their families to seek work in the cities.

Most of the factory jobs employed men, and not women or children. Those women who worked were predominantly single. The passing of new child labour laws prohibited children from working in factories. Instead, new compulsory education laws sent children to schools instead of work, and children could no longer be considered a financial asset to the family.

On the farm, the home and the workplace were one and the same; in the cities they became separate worlds. Men went out to work, while the women stayed at home. No longer did women have a strong economic role to play in the family. At the same time, because men spent most of their day away from the family, they no longer played a strong role in the socialization of their children.

WOMEN IN THE WORKFORCE

Prior to World War II, the nuclear family with the wife at home was the norm. After the war, the nuclear family with the wife working outside the home has become increasingly common.

In 1931 only 3.5% of married women were in the workforce. By 1981 this figure had increased to 50.5%. One of the reasons for this steady increase has been the expansion of the service sector of the economy — sales staff, restaurant workers, cleaning service workers. As jobs became available in this service sector, women filled many of them.

Another factor that has tended to influence the number of women entering the workforce is that women are becoming better educated. With more education, more jobs are open to them. Most of the careers and occupations traditionally held by males are now available to women. The women's liberation movement has helped to bring about this change in attitude by redefining the role of women in Canadian society.

An important reason for the increase of women in the labour force has been economic necessity. As the cost of living rises, one-income families are placed at a disadvantage. More and more women have joined the workforce to supplement the family income.

Margrit Eichler, in her study *Families in Canada*, points out why women would be motivated to seek work outside the family. "It is generally understood that having a paying job confers feelings of status, satisfaction, self-fulfilment, prestige, and a measure of economic independence. If these are factors that weigh heavily for men, why should they not weigh heavily for women?" Married women who were questioned about their reasons for working supported Eichler's conclusions. The findings are in Table 11.2.

TABLE 11.2

Need the money to provide necessities	11.5%
Need the job to improve standard of living	11.8%
Gain personal satisfaction from the job	46.1%
Personal needs and financial necessity	30.6%

THE RISE OF SINGLE-PARENT FAMILIES AND RECONSTITUTED FAMILIES

Another change that has followed World War II is the rising incidence of divorce. This has resulted in an increase in reconstituted families and single-parent families. The following table shows the number and rates of divorces per 100 000 population for specific years.

TABLE 11.3 Increasing Divorce Statistics in Canada

Rate per 100 000 population

1931	700	6.8
1941	2 462	21.4
1951	5 270	37.6
1961	6 563	54.8
1971	29 685	137.6
1982	70 436	285.9

The Reconstituted Family

After a divorce or the death of a spouse many people find that living as a single person is too stressful, and they may begin looking for another partner. If they have children, the emotional or financial pressures of raising them single handed may contribute to the decision to remarry. Sometimes too, the loneliness of living without a partner makes remarriage desirable. Whatever the reasons, many divorced and widowed people marry a second time. In 1962, 12% of all marriages involved at least one previously married partner; by 1982, the figure was 28%.

The new family that is formed through remarriage has been given different names by sociologists — the reconstituted family, the blended family, and the remarriage family are three. There has been insufficient research carried out to evaluate the success of reconstituted families in Canada. Esther Birenzweig, a family counsellor, has observed the strains caused by such issues as discipline, loyalties, and role expectations. Stepparents are often uncertain how to take on a parenting role with children who may view them as an intruder and may resent their attempts at discipline. Children may be unsure how the new parent should fit into their lives, and may feel a conflict of loyalties. One ten-year-old child made a point of ignoring his new stepmother. According to her, "He felt that no matter how nice I was, if only I would go away, mom and dad would get back together. He couldn't afford to risk liking me; he thought his mom would get mad."

The Single-Parent Family

Parents who have become single parents have many decisions to make and many problems to overcome. They are in charge of the family finances, and the health, schooling, and emotional well-being of their children. They must also try to carry out the responsibilities normally split between two people. Women who previously worked in the home may have to look for employment outside the home. Often their income will be insufficient to meet family expenses. Men who are unused to running a household may require housekeeping assistance. Daycare or in-house child care may need to be arranged.

Children of single-parent families also feel stresses. Some feel uncertain about the future. Some may feel different from other children, and retreat into themselves. If they do not have contact with the other parent, they may suffer from not having a suitable gender role model.

One of the major stresses of the single parent is loneliness. There may be no one from whom to get moral or emotional support. The double load of working outside the home and caring for children may not leave time for establishing a social life with other adults. Sometimes single parents are excluded from social gatherings composed of couples. As Benjamin Schlesinger points out in his study *One Parent Families in Canada*, "Isolation from normal community life to some degree is the fate of parents without partners ... They don't seem to fit any of the normal social patterns. They are the fifth wheels of society."

There are now organizations and support groups, such as One Parent Family Association and Parents Without Partners, that are geared toward bringing single parents into contact with each other. Public opinion is much more tolerant of the single-parent family now than it was in the past. Eventually this family structure will be recognized as "normal" along with the nuclear family and the extended family.

Progress Check

1. What changes did industrialization bring to the household in Canada?
2. Give some reasons for the increase in the number of women in the workforce.
3. What are some of the problems experienced by reconstituted families?
4. What problems are associated with single-parent families?
5. What effects might single-parent family life have on children?

11.4 *The Post-Modern Family*

As the twentieth century draws to a close, the Canadian family is undergoing considerable change. More and more, people are living in single-parent or reconstituted families, and the trend shows signs of increasing. What has happened to the traditional nuclear family? Some sociologists feel that it is moving into a new phase which they call the "post-modern family". The changes that are giving rise to this new post-modern family are: 1. the socialization role of the family is weakening; 2. couples are working out more equal roles in their marriage, with responsibilities and decision making being shared by both partners; 3. many family roles of the past are being taken over by non-family agencies.

SOCIALIZATION AND THE FAMILY

Socialization of children has always been one of the most important roles of the family. This role, however, has been weakened by what some sociologists call the "generation gap". Through school, the media, and friends, children are exposed to a wealth of information and ideas. They are encouraged to analyze information and think for themselves. Consequently, by the time they become teenagers, children frequently hold views that differ from their parents. The two generations may disagree on matters such as love, sex, politics, and economics. Holding different views of the world and their place in it does not necessarily imply that the younger and older generations are in conflict, although it can come to that.

In the past children were socialized by the family until they married. Today the peer group has an important role in the socialization of adolescents. Many teenagers are more concerned with how their peers view the world than with how their parents view it. Parental attitudes about good and bad, right and wrong, and parental tastes in clothing and music may seem irrelevant to adolescents. One seventeen-year-old student remarked "I'm old enough to know what I should do, and to see right and wrong. I listen to my parents' advice, but I want to accept only those parts that I choose."

Studies have shown that in 1960 teenagers were influenced primarily by their parents, while the influence of friends ranked third. By 1980 parents were second and friends were first. The influence of grandparents and other relatives also dropped between 1960 and 1980. One of the features of this shift is a growing sense of independence in decision making among adolescents. One girl in grade eleven from St. John, New Brunswick said "If my friends want me to go somewhere with them I will. But if they wanted to do something I felt was wrong, I wouldn't participate. I'd also give them my opinion. I am the kind of person who can make up her mind and make her own decisions."

The results of such studies indicate to some historians and sociologists that the socialization role of the family has been substantially weakened once children reach adolescence. No longer are values and attitudes learned mainly through discussions around the dinner table and in the living room of the nuclear family. The values and attitudes of peers are likely to carry more weight than those of the parents.

CHANGING ROLES OF HUSBANDS AND WIVES

The traditional picture of the wife at home looking after the children, keeping house, and having dinner on the table for her husband when he returns home from a day at work is no longer accurate for most homes these days. More often, both parents return home after a day at work and share the chores associated with running the household. The home was once viewed as a place of comfort and refuge for the male; today it is viewed by many as an area of shared responsibility and shared decision making.

Where once fathers paced hospital corridors waiting for news of their newborn infants, now they participate in the prenatal classes and the birth. Where once care of the infant was entirely the mother's responsibility, now fathers take their turn bathing and diapering their babies. Fathers have benefitted from this increased responsibility by developing a closeness with their small children that their own fathers probably never had.

The role women play in the family has changed even more than the role played by men. Henry A. Bowman and Graham B. Spanier, in their study, *Modern Marriage*, point out "A woman's role is complicated today by the fact that she is caught between the pressure of several forces. On the one hand there is the weight of tradition pushing her in the direction of homemaking and motherhood. On the other hand is the open door of opportunity in the world of gainful employment."

The choice between homemaking/motherhood and career is not clearcut. Most women do not wish to sacrifice having a family in order to have a career. However, the combination of a family and a career is not easy for they must juggle several roles — homemaking, childbearing, child rearing, and employment. If husbands are willing to shoulder some of the household responsibilities and the childcare, then the stress on their wives may not be too great. A realistic division of labour within the home is necessary in today's two breadwinner family.

NON-FAMILY CHILD REARING

As more and more parents with children work outside the home, less of the socialization of young children is being carried out within the family. In extended families the grandparents would often be available to look after the children. However, extended families are a rarity today. For an increasing number of families, the solution is to assign much of the socialization of young children to daycare centres, live-in nannies, or babysitters.

Several choices of daycare exist: cooperative group care (which may be associated with a workplace or neighbourhood), public day care, and private day care. All provinces offer funding toward daycare costs to people whose income does not exceed a certain level (usually around $18 000). Daycare can be expensive, especially if more than one child is involved. The number of daycare spaces available has not yet come close to meeting the needs of working parents. The social issue of adequate inexpensive daycare is one that is receiving a great deal of attention currently.

There has been a lot of research on the effects of daycare on children. One of the problems in conducting studies of this sort is taking into account the range of quality and types of daycare. It appears though that good quality daycare for children from happy, secure homes can be just as good as home care with a loving parent. Not all daycare is good quality, however, and it is necessary for parents to be selective if they want their children cared for well.

In a segment of Israeli society, the **kibbutzim**, or collective farms, the nuclear family is not the basic social unit. Instead,

the entire community forms the social unit. On kibbutzim, children are raised communally. From a very early age they live in a children's centre under the care of a few adults. Kibbutz children spend most of their day with other children in their group, but in the late afternoon or early evening they may visit with their parents. The parents play with them or discuss things with them, and in this way maintain a family bond. The entire community shares in the socialization of the children, and the children view the community as their family. It is rare for young people from the same kibbutz to marry each other, for they tend to view the children with whom they grew up as their brothers and sisters.

Social scientists who have studied the kibbutzim have found few behavioural or emotional differences between children raised communally and those brought up in a traditional nuclear family. If anything, it would appear that kibbutz-raised children have lower rates of emotional disturbances and more positive relations with their parents than children from traditional families.

PROGRESS CHECK

1. Which changes have led to a post-modern family?
2. Why has parental influence on teenagers changed?
3. In what way are some couples moving into a more shared lifestyle? What problems do they face?
4. What are the trends in family life in the latter part of the twentieth century?

OVERVIEW

The family in Canada is in the process of change. No longer can it play all the roles it once did. Instead, in this century, other institutions have taken over some of the traditional family roles. The school system and, in many cases daycare, are now responsible for a good part of the socialization of children; the economy of the family is largely controlled by business and industry; opportunities for recreation and play are provided by various community organizations; and health care is in the hands of the medical profession.

It is expected that the percentage of women in the workforce will continue to increase. The division of labour, with the father going out to work and the mother at home with the children, will be present in fewer homes. Instead, with both parents working outside the home, the responsibility for household tasks and care of the children will be shared. The daytime care of children and their socialization will likely be delegated to agencies such as daycare centres or, for those families who can afford them, to nannies.

The number of single-parent families and reconstituted families is likely to continue to increase, resulting in a variety in family structure. Families will be based on individual emotional and economic needs and less on prescribed structures of the past. Do you think the family as a social institution will survive all these changes and adaptations?

KEY WORDS

Define the following terms, and use each in a sentence that shows its meaning.

reconstituted family   polygamous family   polyandrous
single-parent family   polygynous   kibbutzim

THE FAMILY 337

KEY PERSONALITIES

Give at least one reason for learning more about the following.

Frederick Elkin
G. P. Murdock
Margrit Eichler
Benjamin Schlesinger

DEVELOPING YOUR SKILLS

FOCUS AND ORGANIZE

What questions should students consider when examining and thinking about the following: the origins of the family, the functions of the family, the importance of the family in Canadian society, and the changing nature of the family? Organize your questions under definitions you may develop for "a family".

LOCATE AND RECORD

Divide the class into groups and have each group research one of the following: the nuclear family, the extended family, the reconstituted family, the single-parent family, and the polygamous family. Students can base their research on their readings in the text and on vertical file materials and books in the library. When the research is completed each group will write a report, and present its findings to the class. A one-page summary can be given to each member of the class before the report is presented.

SYNTHESIZE AND CONCLUDE

1. On a sheet of paper list the functions of the family in the order of their importance. The first shall be the most important. In a separate column give your reasons for placing these functions in this particular order.
2. Statistics Canada reported that in 1987 there were 1 200 000 children under the age of sixteen who lived in poverty — more than one out of six children living in Canada. The poverty line for a family of four living in a city was $21 666 a year and $10 651 a year for a single person. Statistics Canada states that an individual or family is below the poverty line if they spend more than 58% of their income to buy basic necessities like food, clothing, and shelter. It reported that

83% of those children living in poverty live in female-headed, single-parent and young two-parent families. In 1987 the average income for a lone female parent was $17 353 compared to $44 919 for two parent families.

What evidence exists to support the idea that many Canadian children are living in poverty. What could be done to reduce the number of Canadian children who live in poverty?

APPLY

1. Divide the class into groups and have each group examine the following Gallup Poll results. What changes are occurring in public opinions on the issues covered in these polls. Have members of each group ask these same questions of community, school, and family members. Do the results of these interviews reflect the opinions of the Gallup Polls? How are they similar and different? Each group should report its conclusions to the class. The class can then speculate about the reasons for any changes in attitudes that are taking place on the issues covered in the polls. They can be listed on the board under the appropriate headings. Class members can then speculate about the future by answering the following questions.
   – Will the attitudes of Canadians on these topics continue to change? In what direction?
   – What changes might occur in the family? Will they be positive or negative?
   – What forces are causing changes to take place in the Canadian family?

Question: If they can, do you think that a couple should live together for a time before deciding to get married or not to get married?

	Yes	No	Undecided
1971	22%	70%	8%
1986	51%	38%	11%

Question: What is the ideal number of children for a Canadian family?

	Four or more	Two or fewer	Other
1945	60%	17%	23%
1987	13%	58%	29%

Question: Do you think married women with young children should work outside the home?

	Yes	No	Undecided
1960	5%	93%	2%
1987	47%	48%	5%

Question: Do you think that having married women in the workforce harms family life?

	Yes	No
1973	73%	27%
1983	62%	38%
1988*	48%	45%

*Breakdown by age of 1988 results of those interviewed

Under 30	40%	60%
30-49	54%	46%
over 50	68%	32%

2. Divide the class into groups for a brainstorming session to determine what family roles of the past have been taken over by non-family agencies. List the agencies and the functions they now perform. Determine why these agencies have taken over traditional family roles and speculate about the effectiveness of these agencies in the performance of their roles.

COMMUNICATE

1. Write a short essay or hold a class debate or discussion on one or all of the following topics.

- the Canadian family is flexible and is adapting itself successfully to economic changes
- the increasing acceptance and tolerance of new and different lifestyles and types of families is undesirable and will have harmful effects on children
- peer groups have more influence on teenagers than do families
- women are trying to perform too many roles at once with negative and harmful effects on themselves and their families
- daycare should be universal and paid for by the government

- all employers should be required by law to provide free daycare on the job site
- it is inevitable that the nuclear family will disappear
- the smaller number of children in families reflects an attitude of extreme selfishness and self-centredness in Canadian society
- polygamy should be made legal in Canada for those who wish to practise it
- child abuse would be greatly reduced if families were required to obtain a licence before they could have a child
- it is the quality, not the quantity of parenting that counts

2. Based on the recent dramatic changes in family structure prevalent in our society, predict the structure you think will be most common in the year 2000. Give well thought out reasons for your predictions.

CHAPTER 12

Love, Marriage, and Divorce

Why is marriage considered a social institution? How has marriage changed? What are the consequences of divorce in our society?

INTRODUCTION If asked to imagine a wedding ceremony, what would you visualize? Possibly it would be a beautiful bride, dressed in a flowing white gown walking up a church aisle on her father's arm to the familiar strains of matrimonial music played on the organ. The groom, handsome in his tuxedo, turns to smile at the bride as she joins him at the front of the church. In the presence of their families and friends the young couple promises to love and honour each other through sickness and health until death do them part.

This standard romantic image of a wedding is reinforced by television, films, newspapers, and literature. Sometimes, when well-known people, such as Lady Diana and Prince Charles, marry, the ceremony is turned into a lavish media event.

"And they lived happily ever after ..." True for fairy tales, but is it possible in real life? With the many pressures to which marriages are subjected, many do not stand the test of time. Some marriages barely last a year; others may fail after twenty-five years. Yet in Canada weddings take place regularly and frequently, and marriage remains one of the most important of our institutions.

In this chapter we will define marriage and examine the role it plays in society. We will also discuss the reasons for getting married and the changes in courtship patterns. Finally we will examine the increasing divorce rate and the effects of divorce.

12.1 *Defining Marriage*

What does it mean to become a married couple? It means two people have formally established themselves as a family. They have, in the presence of witnesses, declared a commitment to each other. They are prepared to share the daily responsibilities of living, to share sexual relations, to share in a social life that includes the friends and families of both of them, and to care for each other emotionally. Their marriage also implies that they are prepared to share the responsibility for any children they have or adopt.

Sociologist, Diane Vaughan, believes marriage is a process whereby two people restructure their lives around each other. Through common friends, belongings, memories, and a common future they shape a joint identity. They attend social functions as a twosome. Mail arrives addressed to both, and the government taxes them jointly. There is a new social status for the two marriage partners — the status of married couple.

The fact that marriage is also the beginning of a legal relationship is often not apparent. Both partners are committed to fulfilling the duties and obligations of marriage. However, if when a relationship is weakening, one or both partners fail to honour the obligations, there are social and legal sanctions that can be called upon.

Some couples are spelling out the legal responsibilities of their marital relationships by drawing up a marriage contract. These contracts, now legal in all provinces, may define how household tasks will be shared, how the partners will contribute financially to the marriage, or how possessions will be divided should the relationship be dissolved. The only area that cannot be decided by contract is custody of the children. Shirley Greenberg, an Ottawa lawyer, says: "Because women are so vulnerable in marriage, contracts are really important. They help everyone clarify expectations."

Family law in Canada was reformed after the Murdoch case in Alberta. Irene Murdoch, the wife of an Alberta rancher, was told by the Supreme Court of Canada that a wife did not share ownership of property registered in her husband's name merely because she assisted him in acquiring and developing the property. Many people were appalled by this decision and worked to have the law changed. On June 1, 1986, the new federal Divorce Laws came into force.

Today Canadian courts recognize child care, household management, and financial provisions as joint responsibilities. Since both spouses contribute to fulfilling these duties, each is entitled to an equal share of the family assets. Marriage is thus seen as a partnership in which the contribution of a spouse's unpaid labour within the family is equal in value to the contribution of money.

12.2 Arranged Marriages

Marriage in all societies is important because it indicates the creation of a new family and the potential addition of new members to the society. Most marriages in North America and Western Europe result from two people meeting, being attracted, and deciding after a period of courtship that they love each other and wish to live together. Whether or not to marry, when to marry, and whom to marry are an individual's decisions in our society. Relatives and friends may give advice, but it is the couple who will make the final decision of whether to marry or not.

In many societies, especially those in which there is a substantial transfer of wealth with marriages, the decision who will marry whom is felt to be too important to be left up to the couple. Instead, the parents of the couple, a religious leader, or a group of elders may be the ones to decide. Sometimes the couple may not meet until the day of the wedding. In some cases they may have been promised in marriage at birth. Once the couple marry and get to know each other, love and affection may develop, but these feelings are not deemed necessary to the success of the marriage. Mutual respect and fulfilment of responsibilities, including childbearing, are often judged to be more important.

In some societies, a father gives his daughter in marriage to the son of an important man in a neighbouring community or tribe. The arranged marriage, it is hoped, will cement the bond between the two tribes. It also may increase the status of the poorer or less influential family.

There is no way of telling which type of marriage is more successful — the arranged marriage or the marriage agreed to by the partners. We do know, however, that our North American custom of courtship leading to marriage does not ensure that the marriage will last.

12.3 Developing Marriage Customs in Canada

In eighteenth and nineteenth century Canada, marriage was not so much a love bond as an economic necessity. Husbands and wives married in order to become a successful economic unit producing life's necessities as well as children. The family and the business of earning a living were closely related. Often the home was the centre of the family business, and both men and women shared the economic activities. In eighteenth century New France (Quebec), for example, the wives of craftsmen opened taverns beside their husband's shops. On the farms wives worked in the fields beside their husbands. Since women in New France were usually better educated than women in France, they also kept the accounts and managed the finances.

Marriage in those days was an important business partnership. Numerous relatives signed the marriage contract, and the property rights of both parties were protected under law. Wives bringing family property into the marriage kept a legal right to it. Marriage was also an important means by which the colony of New France could expand its population. The government encouraged a high birth rate with financial bonuses and heightened social status. Thus children, one of the main components of the family, also became central to the political and economic continuance of the colony.

In Upper Canada (Ontario) as well, the basis for marriage was strongly economic. If a man's wife died, she had to be replaced as soon as possible so that the work could carry on. On the frontier, young men, who had emigrated to take advantage of the cheap land offered in Canada, found the life exhausting and lonely. The Canadian government undertook to import young British servants to become future pioneer wives.

Parents in Upper Canada wishing to find a good match for their offspring used to bring them to York (Toronto). Specially arranged dinner parties and other social engagements provided an opportunity for well-bred, but often isolated, young people to meet each other. They were encouraged to marry for wealth and position rather than for love.

During the nineteenth and early twentieth century young people met at church dances, at community functions, at work, or through introductions by a member of the family. Courtship was well organized with rules governing the behaviour of the couple. Although arranged marriages were not generally the rule, women usually married men with similar social and economic backgrounds.

In early Canada most girls were married between the ages of twelve and sixteen, while men usually married in their twenties. By the nineteenth century the average age of marriage for both men and women had climbed to the mid to late twenties. Since the middle of this century the average marriage age for females has hovered around twenty-two, and for males around twenty-four to twenty-five.

PROGRESS CHECK

1. What is marriage?
2. Looking at various cultures, give some of the different reasons for marrying a person.
3. What is an arranged marriage?
4. How was marriage in early Canada "an economic necessity"?

12.4 Love and Marriage

In our society "love" and "marriage" are words that are frequently used together. As the song says, they "go together like a horse and carriage". But this has not always been so. In the past, love was not thought to have much at all to do with marriage.

During the twelfth century a phenomenon called "courtly love" developed in Europe. Troubadours sang about tender, flirtatious relationships that took place outside of marriage. Courtly love was considered a noble and virtuous activity, and was called "pure" because it was unconsummated by sexual relations. Courtly love gave married and single couples an opportunity for adventure and romance.

During the late eighteenth century a different type of love emerged. Unlike courtly love, this type of love — romantic love — was not based on celibacy (on not being married). People began to talk of "lovers" and "mistresses", and marriages founded on romantic love began to occur. Young people thinking about marriage began to pay more attention to inner feelings than to status, property, and parental wishes. But, status and property still played a role in mate selection, because most people still married within their social class.

The modern version of love — romantic love and falling in love — is a more recent phenomenon. Its roots can be traced to industrialization. People who worked in the factories, had little or no property. They did not have to worry about maintaining their social status by marrying someone with similar resources. They were free to marry for more personal reasons. As well, when young people left the farms and villages to seek their fortunes in the cities, they were no longer under the watchful eyes of their parents. Consequently, they were less restricted by the traditional marriage customs imposed by the family and community. As courting couples gradually gained more privacy from family and peers, there was an increase in sexual contact, and affection and love became the principal reasons for marriage.

By the twentieth century, romantic love became the main basis for choosing a marriage partner in most of Western Europe and North America. Young people chose mates based on their feelings rather than on their parents' wishes. They wanted more from their marriage than merely the satisfaction of fulfilling family responsibilities. In addition they wanted a more emotional and affectionate relationship. Personal happiness, it was felt, could be achieved through a loving relationship with someone of the opposite sex.

THE PRIVATIZATION OF COURTSHIP
Traditional courtship situations which brought boys and girls together for the first time used to be monitored by the community or by a couple's peers. Girls and boys were usually not allowed to encounter each other without a chaperone. However, by the mid nineteenth century, courtship practices had changed. Courting couples were able to see each other in private. According to social historian Edward Shorter "courtship had become privatized".

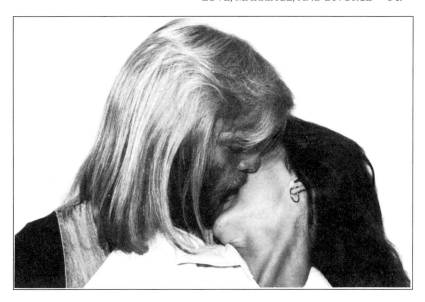

Courtship Today

One of the most important factors in the evolution of courtship in North America has been the automobile. With the automobile, young people could escape parental restrictions, and develop their own individualized courtship rituals. Today young people meet in much the same way as did their parents — at high school, at university, at parties, and on the job. Many are introduced by friends. But some things have changed. Many new avenues to bring compatible people together have been opened up.

Dating services to which individuals pay a fee to be introduced are popular. Video-dating in which individuals make a short video-tape about themselves has also become popular. Companion ads in newspapers advertising one's availability and attributes flourish throughout North America. Singles bars, clubs, dances, apartments, and ship cruises also bring people together. Some super-markets have singles nights during which only singles show up to shop, mingle, and possibly find a mate.

Toward the end of the 1950's the conduct of young people during courtship became their own "affair" and sex played a larger role. Sexual contact became more frequent and was seen as less binding than ever before. Couples who had sex were no longer expected to marry, as had been the case previously. Couples began to link the intimacy of sex with the strength of the love bond.

A change of attitude has accompanied these changes in courtship practices. There has been a steady increase in the percentage of Canadians who approve of premarital sex, according to surveys taken in the 1970's and 1980's. Religious affiliation has some effect on attitudes towards sex. Young people who identified themselves as Protestant or Catholic and who went to church regularly were more conservative than those with no religious affiliation.

TABLE 12.1

	Males	Females	Protestant	Catholic	No Religious Affiliation
"Do you agree that sex before marriage is all right when people love each other?"	84%	77%	75%	81%	94%

Edward Herold, a sociologist at the University of Guelph, found a difference in the attitudes of males and females toward premarital sex. Young Canadian females more frequently felt sexual involvment should be preceded by a meaningful relationship. Sixty percent felt love should precede sex. "Women have been conditioned" according to Herold "to associate love with sex and are less willing than men to engage in sexual relations solely for obtaining physical pleasure".

Unfortunately, the increase in premarital sexuality has also been associated with problems such as unwanted pregnancies. Approximately 13% of all children born in Canada are born to unmarried mothers. Between 1960 and 1980 the number of children born to unmarried mothers almost tripled. Another problem faced by young people engaging in premarital sex is that of sexually transmitted diseases, including the recent threat of AIDS. Although 50% of young people experiment with sex, polls indicate that only half that group used contraception. As sociologists Reginald W. Bibby and Donald C. Posterski point out "It seems apparent that many adolescents are sexually active but sexually illiterate".

PROGRESS CHECK

1. How had courtship changed by the mid 19th century?
2. How are courtship practices today similar and different from those practised a generation ago?
3. What are current attitudes toward premarital sex?

MARRIAGE: A WEAKENING INSTITUTION?

A report in the *Journal of Marriage and the Family* concludes that marriage is a "weakened and declining institution". The reason cited for this decline is that women are getting less from it. Sociologists Norval Glenn and Charles Weaver who wrote the report, found that women are less happy in marriage now because of an increased load of responsibility.

In the past, in many families there was a traditional division of labour. Husbands worked outside of the house while their wives provided what economists refer to as "home production". Many women worked on a part-time basis but their primary role was homemaker. This role included cooking, cleaning, caring for children, and performing other household duties. Today this division of labour has disappeared in many marriages. The "gains from marriage for women have declined" according to *Cosmopolitan* magazine, and many women are postponing marriage or avoiding it altogether. They may not be able to be the caring wife, the nurturing parent, and dutiful housekeeper, as well as maintain a successful career.

But women are not the only ones who complain about the direction relationships between the sexes has taken. "A lot of men are still making adjustments to the fact that their wife is not just like "dear old Mom" according to Fred Rhodes of the Texas Children's Hospital. Many psychologists and sociologists have come to the defence of "post women's liberation men". They believe many men have become more sensitive and responsive to their partner's needs. Psychiatrist E. James Lieberman says " we are in the middle of a revolution, and it's getting better".

THE CONTROVERSY OVER ROMANTIC LOVE

Today romantic love is a controversial and much analyzed issue. M. Scott Peck tells us in his book, *The Road Less Travelled*, that romantic love is a "trick that our genes pull on our otherwise perceptive mind to hoodwink or trap us into marriage". Of the women and men questioned by Dorothy Tennov in her pioneering study *Love and Limerence: The Experience of Being in Love*, 32% said they had been "foolish" for falling in love. Twenty-six percent even "hated" themselves for being so helpless and dependent.

Author Bonnie Kreps, writing on the subject of romance, believes "for all our involvement with romantic love in our culture, we live with a curious ignorance about its nature. It's time we examined romantic love more closely. Most people know when it begins but few know why.... The truth is, romantic love . . . has a built in self-destruct." Kreps believes there is an important difference between romance and "the more real, mutual regard and caring that is fostered by intimacy and time and that more properly can be called love".

Krebs feels romantic love has outworn its use. She believes that having "so transitory [short-lived] and self-indulgent a state institutionalized at the core of our society is dangerous folly". There is, according to Krebs, an important difference between romantic passion and love. "The romantic passion" she says, "may be thrilling and delirious, but it is not true". Real love goes beyond infatutation and thrill seeking. It is rooted in respect and long-term commitment, and does not self-destruct after a brief time.

When we fall in love, according to Krebs, it is not so much with a person as with a feeling. The desire for the experience of falling in love is an important motivating factor. Often falling in love means that "you relinquish control and deliver yourself over, not only to the unknown, but to an unknown person". Krebs believes that this process is risky and produces anxiety, based on fear of the unknown and vulnerability. But it is precisely this "adrenalin rush" that we seek when we fall in love.

Today people are beginning to reflect on the process of falling in love. Since the majority of people getting married do so for love it may be time to re-evaluate the quality and strength of the love bond in the face of rising divorce rates.

In *Women and Love, A Cultural Revolution in Progress*, Shere Hite is encouraged by the fact that both men and women are in the process of "redefining" love. Instead of accepting cultural norms of what love should be or what love used to be, Hite says people are "demonstrating an ability to see and analyze what is going on inside their relationships".

Progress Check

1. Why are women postponing and avoiding marriage?
2. What complaints do men have about the direction relationships are taking?
3. According to many social scientists, how have men's attitudes towards women changed?
4. What does Krebs say about romantic love?
5. What is an important motivating factor for falling in love?

CASE STUDY

The Importance of Love in a Relationship

Love is a word that is frequently used — in songs, conversations, and in our thoughts. It is a constant theme in magazines, books, movies, television programs, and videos. It can be used to describe desires, intense feelings, and strong preferences for food, music, entertainment, literature, interests, or sports.

Most people have a real need to belong to a group, and to have strong emotional attachments with certain people. One type of emotional attachment is an intense feeling for and attraction toward another person. This kind of emotional attachment usually involves heterosexuals, people of the opposite sex.

During adolescence most people experience intense feelings and attractions for particular people, but usually for only short periods of time. These feelings are sometimes referred to as "crushes" or "puppy love". Sometimes teenagers are overwhelmed by their feelings for another person.. They might find it difficult to deal with their romantic fantasies or to cope with the pain of a relationship that does not work out as expected.

Dr. Aric Schichor believes that it is important for parents to encourage their teenage children to talk about their feelings and experiences. They should try to understand that their children are exploring new areas in their lives and may be experiencing doubt, inadequacy, and confusion. Parents should not expect teenagers to act as adults when they encounter their first love relationships.

Dr. Schichor states that most teenagers are successful in working out their feelings and reactions to their first romantic experiences; a small minority are not. Those teenagers who are not successful, may develop eating disorders or drug problems, or they may resort to irresponsible behaviour such as reckless driving.

Such people usually need professional help in order to deal with the roots of their problems.

There are many characteristics that are associated with romantic love. Usually lovers are preoccupied with each other and are constantly on each other's minds. They wish to be with each other, and want to be mentally, emotionally, and physically involved. They receive pleasure from doing things for their loved one — even if it requires personal sacrifice. They experience feelings of elation, and life becomes fresh and exciting.

Often the rule of **homogamy**, *the tendency to seek a love relationship with someone similar to oneself, applies. Despite the common phrase "opposites attract", people who fall in love often share many qualities. They might have the same level of eduation, the same ethnic background or religion, or belong to the same social class and age group. Frequently they have similar attitudes, values, mental outlooks, interests, and intelligence levels.*

Psychologist Robert Sternberg believes that a mature love relationship is made up of passion, intimacy or closeness, and a commitment to each other. He states that passion develops most quickly in a relationship, but it also fades most quickly. The feeling of intimacy develops more slowly, but in a successful relationship it becomes increasingly important. Through intimacy, the couple learns to support each other and come to understand each others' needs and feelings. Strong emotions of intimacy strengthen the sense of commitment that lovers have for each other. Sternberg states that both partners have to work constantly on the relationship to keep it healthy and ongoing. A love relationship cannot be taken for granted.

1. *What evidence is there that love is a popular theme in Canadian society?*
2. *Why do teenagers sometimes experience problems in romantic relationships? What can be done to help them?*
3. *How does one know that one is in love?*
4. *What attracts people to each other?*
5. *According to Robert Sternberg, what are the components of a long-term love relationship?*

12.5 Divorce in the Twentieth Century

In North America and Europe married life has increasingly become unstable, as reflected by the dramatic rise in divorce rates over the past thirty years. In Canada, from a divorce rate of 39.1 per 100 000 in 1960, the rate has climbed to 259.1 per 100 000 in 1980. If you conducted a quick survey in your class you might discover that the parents of at least one-third of your classmates are divorced.

Some people dispute the accuracy of the divorce statistics. Louis Harris, who administers the highly regarded Harris Poll, claims that U.S. government figures and his own surveys indicate only one out of eight marriages will end in divorce. Harris says the major finding of his survey of American family life was "that the American family is surviving under enormous pressure... and the burden is on women far more than it is on men". The misconception about the high divorce rate, Harris believes, was caused by the 1981 report of the U.S. National Centre for Health Statistics that reported 2.4 million new marriages and 1.2 million divorces during the year. Harris feels there was an important element left out of the equation — "the much, much bigger 54 million marriages just kept flowing along like Old Man River".

Along with the continuing number of marriages, there is an increase in wedding ceremonies that reflect the interests of the couple. A recent wedding was solemnized in the setting of Algonquin Park.

Statistics Canada divorce figures indicate that between 1968 and 1982 the divorce rate per 100 000 population increased every year. However, the rate of divorce in 1985 dropped 4.9% from the previous year. In 1985 61 980 couples in Canada divorced.

SEVEN HYPOTHESES TO EXPLAIN THE INCREASE IN DIVORCE

Sociologists and psychologists have developed many different hypotheses to explain the high incidence of divorce in today's society. Following are seven of these hypotheses.

(1) The liberalization of Canadian divorce laws in 1968 and 1986 has increased divorce rates. This has certainly played an important role. The change in the law was a reflection of a social need. Before the new laws, many married people lived apart or had unsatisfactory relationships, but were still counted as being married.

(2) Marriage as an institution is no longer popular. This hypothesis has less merit. Marriage rates have not declined; people are still being married in large numbers. Between 1968 and 1975 for example, the marriage rate remained steady with over 8 in every 1 000 people marrying each year. People who divorce often quickly remarry.

(3) Couples marry less for commitment to each other but more for personal satisfaction and gratification. This hypothesis may have considerable merit. People may be in love with the feeling of being in love. When this feeling lessens, a new relationship is sought.

(4) People entering marriage do so with the attitude that it is not a lifetime commitment. There has been a dramatic change in attitude toward marriage in the past fifty years. Most of our grandparents saw marriage as a permanent bond until the death of one spouse. People had to make it work because there was no acceptable way out. This attitude has changed: today, marital breakup is no longer considered disgraceful or unacceptable.

(5) Women are becoming more independent economically and emotionally. Today a woman's economic survival is no

longer tied to her husband. Many women are educated and pursue careers outside the home. Just as important, women are growing more emotionally independent of men. The need "to have a man" for the sake of emotional security or social reasons has decreased. Women are also more aware of their rights and they have greater self-awareness.

(6) The influence of the media and the market place has led to higher divorce rates. It is not possible to prove this hypothesis, but it is worth considering. We are all consumers with choices all around us. Every year new models and designs of houses, cars, and appliances are presented. This emphasis on newness and glamour may carry over into other areas of our lives, including marriage. Life-long monogamy may not satisfy the media-created need for periodic change.

(7) Divorce is on the increase because of a bandwagon effect. This hypothesis may have considerable merit. In Canada the divorce rate has skyrocketed since the early 1960's. Divorce has become an acceptable solution to marital conflict. All social classes are well represented in the divorce statistics. Married couples see many examples of divorce around them; instead of working out their problems they may follow the example of others.

CASE STUDY

A Marriage Break-up

Peggy is a thirty-eight year-old mother of five. Her husband's heavy drinking bouts over a period of several years caused the family a great deal of pain, and it was a struggle to keep the family together. The couple fought frequently. One night after another fight with her husband she bundled baby David and four-year-old Laurie into snowsuits and left. The other three children were out, so she phoned them and told them to go to their assistant minister's house. The family stayed there for four days until they could find other accommodation. Peggy's marriage was finally over.

It took some time for the children to accept the new reality. Her seventeen-year-old, Adam, had been misbehaving before the separation. Now his behaviour grew worse as he tried to draw attention to himself in the hope that this might bring his parents back together. Although the second oldest, Steven, loved his dad he was angry. "Doesn't he see what he is doing to us?" he asked. The oldest daughter, Cassie, was her father's favorite. He had always called her his "Princess". Once the family split up she ignored him and withdrew into her own grief. She felt betrayed by her parents. The two youngest children strongly believed their father would eventually come home. They would cry, "Why can't daddy sleep here tonight?" or "I want daddy to stay".

The problems of dealing with her children's pain, her own pain, and the tightly stretched family finances were overwhelming for Peggy. When Adam had a scrape with the law and ended up in a group home for three months, she sought family counselling. She was directed to the Families in Transition program (FIT), a service developed to help families like this one cope with divorce.

With the help of family counselling, Peggy and her children are adjusting to being a single-parent family. The three older children are learning to deal with the anger and blame they have felt since the separation, and learning how to pull together as a family.

After six months of living alone, Peggy's husband began attending Alcoholics Anonymous meetings. He now sees the children regularly and contributes toward their support. The two younger children are happy to see him, and the older ones are developing a greater respect and understanding of him, now that he is meeting his problem head on. Although it is doubtful that Peggy and her husband will ever be reunited, they are managing to treat each other with consideration, and without the hostility that formerly characterized their relationship together. As Steven points out two years after the separation "They are both nicer to be with now that they are not spending so much time fighting. I think the separation was for the best."

Rhonda Freeman, director of FIT, believes professionals are now more aware of the social and psychological problems associated with divorce. But she warns "there is a danger in thinking it is not such a big deal anymore. No matter how common divorce is, it is still awful. There isn't safety in numbers. It's still devastating".

Do you think this is a fairly typical case study of family break-up? Why or why not?

THE EFFECTS OF DIVORCE ON CHILDREN

There are many factors involved when researchers study how divorce affects children. What was the relationship between the parents prior to separation? How old are the children? Do the parents continue to battle after they have separated? Are the children drawn into these battles? Do the children continue to have a close relationship with both parents after the separation? Are the children kept ignorant of any problems until the separation becomes a fact? These are just a few of the factors that will affect children when their parents divorce.

Sociologist Teri Kay believes that while children may wish that their parents would stop fighting, they do not anticipate that they will break up. The separation usually comes as a shock. The initial shock may be followed by feelings of shame or anger and by the fear of being alone. According to a study by Dr. Claire Lowry of the Child and Family Centre of Kitchener-Waterloo Hospital, children find it easier to cope with a death in the family than with a family break-up.

Martin Richards and Margaret Dyson, two social psychology professors at Cambridge University, have studied the effects of divorce on children. They found that each age group reacts differently to divorce. Preschool children may withdraw into a fantasy world, often regressing into baby talk and becoming excessively clingy. They may be fearful of being left alone under any circumstances. Children between six and eight often feel sadness and anger. Torn between their mother and father, they often compensate by being materialistic and asking for special compensation, such as toys, from their parents. They can also be aggressive or withdrawn and they may do poorly in school. Children between nine and twelve years often try to act "cool" while underneath they may really feel embarrassed and lonely. A boy whose father has left may try to take on a protective father role in the family.

Many recent studies indicate that the effects on children are not always so damaging as traditionally portrayed. Margrit Eichler summarizes many of these studies in *Families in Canada Today*. She believes "there is by now increasing evidence that divorce per se may not be half as significant for the children as had been thought, what matters are other factors. . . such as the relationship with both parents, poverty, mental health of the custodial parent and others".

Martin Levin, an American sociologist, concludes from his research that "divorce does not necessarily produce psychological damage in the child in every case. Other studies show that a happy one-parent family is less stress-producing for a child than an unhappy two-parent one . . . children who appear to be distressed most by divorce are those with parents who continue to have turbulent relations following the divorce, or those who have been forbidden contact with the parent not in the home after the divorce."

Child psychiatrist, Dr. Louise J. Despert, maintains that the emotional situation in the home is what determines the child's adjustment. If the relationship between the parents is very disturbed then the child is very disturbed. She believes that divorce is not necessarily destructive; it may be a cleansing and healing experience.

TEENAGERS AND DIVORCE

Teenagers often feel resentment toward their parents for undermining their security at a critical time in their lives. In *The Emerging Generation — An Inside Look At Canada's Teenagers* some of the conclusions reached contradict earlier beliefs about children of divorce. Young people whose parents are divorced differ little from others in the value they place on love, friendship, and family life. Adolescents who live in one-parent families do not indicate that loneliness is a major problem. They usually receive enjoyment from both of their parents although the two are physically separated. However, they show a slightly greater tendency to turn to friends and to the parent they are living with (usually the mother), and tend to be alienated from the parent who is gone (usually the father). These children from separated families "are somewhat more skeptical about the future of the family generally" although they value their family.

Author and broadcaster Warner Troyer offers us some helpful insights in his book *Divorced Kids*. After interviewing hundreds of children of divorce he reached a number of generalizations about how children of divorce feel about marriage, divorce, parents, and relationships.

On the whole they give more thought to marriage, separation, and divorce than do most other children. Rita, a university student whose parents divorced when she was fourteen states "Well I'm just not sure about marriage; not a bit sure..." Alex, at twelve, shares these feelings about marriage. "I'll want to get married, someday, I think." Sherry who is eleven says "I'll get married when I grow up . . . But I'll be real careful. That's most important. Kids would be more careful. But adults they don't think about the future — just about then, and what they want right then."

How do children of divorce feel about divorce? Almost all those under the age of ten believed divorce was "wrong". They felt divorce should be harder to get. "Parents should have to stay with their kids" was repeated frequently. Teenagers' views differed from their younger counterparts. Most were unanimous about divorce: if it was going to happen it should be quick, cheap, and painless with no courtroom struggles. "No fault" divorce, whereby neither side is held responsible for the break-up was suggested by many of the teenagers. How do children of divorce feel about parents? Many feel parents are doing a basic minimum for them. "They're just too busy to talk to me, I guess" said one.

Many children complained that their fathers were absent too long, and unreliable about their visits. A third complaint was that they never "have a chance to really visit with Dad". Dad was always taking them to some exotic location to entertain them. Or Dad was courting a potential wife and may not have had much time or patience with the kids. Most teenagers were usually not concerned about financial problems which their mothers may have encountered. In general, they were content to live with lowered spending expectations, but most resented their mother's complaints about lack of money. They also disliked the persistent criticism of their fathers voiced by their mothers. Mothers who used their children to spy on Dad or to take messages to him were also sharply criticised by the children interviewed.

Many of the children interviewed were positive about their mother. They said "She's prettier now" or "she laughs a lot more" or "Mom experiments more now; she tries cooking fun stuff more and that's better for us". Once their mother took a job outside the home, children often felt "She's more interesting now that she's working, takes better care of herself, has a more interesting life and is better to be with than when she was at home". The young people who were most hurt by divorce, according to Troyer, were girls in their late teens. They were often "deeply involved in a love relationship with someone just before their parents split up". It may be that at the moment in their lives when they were making their own, first serious emotional commitment, says Troyer "they were doubly injured by the shock of failed love."

THE EFFECTS OF DIVORCE ON WOMEN

The majority of one-parent families are headed by women. Nurturing children, managing a household, and holding down a job constitute a heavy load. Add to this load the emotional strain of dealing with a separation and divorce, and you have a situation of great stress. A recent American study by Robin Akert indicated that women feel more guilty, unhappy, lonely, depressed, and angry than do men when their marriages break up.

Michael Wheeler, a researcher with the School of Social Work at McMaster University, found that the ability to cope with divorce was closely linked with a woman's age at the time of her marriage. Those who married before the age of twenty suffered from depression to a much greater extent after the separation than did those who married at an older age. Wheeler believes that women who marry very young have not developed their identities. Often they expect marriage to provide them with a role to play and a meaning to life, so when the marriage fails, they lose their purpose.

Some women find divorce to be a positive influence on their lives. Gail Gardner, a 36 year old mother of two, claims "now I have accomplished an entire new life. Divorce gave me an ego, which I didn't have before."

Other women find the problems of loneliness and financial insecurity to be overwhelming. One woman who earned $20 000 in 1986 as a secretary, supplemented by child support of $50 a week for each of her three children says "I can't remember one happy moment in marriage, yet now that I'm finally on my own, I'm starting to go through all the trauma of being a divorcee. It's loneliness; it's financial - that's the killer; that hits you in the pit of your stomach. I wish I were married to anybody, I don't care. It's so lonely. I see so many unhappy marriages, so many people coping with marriages that are lousy, absolutely lousy but they've got financial security — I'm not saying rich, I'm saying financial security. They have somebody to come home to, somebody to take over with the kids, somebody to hammer a nail into the wall, somebody to yell at, somebody to look at across the table even if they hate this person. I'm really feeling what it's like to be part of the new statistics. The new poor. I'm the new poor."

Poverty is one of the biggest problems facing divorced women. In 80% of cases women get custody of the children and in a majority of these cases the family support orders are ignored by husbands. Courts have also set limits on wife support encouraging women to become self-sufficient. "Women take a nosedive in financial security", says Teri Kay of the Jewish Family Services Association. "They have economic problems and they have career problems at a time when they are probably feeling the most vulnerable, when their self-esteem goes down, when they are questioning their worthiness, their compatability. And they also have to manage a home and children at a time when they are the most depleted."

"A young woman who neglects her own education and job training because she thinks her future husband will support her for life is running a grave risk of ending up alone and in poverty."

Louise Dulude, lawyer

A substantial number of women are left without adequate financial support or earning power since most working women still occupy the lowest rungs of the job ladder and many are paid less than equally or less qualified males. One sociologist, Artie Russel Hochschild, has said that what has happened is not so much a rise in the status of women as a feminization of poverty.

THE EFFECTS OF DIVORCE ON MEN

Divorce also has a strong impact on men, according to Joe Rich, a social worker who runs counselling groups for men. "The man feels all the emotions (sadness, anger, loss) associated with ending the relationship. And if he moved out he will also see his children less frequently. He's depressed and he's lonely," says Rich. "He's afraid of what his relationship with the kids will be like. There is no other relationship that is like that of the visiting parent relationship; no feeling can match being a parent and visiting a child." Rich believes most men feel remorse and regret but not the sense of failure may women have. If there are children, men sometimes feel they have lost control over them.

Men may need as much help after divorce as women. They may suffer from the conflict between natural emotions such as guilt and loneliness and social conditioning which requires them to be tough or "macho". Stephen Carpenter is a therapist who runs a separation and divorce service for men. "Men are socialized to sacrifice, to be the father, protector, and provider and to keep things inside. It's okay for women to cry but men aren't supposed to."

One of the strongest feelings encountered by men as well as women is the sense of loss. Therapists sometimes compare the reactions to divorce to the grief that follows the death of a spouse. Many men do not want to talk about intimate feelings related to their grief. They have a tendency to suffer in silence and "tough it out". Women, on the other hand, have a tendency to rely more on close friends for emotional support in times of personal crisis. For both men and women, many friendships dissolve following a separation. Most of their friends may have been couples who now avoid the separated couple because they may not want to take sides. This loss of friends and the feeling of isolation it brings can be hurtful.

"Divorce is one of the toughest, emotional experiences anybody has to go through" says Dr. Robert Segraves, a family therapist and professor of psychiatry at Tulane University Medial Centre in New Orleans. "Instead of asking for help, though, to sort out some of their feelings, unlike women, men tend to feel they can work it out with their own resources. The tough-guy image, Marlboro man image is still with us. And it takes its toll." If and when men do reach a point where they want help, there are a number of support groups to help them cope with the stresses of divorce.

PROGRESS CHECK

1. In chart form list the seven hypotheses for divorce and indicate why they do or do not account for the increasing divorce rates.
2. In chart form list the effects of divorce on women, men, children, and teenagers.
3. What conclusions do the authors of *The Emerging Generation* reach about teenagers and divorce?
4. According to Warner Troyer how did children of divorce feel about marriage, divorce, parents, and relationships?

OVERVIEW

Marriage remains an important institution in our society despite the high failure rate we are seeing from the 1960's on. The roles in marriage are changing as more women work outside the home. Many women are finding that their responsibilities have multiplied while the benefits have declined. Men are having to adjust to the changing roles as well. They are having to contribute more time and effort to household tasks and caring for the children, while at the same time having to adjust to the fact that their wives are not going to be able to play the same role in the home that their mothers did.

Our notion that "falling in love" is a realistic basis for getting married is also coming under scrutiny. Some people feel that this kind of romantic love is a state that cannot last, and for that reason it is a poor basis for forming a long-term relationship.

With the increasing numbers of split-up families, many researchers are trying to find the causes of break down and trying to understand the effects on the families. Determining the negative effects on children, and learning how to minimize them is especially important. Recent studies indicate that the parents can do a great deal toward easing their children through the stressful times. By allowing the children to maintain contact with both parents, and by taking care not to involve them in their battles, divorce can be less distressing for children. Many experts feel that a happy one-parent home is preferable to an unhappy two-parent home for children.

For divorced parents loneliness and financial strain are major problems. For the parent who does not have custody of the children, there is the stress of maintaining a good relationship with them while living apart.

Many people wonder if this is a period of social upheaval and change we are currently experiencing, and if the time will soon come that male/female roles and relationships will be worked out successfully. They also wonder how society will be changed as a result of the changing roles men and women are playing.

KEY WORDS

Define the following term, and use it in a sentence that shows its meaning.

homogamy

KEY PERSONALITIES

Give at least one reason for learning more about the following.

Irene Murdoch
Edward Herold
Bonnie Krebs
Shere Hite
Louis Harris
Margrit Eichler
Warner Troyer

DEVELOPING YOUR SKILLS

ORGANIZE AND FOCUS

1. (a) Create an organizer that summarizes the functions of marriage.
   (b) What questions should we ask about the changes that have taken place in people's attitudes toward the social institution of marriage?
2. (a) Design an organizer that traces the relationship between love and marriage over the years.
   (b) What questions can we ask about that relationship?
3. (a) Develop an organizer that lists the possible reasons for the increase in the divorce rates in Canada.
   (b) What questions can we ask about the causes of divorce?

LOCATE AND RECORD

1. Divide the class into groups and have each group investigate and research one of the following:
   – the effects of divorce on children
   – the effects of divorce on teenagers
   – the effects of divorce on women
   – the effects of divorce on men

   The individual groups can refer to the information contained in this chapter, and vertical files and books in the library. Share and compare your conclusions with the rest of the class. How are the effects of divorce similar and different on each group studied?
2. (a) Divide the class into pairs and have each pair discuss and list the positive qualities of the person whom they would like to love and/or marry. Are the qualities similar or different in both relationships?
   (b) Discuss the situations in a love or marriage relationship that you would consider unacceptable.

EVALUATE AND ASSESS

Summarize the results of the following chart on satisfaction with romantic relationship. Which groups are the most satisfied and which are the least satisfied? Assess the accuracy of these results — do the results seem logical? Are any of the results surprising?

Satisfaction With Romantic Relationship

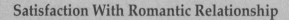
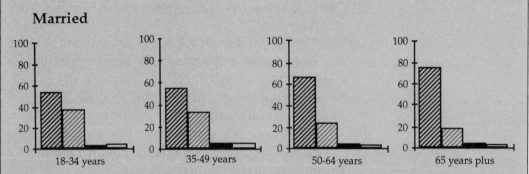

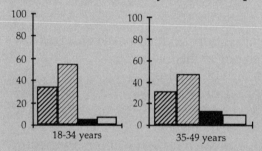

*older groups not included because of too few respondents to the survey

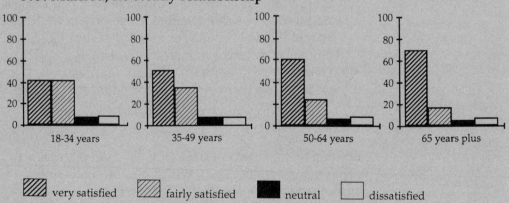

SYNTHESIZE AND CONCLUDE

Analyze the following chart by determining the most satisfying and least satisfying points in the family life cycle for both wives and husbands. What do you think were the causes for their satisfaction and dissatisfaction during the various phases in their marriages? Do you find any of these results surprising? If so, why?

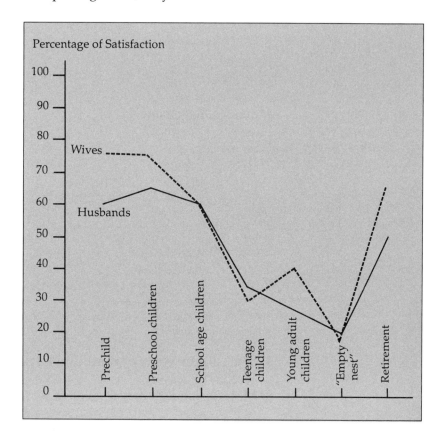

APPLY

A study was done over five years by a Canadian research team with 63 couples on areas of disagreement before they were married and at various times during their marriages. (This study was done by Edward Bader, Robert Riddle, and Carole Sinclair.) Analyze the results of the following summary by answering the following questions.

(a) Which areas of disagreement were most important during the entire five year study? Which were least important?

(b) Which areas of disagreement became less important as the marriages lengthened? Which became more important?
(c) Speculate about the reasons for the areas of disagreement. Why do they remain the same or become more important or less important over the years?

Areas of Disagreement	Ranking in order of most disagreement to least			
	Pre-marriage	Six Months	One Year	Five Years
Husband's job	1	4	6	4
Wife's job	8	6	10	10
Household tasks	3	1	1	1
Money management	6	2	3	6
Husband's relatives	11	10	7	6
Wife's relatives	5	9	4	6
Husband's friends	7	6	12	11
Wife's friends	10	12	13	12
Affection	8	5	7	4
Children	12	13	11	12
Religion	14	14	14	14
Social activities	3	11	4	9
Time and attention	1	3	2	1
Sex	13	8	9	3

COMMUNICATE

1. Write a short essay or debate and discuss one of the following topics.
- Marriage and divorce laws are now fairer than they were in the past.
- The rule of homogamy applies in most love relationships.
- Couples should be required to live together for a certain period of time before they can be legally married.
- Monogamy is still the only acceptable and workable form of marriage arrangement possible in Canadian society.
- Romantic love is overemphasized in Canadian society, and can lead to serious problems in relationships.
- Divorce is more devastating to young children than to other family members.
- By the year 2050, there will be no marriages and, therefore, no divorce.

CHAPTER 13

Schools: Looking Back and Looking Ahead

Why is education a social institution? What have been some recent reforms? Are there alternatives to the present educational system?

INTRODUCTION When Canadian high school students were surveyed by sociologist, Reginald Bibby, about what made them happy, school came 15th out of 17 choices they were given. Friendship, music, and boy and girl friends were at the top. For most of them, relationships with other students were the best thing about school. Yet schools have their own intentions for students which go far beyond personal desires and choices. This chapter is mostly about the purpose of schools, and how effective they have been. It examines the history of education in this country, and how the schools we have today came about.

13.1 *The Purposes of Education*

Your life at school and your friends are the personal aspect of education. The choices you make for courses and decisions about where you want your education to take you are crucial for building your life.

The other side of schooling is education as an institution. Although it is responsible for helping individuals learn, it has several other purposes as well. You can be helped in preparing for a career. Your social status can be improved through achievement. All students are made aware of their roles in a democratic society. A common cultural identity is fostered by schooling while your individual potential is also developed.

Surveying students during outdoor class, Ottawa.

EDUCATION AND JOBS

In order to survive, we must support ourselves. Education helps people acquire the skills necessary to obtain a job. In order for society to continue, people must constantly be supplied for the labour market. The schools' primary purpose is to prepare students for eventual careers. A **curriculum** is an organized course of study, that may be a combination of required and optional subjects or courses. At each stage of a student's development, new skills need to be mastered.

Two Kinds of Curriculum

The obvious kind of curriculum combines all the subjects students have to learn and the varied ways of thinking these courses demand. The less noticeable curriculum does not have a course of study, but it has to be learned, all the same. It is made up of attitudes and behaviour, like neatness, politeness, and punctuality. These prepare students for relating to people in the work world, and for relating socially. By behaving the way the school wants them to, the students create an orderly environment for learning.

EDUCATION AND STATUS

Canadian society is democratic — everyone is supposed to have equal opportunity. It is apparent, however, that there are various levels in our society. A person's status is considered to depend on ability, aptitude, or achievement. The ideal is that, no matter whether your family is poor or rich, you should be able to better yourself through education. Some people say that children of millions of Canadians have a higher status in society than their parents, because they have reached a higher level in school or university. The opposing argument is that it takes more education now to keep you in the same place in society as your parents.

The schools believe they have a part in improving and maintaining people's social status. Their job is to identify ability in their students and support it. This is the education system's second purpose.

EDUCATION AND THE SOCIAL ORDER

Learning to behave in an orderly school system is a skill that is meant to carry over into social and political life. It is considered Canadian to be a responsible citizen. This means staying informed on public affairs, and understanding how the political system works. The issues of right and wrong were thought of as religious questions in days gone by. More and more, they are thought of as social values, or **ethics**, and they are built into courses of study.

EDUCATION AND THE TRANSFER OF CULTURE

A fourth purpose of education is to bring people together into a common cultural identity. Canadians have a variety of attachments — ethnic, religious, and regional. But the schools set out to supply their pupils with a single cultural heritage. We learn in school that Canadians are bound together by a common tradition and two official languages. Schools give us our second identity, which we share with other members of society. It comes, not just from shared traditions, but from values, attitudes, and specific ways of doing things.

EDUCATION AND THE NEED FOR PERSONAL FULFILLMENT

A fifth function of education has become more dominant in the recent past. Students are encouraged to achieve personal fulfillment through exposure to art, music, sports, and theatre.

A high school student, enrolled in a summer field school, helps with the archeological excavation of a nineteenth century inn.

13.2 *The History of Schools in Canada*

Each province has its own system of education. Though some goals, like forming a common national identity, are shared across the country, the ten provinces set up and control their own educational systems. The provinces received the right to do this through the Constitution Act of 1867.

COMPULSORY EDUCATION

By law, throughout Canada, all children must attend school; this is **compulsory education**. They may attend schools in the public system, private schools, or religious separate schools. In the late 1800's schooling became compulsory, not only to teach children, but to get them off the streets and out of the factories. Children had become a cheap source of labour in factories. Putting children in school opened the factory jobs to adults.

CAREER PROFILE
WALLACE LAMBERT

Canada is internationally renowned as a pioneer in bilingual education. That reputation is in large measure because of the work of Wallace Lambert.

Lambert was born in Amherst, Nova Scotia in 1922, but was raised and educated in the United States. After army service in World War II, he studied at universities in the U.S., Britain, and France. He came to Montreal's McGill University in 1954, where he has taught ever since. His wife, a French Canadian and he raised two children who are perfectly at ease in both languages and cultures.

Lambert's most famous work is probably the 'St. Lambert Experiment'. A Montreal suburb, St. Lambert was, in the mid-sixties, the site of an exciting innovation in Canadian education. Elementary-school, English-speaking children were placed in a variety of French immersion classroom situations. Their educational progress, especially their development in the two languages, was carefully monitored and tested by Lambert and his team of researchers over a number of years. The evidence gathered from the experiment showed that second-language immersion, combined with some education in the mother tongue, was not only feasible but successful. The students in the St. Lambert group, often jokingly referred to as the world's most-tested students, showed a remarkable ability to perfrom effectively in both French and English. Since then, the St. Lambert experiment has served as a model for immersion and bilingual education across Canada and abroad.

Lambert has had several other related interests in psychology and education. Among them are comparisons of the social and cultural values of different linguistic groups, in Canada and elsewhere. His theory about what motivates people to learn a second language is widely-accepted. According to Lambert, the more closely a second-language learner is attracted to the community that speaks that language, the more successfully the language is learned.

In the early years, educators felt children were like vessels which could be filled with knowledge. What was poured into children's minds was regulated through a network of administration. At the top were elected ministers of education, with a large support system of civil servants. At the levels concerned directly with students, there were teachers, principals, and classroom inspectors. Teachers were required to teach certain concepts in particular ways, in a structured and regulated curriculum. Students were expected to learn the required skills, work hard, and obey authority.

THE EXPANSION OF THE SYSTEM

The education system expanded with the large influx of immigrants from Europe after World War II, and the baby boom of the forties and fifties. Between 1951 and 1971, the enrolment from Kindergarten to Grade 8 increased from just over 2 million to over 4 million across the country.

Thousands of new schools were built, and there was a great demand for teachers. "Stay in school" became the slogan of governments, employers, parents, and students.

ENROLMENT IN SCHOOLS

Schooling in Canada has changed to meet the needs of a changing society. At one time formal education took ten years. A person would then enter the work force with skills and knowledge that were considered to be adequate.

Nowadays, for many people, education has become a lifelong process. There are nearly 6 million full-time students enrolled in Canada's 15 750 schools, colleges, and universities. In addition, there are 3 million adults taking part-time courses. Most people now feel that institutions of education are valuable to them throughout their lives.

PROGRESS CHECK

1. Of the five purposes of education which one seems most important in your school? How can you tell?
2. Which purpose of education is most important, in your own view? Explain.
3. Why were many new schools built in the 1950's and 1960's?
4. Why has education become "a life-long process"?

13.3 The Challenge to the Schools

The newly expanded school system began to be criticized. The things that educators thought were their job to do were seen by some people outside the system as not their job at all. To make this point, the educational theorist, Ivan Illich, said Canada should abolish schools. He said that most learning occurs outside school: "We learn to speak, to think, to love, to feel, to play, to engage in politics and to work without schools or teachers."

If we seemed to have learned a lot from school, Illich thought, it was only because we had spent so much time there. Not all children learn best in the structured setting of the school. We all learn from experiences all around us, from each other, from television, and just from growing up.

Once open discussion of the schools began, new models were found for our system. Summerhill, a private school in England, was discussed as an example of a place where students could learn at their own pace and where the students made many of the rules about their rights and obligations.

CASE STUDY

Summerhill

Joshua Popenoe, a graduate of Summerhill, wrote down his impressions of the school in a book called Inside Summerhill. *The following is a summary of some of his impressions.*

"The bell for the first class rang at 9:30 a.m. Those students who wanted to go did so. Those who did not, did something else. A majority of students did not attend classes regularly. Each day was divided into seven periods of forty minutes. Five periods were scheduled in the morning and two in the afternoon. The average class was composed of six students. They learned at their own pace. Many Summerhill students took only two years to prepare for national exams; three or four years was normal in other English schools.

However, Summerhill was not a school without rules. General meetings, at which all school business was discussed and voted on, were held every Saturday evening. Rules were proposed by both students and teachers, and each proposed rule was submitted to a vote. All those present, students and teachers alike, had one vote.

One of the most important rules made was that one person should not infringe on another person's rights. Another rule was that students under a certain age could not carry knives or matches. Most students attended these weekly meetings because they did not want rules passed with which they were not comfortable."

1. If Summerhill students did not attend classes regularly why were they able to write their national exams sooner than other students?
2. Why, do you think, were rules discussed at general meetings?

ALTERNATIVES TO THE SYSTEM

A Canadian historian, Robert M. Stamp, wrote a book called *About School: What Every Canadian Parent Should Know*. He described six possible school styles. Some were in great contrast to each other; others looked much alike.

School A is a traditional school, emphasizing academic achievement. Teachers instruct and direct. Students are required to adjust.

School B is nontraditional and nongraded. Materials are made available for work and play. Teachers assist and guide rather than direct and instruct. Students do not all do the same things at the same time, and many of their learning activities take place outside the school building.

School C is called a "free" school. It is patterned after Summerhill, with students taking responsibility for selecting and planning activities. Much of their learning takes place in the community.

School D is oriented to technology. Here computers are used extensively for learning and for diagnosing individual needs and abilities. The library is stocked with learning tapes and video tapes through which students and teachers can concentrate on specific areas of inquiry. Students can use television cameras for making their own programs.

School E is a community school. Here, not only young people but anyone else who wants an education is welcome. In the late afternoon there are activities for children, and, in the evening, adult classes are held.

School F is patterned after the Montessori schools, which originated in Italy. Each classroom contains various learning areas with a wide choice of materials and specialized learning equipment. Teachers help only when required. For the most part, students do not do the same things at the same time. Also, many of their learning activities take place outside the school building.

GOVERNMENT REPORTS ON EDUCATION

Criticism of school systems caused provincial governments to call for reports on education. In Ontario, a report called *Living and Learning* (1964) said schools were out of date. Students should be given more options to choose different courses to satisfy their own needs. The report said that schools should be less structured in their programs to allow for these personal choices.

In Alberta, the report called *A Choice of Futures* made similar recommendations. Schools were described as "cocoons" which isolated youth and kept them from playing effective roles in society.

PROGRESS CHECK

1. Which of Robert Stamp's six schools does your school resemble, or is it a combination of more than one style? Give examples.
2. Which school style do you think most helpful for your own achievement?
3. What would be the result if schools were abolished?

13.4 *Educational Reform in the 1970's*

By the early 1970's, most Canadian educational systems had changed. "Open concept" schools were built in some places, with no walls separating the classrooms. Under the option system, students could choose from a large list of courses, and few compulsory subjects were required. Exams were eliminated in many schools, and teaching strategies changed. Student-centred approaches such as presentations, debates, and individual research projects became popular. Student dress codes were almost completely discarded. Self-expression and self-realization became two principal educational goals. Schools were expected to help students clarify their own values, and not merely impose society's values on them. Failure was seldom acknowledged. Students were sometimes promoted whether they had mastered the courses or not.

An open concept classroom in Edso, Northwest Territories.

A LACK OF SKILLS

This educational experimentation was bound to place considerable stress on everyone concerned. Some teachers and students were lost along the way. New methods encouraged learning, but there was also frustration and heated debate. By the late 1970's, many educators and parents wanted a change back to compulsory courses in English, Mathematics, and History. There were worries that students were graduating without marketable skills. Business and industry wanted the schools to prepare students for their special demands. If students were given a false sense of achievement, they would be unprepared for setbacks in the "real world" of work.

INEQUALITY IN SCHOOLS

Besides these criticisms, there were other problems. As schools spent less time on skills and work habits, working-class children suffered the most. They had the highest dropout rate. Well-to-do parents could pay for tutors or special camps to teach their children the skills the schools should have taught.

Children whose language at home was not English depended on the school to provide them with literacy skills. If literacy was not emphasized, they would have less chance of future success.

13.5　A New Call for Reform in the 1980's

Several billion dollars of tax money was spent to operate schools in Ontario in 1986. With so much money being spent on the system, citizens were anxious to see results from schooling. Some people wanted all students to write province-wide exams, instead of teachers giving their own tests. Ministries of Education could then compare students' performance across the province, and see if schools and teachers measured up to a standard.

Alberta and Saskatchewan set grade 12 diploma exams. Quebec announced examinations for the senior year of high school. In New Brunswick, grade 11 and 12 students had to pass at least one province-wide math exam.

Professors of education took sides on the issue of province-wide exams. At the Ontario Institute for Studies in Education, one professor, Mark Holmes, said students needed exams to know where they stood. "We need uniform, external examinations to re-instill a spirit of purpose in the educational system" said Holmes. But another professor, Les McLean, disagreed. He said that teachers would teach to what they were sure would be on the exam; courses would be less enriching. "(Exams) wouldn't ruin the system, but they would consume large amounts of time and a substantial amount of money that could be better spent elsewhere ... In the end, teachers will teach to the specifics of what will be on the exam and the kind of education in the classroom will narrow."

But exams were not the only concern. For example, Ontario students were found to be poorer in science than students in the western provinces. Grade 8 students in Ontario were slightly better in math than British or American students, but lagged behind students in British Columbia and such countries as Hungary, France, and Japan. When students were found to be doing better in math than was formerly the case, was that only because standards had dropped?

RE-EVALUATING THE SCHOOLS

A survey carried out in April of 1987 in Ontario interviewed parents, teachers, educational experts, and government and business leaders. Everyone asked agreed that the schools should raise their standards.

Yet more than 75% of the students interviewed later said they were getting a "good" or even "excellent" education. Most felt they were being well prepared for the world of work. Employers had strong doubts about this. They said that high school students had problems thinking logically and communicating, and that they lacked self-esteem. University and community college teachers thought that high school graduates had poor work habits, and little self-discipline or commitment.

CASE STUDY

Report Card on Our Schools

When 1150 parents, teachers, students, and employers were interviewed, some strong opinions on Ontario's education system became apparent. Four of five Ontario residents agreed that schools evaluate their students differently. Many felt that standardized tests would be useful tools to measure student performance.

Employers and post-secondary teachers were strongest in a call for standardized tests. Eighty-three percent of community college and university instructors wanted province-wide tests so that students could be evaluated by the same standards.

"I agree that province-wide standards or tests of skill should be established in certain subject areas."

Parents	Elementary/ Secondary teachers	College/ University teachers	Secondary school students	College/ University students	Employers
69%	54%	83%	69%	59%	75%

Most of those in favour of some standardization listed Mathematics and English as the subjects to receive priority. "A lot of students cannot do simple multiplication and division. In writing reports, the need for good grammar is essential, but many students just do not exhibit that talent", said one community college teacher. One parent said that external tests "would give all parties — students, parents, teachers, schools, and universities — a measure of how well the system is performing." Another parent said, "What we need is a system that makes sure that a mark in an Ottawa school is the same as that mark in a Hamilton school."

In another part of the survey, parents and employers criticized the school curriculum for not having focus, for not teaching the basics, and for wasting time on irrelevance. What should curriculum do? At the very least, it should pass on information as well as train students to communicate, understand, and assess, said most of those surveyed. In other words, the curriculum should teach basic facts and also teach how to solve new problems.

The growing number of dropouts concerns educators who feel that some young people may be moving into adult life without the necessary skills. "Education has to be seen as relevant — only if schools make courses relevant can they hope to motivate students to learn and to continue learning.", said a Ministry of Education official when questioned about the survey results.

In general, the survey gave the Ontario education system failing grades on its report card. A former dean of the Faculty of Education at the University of Toronto lays the blame for this poor report card at everyone's doorstep. John Ricker says that we have burdened the school system with an ever-increasing collection of functions. "In addition to their traditional academic role, schools are expected to act as nursemaid, baby sitter, welfare worker, police officer, and informed advisor on health, law, economics — anything that pops into the minds of the disparate groups of a multicultural society. Of course the system lacks focus." He adds that "there are some encouraging signs. There is a general conviction that our education system needs serious rethinking."

Ricker believes that schools should concentrate on two main objectives. The first is a "content" objective — to ensure basic, minimal knowledge and understanding of our intellectual and cultural heritage. To attain this objective means basic literacy in language and literature, mathematics and science, and history. The second objective is a "mind-training" one. Students should be trained to think analytically and critically, to gather and assess information from many sources, and to determine what is and is not relevant in the solution of a problem.

Ricker goes on to say, "My concern with the present curriculum is that some students may get caught up in its clutter and leave school not only lacking basic literacy but with such an adulterated stock of ideas and information that future learning will be very difficult, perhaps impossible. Since the ability to learn continuously will have a large bearing on their future welfare, we should at least make sure that the curriculum itself is not an obstacle to learning. In the long run, the student's most important safeguard is the teacher, the best of whom have always been able to surmount most curriculum difficulties."

(based on survey conducted for, and articles in the *Toronto Star*, April/May, 1987.)

1. What do you think is relevant information for the schools to offer?.
2. Do you agree with John Ricker that we expect the school system to do too much? Why or why not?
3. Discuss the statement, "In the long run, the student's most important safeguard is the teacher...".
4. Research another province's curriculum guidelines. How does that curriculum compare with Ontario's? What sort of report card would that province receive for its education system?

Progress Check

1. What is the value of the option system of education? What are its drawbacks?
2. Is literacy the job just of the schools? Explain your opinion.
3. Why do working-class children have the highest dropout rate? What can be done to correct this situation?
4. Explain (a) the value of province-wide examinations
   (b) the drawbacks to such exams
5. Should the schools or business and industry prepare students for jobs? Explain.

THE SCHOOLS AND THE FUTURE

The school system in Canada was set up in the 19th century, at a time when it took over a century to double society's body of knowledge. Today the sum of knowledge doubles every two years. Can schools designed for the past fit students for the future?

Ann Jamieson, who studies how children learn, says: "A lot depends on the individual attention kids are getting. If we had the money to do everything on a small scale my guess is we could do everything a lot better."

New or Old Alternatives: The Japanese Example

Canadians have often looked to other school systems for new ideas. At one time we copied England or the United States. Today, some of our attention is drawn to Japan. The

Japanese are admired for their great economic success. Moreover, 90% of Japanese students finish high school. The illiteracy rate for Japan is 0.7% compared to 20% in the U.S.A. In Canada there are at least 4.5 million people who are classified as illiterate or functionally illiterate, about 22% of the population.

Merry White, who wrote *The Japanese Educational Challenge*, summarizes Japanese objectives for school children: "It is desirable that, in the lower grades, one should learn to bear hardship, and in the middle grades, to persist to the end with patience, and in the upper grades to be steadfast and accomplish goals, undaunted by obstacles or failure."

School in Japan is difficult, and parents tell their children "Gambatte", which means "You can do it. Tough it out!" There is a single national curriculum that all students take in each grade. The school day is seven hours long, and the school week includes a half day on Saturday. The Japanese school year is 240 days as compared to about 180 days in Canada. Class size in Japan is 45 to 50 students.

There are rigid rules of behaviour in Japanese schools that include styles of haircuts and the number of pleats in the school uniform skirts. It is common practice to stand to answer a teacher's question. There is an emphasis on memorizing facts and preparing for examinations. By the age of 10, students are preparing for university entrance exams which will be taken at age 17.

The national attitude in Japan is that a student's education is the most important consideration. The education received will determine the student's career and social status. However, some Japanese parents are admitting that the high level of competition for university acceptance is becoming too difficult. There is concern that the educational system is too rigid, and that there is no room for creativity.

TRENDS FOR THE 1990's
Japanese educational values may be hard to establish in Canada. Some may even be wrong for us. In Canada, more young people attend school from Kindergarten to Grade 12 than in most other countries. Our schools must deal with every level of ability, emotional state, and social background.

TABLE 13.1 Quality of Public Education

In February, 1988, the Toronto *Globe* published the results of a survey of 2 045 adult Canadians who answered the question: How satisfied are you with the quality of public elementary and secondary education in your province?

	Total %	Atlantic %	Quebec %	Ontario %	Manitoba %	Saskat. %	Alberta %	British Columbia %
Very satisfied	17	24	7	21	26	29	17	11
Somewhat satisfied	42	60	33	42	48	40	47	44
Somewhat dissatisfied	22	9	34	20	15	15	16	22
Very dissatisfied	11	3	17	9	4	5	11	15
No opinion	8	5	8	8	7	11	9	8

Everyone does not learn at the same speed. Canadian schools have tried to develop programs to meet the needs of every learner. Today, students who once would not have reached Grade 10 go on learning at their own level. Eventually, they exhibit skills that were not apparent earlier.

Meanwhile, however, almost five million Canadian adults cannot read or write or use numbers well enough to meet the literacy demands of today's society. One-third of these Canadians are high school graduates. This may be hard to grasp, considering the billions of dollars spent on education in Canada. One of the biggest tasks for the system will be to educate these adults and to make sure none of today's school children suffer the same fate.

Besides, the old phrase about having "finished school" no longer applies. Our "high tech" society demands that we keep learning throughout our lifetime. Futurist, Alvin Toffler, points out: "Today we need to combine learning with work, political struggle, community service and even play. All our conventional assumptions about education need to be re-examined."

PROGRESS CHECK

1. Is individual attention as important as Ann Jamieson says it is?
2. How does the Japanese school system differ from the Canadian school system to which you belong? Are there ways in which they are similar?

3. Which system is preferable in your view? Why?
4. Do you think Toffler's statement on the future of education is accurate? Have we reached that stage now?

OVERVIEW

Education is one of society's most important institutions. Through education we acquire the skills, knowledge, values, and attitudes that will allow us to take part in society. Education helps people take their place in the job market; it acquaints people with the political and legal system; it helps to give us a cultural identity; and it helps us find personal fulfillment.

Provincial educational systems expanded greatly in the 1950's and 1960's. It was hoped that a better educated workforce would be a more productive one. Also, places had to be found in schools for the children of immigrants and the children of the post-war baby boom. By 1970, more Canadians than ever before were attending school, from kindergarten to university. Parents wanted their children to have more opportunities and a better lifestyle than they had had.

As the school system grew, there was much discussion about methods of education. Critics like Ivan Illich thought schools were unnecessary, that most learning took place outside school. A variety of alternative schools were offered which gave children more choice in what they would learn, and how. By the early 1970's, open schools, the option system, and student-centred approaches were widespread. But some teachers and students could not cope in this largely unstructured system. The school system came to be criticized once again. Many people wanted a return to the learning of basic skills and standardized testing. There was widespread agreement that schools should become more demanding. However, Canadian schools would still be expected to deliver more than just "the facts". Students need to know how to learn and to become good communicators.

Schools of the future will have to deal with each learner's needs, as school becomes more accessible. Even for the most successful in our schools, education is a lifelong task.

KEY WORDS	Define these terms, and use each in a sentence that shows its meaning.

curriculum
ethics
compulsory education

KEY PERSONALITIES	Give at least one reason for learning more about the following.

Reginald Bibby   Les McLean
Ivan Illich   Ann Jamieson
Robert M. Stamp   Merry White
Mark Holmes   Alvin Toffler

DEVELOPING YOUR SKILLS

FOCUS

The focus of this chapter is education. Parents, educators, students, and businesses are among those who are examining the present school systems. They are making suggestions to try to improve the present quality of education. They often disagree among themselves about the various ways in which schools can be changed to improve the quality of education.

Divide the class into groups and brainstorm a list of questions that can be asked about the purposes and functions of education. Each group should write down the questions and ideas that are suggested by group members. At the conclusion of the brainstorming session each group can select the best questions. One question might be "What types of schools are considered to be the most effective? Least effective?" Each question will bring out different points of view and information that will reveal something new about education and its delivery. Answering these questions through discussion, reading, research, and writing will lead students to a better understanding of education and the issues surrounding this important institution in Canadian society.

ORGANIZE

1. In an organizer, describe the changes that have taken place in Canada's education system over the years, and their effects. In a third column list the criticisms that have been made of Canada's schools.
2. Unit Five has looked at several social institutions; the family, marriage, and education. There are other social institutions; for example we are affected by laws and the court system. Brainstorm with other members of your class to think of as many social institutions as you can. In an organizer, list the social institutions in the first column. In a second column, list the ways in which each can affect a person.

LOCATE AND RECORD

1. Divide the class into groups to investigate and gain more information about the recommendations and criticisms that have been made to improve the delivery and quality of education in Canada. Each group can refer to one or more of the government reports described in the text and do additional research on these education reports in the library. You can also refer to the vertical files in the library

under the topic of education. Your group can also further examine alternative schools like the Summerhill and Montessori schools. Each group can refer to one or more of the following books (or additional books) during its research activities. Record your findings in your own words and share the discoveries and findings of your group with the rest of the class.

Coleman.	*Adolescents and the schools*
Goodman.	*Compulsory Mis-education*
Herndon.	*The way it sposed to be*
Holt.	*How children fail*
Kent.	*Ryerson Cake with Dewey Icing*
Kozol.	*Death at an early age*
Lipset.	*Students in revolt*
Lipset.	*The Berkely student revolt*
McGuigan.	*Student protest*
National Education Association.	*Schools for the sixties*
National Education Association.	*Education in a Changing Society*
Neill.	*Summerhill: A Radical Approach to Child Rearing*
Reid.	*Student Power and the Canadian Campus*
Toffler.	*The Schoolhouse in the City*

2. Arrange to interview someone who went to school in your school system thirty to forty years ago. Your school might have records of the names and addresses of some former students. Prepare five questions to draw out this person's opinions and experiences while at school. Present your findings to the class.

EVALUATE AND ASSESS

1. What is the point of view of each of the following authors on the subject of schools? Assess the authors' bias and accuracy.

   "In such a school (a traditional one), the children, like butterflies mounted on pins, are fastened each to his desk, spreading the useless wings of barren and meaningless knowledge which they have acquired."
   Maria Montessori, *The Absorbent Mind*

"Teenagers are captive in a school system. Their actions and behaviour are observed and recorded. Unlike their elders they are compelled by law to be captive. They cannot protect themselves from such systematic scrutiny. "The system" is put ahead of the rights of the individual."
Tim Reid, *Student Power and the Canadian Campus*.

"Dr. Wilder Penfield has said that you can become educated by reading, travel, and talking with intelligent, experienced people. I would go right along with this; you must attend university if you want a degree, but if you simply want an education, there's no necessity to go near one. Go instead to the public library, where you'll find all the truly great teachers who ever lived. Hit the road for a year and find out something about yourself. Talk to older people who've had some experience of life. Get a job, any job, and learn on it."
Richard J. Needham, columnist, Toronto *Globe and Mail*

"Canadians appeared to believe, more emphatically than did Americans, that the public school should serve the individual; Americans believed, on the other hand, that it should serve society. Canadians, as a group, assigned considerably higher priority than did Americans to knowledge, scholarly attitudes, creative skills, aesthetic appreciation, and morality, as outcomes of schooling. Americans emphasized physical development, citizenship, patriotism, social skills, and family living much more than did Canadians."
Lawrence W. Downey, *Canadian Society: Sociological Perspectives*

2. Complete each statement in one hundred words:
    A good teacher is ...
    A good school is ...
    A good student is ...
    Exchange your completed statements with two other members of the class and discuss. Be prepared to report back to the class.

SYNTHESIZE AND CONCLUDE

What conclusions can be made from the following data about university enrolments in Canada?

"Roughly 18-20% of high school graduates continue directly to university. Statistics Canada says one-quarter of first-year students don't come back for a second year."

"Preliminary estimates also show total full-time enrolment, both graduate and undergraduate, increasing again this year by perhaps 10 000 to roughly 470 000. This increase is despite a decline of 82 000 in the number of 18-to-21-year-olds in Canada."

"In 1971, Statistics Canada said, 2.9% of the female population and 6.2% of the male population over age 15 had a university degree. By 1983, 7.7% of the female population and 11.3% of the male population had degrees, so the gap remained almost the same."

APPLY

1. Divide the class into groups and read the following article on age grading. Analyze age grading by discussing the following questions. Have a member of the group record the opinions and ideas discussed by the group. Share your findings with the rest of the class.
   – What are the purposes and importance of age grading?
   – Analyze the relationship and importance of skill, emotional, and social development in the age grading process. What could happen if a very bright student was skipped several school grades and placed with older children?
   – Speculate on what would happen if schools abandoned age grading.

"An important type of social time in the socialization of children is age grading. Children of the same age are expected to attend classes together and learn similar kinds of materials. Age grading is important because children of similar ages have approximately the same types of learning abilities. Students in a particular grade can be expected to perform certain skills and solve problems that are not too easy or too difficult for them. Teachers at certain grade levels generally know what stage of development students have reached.

Age grading is also important because students in a particular grade usually are at the same emotional and social levels of development. They can relate and mix with each other, and this permits them to learn important social skills. They find out how to get along and interact with each other. These are important and necessary skills in any social situation. Each higher grade level expands on the learned abilities of young people. When students graduate from high school, society expects them to have reached a certain level of mental, emotional, and social skills. These skills will permit them to enter university, college, or employment. They will also be able to engage in social and emotional relationships with a certain degree of confidence and ability."

2. Class Discussion: Examine some of the types of schools that are listed on page 376 of the text. Which type of school most closely resembles your school? What changes could be made in your school to create a richer and more effective learning environment?

COMMUNICATE

Write a short essay or debate one of the following.

- schools require stricter discipline and part of this could be achieved if there was greater use of the strap
- schools are asked to perform the tasks that were once the responsibilities of other institutions; as a result, there is less emphasis on skill development
- sex education should be the responsibility of the parents, not the schools
- students are performing an important job for themselves and for society; they should be paid a salary by the government while attending high school and university. A good education guarantees a high social status and a well-paying job
- fewer people would be left illiterate if schools had smaller classes

2. Read one of the following books, and write a critical report on it.
   Myrna Kastash: *No Kidding: Inside the World of Teenage Girls*
   David Elkind: *All Grown Up and No Place to Go*
   Gordon Karman: *Don't Care High*
3. After your class has been divided into groups, each group will brainstorm the following questions, presenting the group's combined predictions to the other groups: how will Canadian schools in 2050 differ from schools today?
4. In this unit you have concentrated your efforts on an analysis of the family and schools as social institutions. Identify with your classmates other institutions that affect you in your daily life (or that could affect you). Together, produce a chart that has three columns entitled: Social Institution /Positive Effects on my Life/ Negative Effects on my Life. For example, the social institution might be *Law Enforcement*. A positive effect would be that the police caught a person who stole your wallet. There then may be a negative effect you think appropriate for the third column.

   After you have listed the social institutions that you and your classmates have thought of, add two more columns and consider positive and negative effects on the two following groups, for each institution listed: People Over the Age of 65/ People of a Minority Group.

   Together, summarize your feelings about ways you like the social institutions as they presently exist, and suggest ways that you'd like to see them change.

UNIT

6

YOU AND YOUR SOCIETY

CHAPTER 14

Behaviour Patterns

What is mental health?
Where is the fine line between normal and abnormal behaviours?
How can mental disorders be treated?

INTRODUCTION What happens when a parent or friend tells you to "act normal?" Usually, you evaluate the circumstances and meaning of the comment and act accordingly. In your mind, you have a clear idea of what is considered normal behaviour. Do you walk more quickly on a dark street at night? Would you break out into a cold sweat if you heard footsteps behind you? Many of us would because we would believe that there was a threat to us. This threat creates anxiety and, to relieve the anxiety, we walk more quickly to get away from the threat. Because many of us would act in this fashion as a way of dealing with an anxiety, this behaviour would be considered normal. If we always walked quickly because we thought that the devil was chasing us, our behaviour would be considered abnormal.

In this chapter, we will explore the various concepts of normal and abnormal behaviour, and consider what it means to be in good mental health. We will consider the various ways in which behaviour is adjusted in response to certain events and examine some forms of mental disorder and the methods used to treat these disorders. Much of our discussion will centre around the discipline of psychology as it is used in psychiatry.

14.1 *Defining Normal and Abnormal Behaviour*

How do we decide what is normal and abnormal behaviour? Most psychiatrists would probably agree that the dividing line between normal and abnormal is determined by the choice of methods we use to cope with certain situations in our lives. Some of those methods are chosen consciously, meaning that we are aware that we control them. Some of them are unconscious, meaning that they are "chosen for us" by our mind to help it adapt to those situations that present anxieties.

We all have fears, doubts, and anxieties about ourselves, certain events, or the future, and we act in certain ways as a reaction to those feelings and mental states. Speaking before a large crowd or taking a test may cause many people to be fearful or anxious and to act differently than usual — impatient, nervous, unable to concentrate, or even to speak! But usually, these feelings and behaviour are temporary. They are balanced by other feelings that allow us to go about our daily activities and feel that we are in control.

When fears and anxieties dominate our thoughts and many of our activities, and are not balanced by feelings of confidence, satisfaction, or control, our behaviour changes. If this continues for a long time, many people lose their ability to adapt to anxiety-causing situations; some people may lose touch with what is real and normal. The more we are out of touch with reality, the more our behaviour becomes abnormal, and the closer we may be to suffering from a mental disorder.

DEFINITIONS OF ABNORMALITY

In the field of psychology, defining abnormal behaviour sometimes depends on which model or approach the psychologist takes. Those who use the **statistical model** suggest that abnormal behaviour is any behaviour that is not within the average for a certain population. For instance, the majority of people do not sit in the middle of a busy street. Persons who do are considered abnormal because their behaviour is not within the average for the population.

The **clinical model** suggests that abnormal behaviour is any behaviour caused by a mental disorder or disease condition that can be diagnosed through an examination of that person's behaviour or physical state. This can be done by the person taking an active role in the examination through conversation or through certain medical tests. For instance, blood tests can show an under-active thyroid that is producing abnormal behaviour.

The **normative model** suggests that abnormal behaviour is any behaviour that threatens social norms. The behaviour of the hippies of the 1960's or of the punkers of the 1980's was considered abnormal because it challenged the current beliefs of the dominant group in society. But norms change and are adaptable; as the abnormal behaviour is tolerated, some aspects of it become part of the norm. At one time, the wearing of jewellery by men was considered abnormal; it is now acceptable.

It should be stressed that these models are not exact nor are they the only ways in which abnormal behaviour can be defined. Categories of abnormal behaviour have changed over the years, and continue to do so.

14.2 Defining Mental Health

When can it be said that a person is in good mental health? It is generally believed that a person who enjoys life, is self-confident at least some of the time, and is hopeful about the future is in good mental health. A person in good mental health is capable of maintaining close and satisfying relationships with other people, and can resolve conflicts and difficulties most of the time. Feelings of discouragement and self-doubt do not control daily living.

People with good mental health can usually identify the causes of any unhappiness and can cope with problems as they occur without being overwhelmed by them. They do not usually run away from their problems, ask others to bear their problems, or make life difficult for other people because of their own setbacks.

Although they have fears and concerns about people or situations, they do not have irrational fears that control their daily lives. They are concerned about their health without being overconcerned. Health problems are not "created" to compensate for feelings of anxiety, frustration, or loneliness. People with good mental health have a well-developed conscience which guides their behaviour. However, they do not allow their conscience to rule them to such a degree that constant guilt and anxiety result. Anger can be expressed when it is justified, but does not linger.

Progress Check

1. List three of the approaches psychologists might use to define abnormal behaviour.
2. Give five general characteristics of a person who is mentally healthy.

14.3 *Defence Mechanisms*

For most people, disappointment, frustration, and anxiety can cause emotional stress. Stress occurs when there is a difference between what we desire or expect and what we receive or experience; the greater the difference, the greater the stress. These desires may be conscious or unconscious. The characteristics that make up our personality determine to some degree what behaviour we show in response to a stressful situation.

For instance, each of us requires recognition from our friends, family, and co-workers. We want to be considered worthwhile in the eyes of others. We try to improve our image, our sense of self-worth, by adopting socially acceptable behaviour, attitudes, or dress. Sometimes events occur which challenge our sense of self-worth. We may suddenly feel inadequate or less than what we thought we were.

To avoid a sense of failure and to eliminate the emotional stress caused by that sense of failure, it is believed that we use certain strategies.

These strategies, called **defence mechanisms**, are a way to adapt to particular situations for our own protection. They are not considered abnormal or uncommon. Defence mechanisms can be compared to the immune system which protects our bodies from infection and disease.

TYPES OF DEFENCE MECHANISMS

Most of us use defence mechanisms to some degree. After failing a test for example, do you ever say, "Oh, I didn't really study very much" or "It doesn't matter because this is not an important subject"? Or, have you ever blamed something or someone else for your mistake? "The teacher failed me." Most of us can answer yes to these questions. Most of us use a defence mechanism to lessen the stress caused by such socially-embarrassing situations. We will examine four of these defence mechanisms: rationalization, compensation, displacement, and repression.

Rationalization

Rationalization is the mechanism which invents a reason to explain a certain behaviour or attitude. This happens when we cannot face or accept the true reason which caused the action, behaviour, or attitude. These "chosen" explanations or rationalizations are usually very good and reasonable ones. We convince ourselves that they are also true, rather than face the real reason for the behaviour. The various forms of rationalization can be called "saving face", "sour grapes", and "sweet lemon".

Saving face: We all have a certain image of ourselves that we like to believe in and protect. Sometimes situations arise that challenge this self-image and create anxiety or stress.

Steve brought his report card home with a failure in French; his mother expressed disappointment. Steve told his mother that French was not really an important subject for him because he wanted to study mathematics in university. The real reason for his failure was that he did not understand French as well as his other subjects. Yet, he had not studied very hard for his exam. Rather than accept the reason for his failure, which would challenge his image of himself as a fairly good student, or admit to his mother that he had difficulties with French, Steve rationalized. He found an explanation that allowed him to feel less embarassed and uncomfortable in dealing with his mother's disappointment. He had "saved face", both with his mother and himself.

Sour grapes: There are many different ways of showing this particular form of rationalization, but all are designed to lessen the anxiety caused by certain situations with which we cannot deal easily. The name comes from a fable by Aesop, written a long time ago.

"A fox was particularly fond of grapes. One day when he was very hungry, he spied some beautiful, luscious grapes hanging high on a vine. He wanted some but found they were above his reach. He jumped and jumped and jumped, but was unable to get any. Finally, he gave up in despair. As he walked away from the vineyard, he said, "They were probably sour grapes anyway. Anyone can have them for all I care. Who wants sour grapes. Certainly not I."

You are dating someone and believe that all is going well in the relationship. Suddenly, without warning, your friend wants to end the relationship. What would be your immediate reaction? It would likely be considerable anxiety and self-doubt. You may start to look for flaws in your appearance or errors in your behaviour that would explain your friend's decision.

Finally, you decide that the relationship was not that desirable after all. You might claim that your friend was not all that attractive anyway. You were only dating until someone you liked more came along. You have lessened your anxieties and protected your self-image by using a "sour grapes" rationalization.

Sweet Lemon: Sometimes, if events do not turn out the way we want them to, we find a soothing or a "sweet" rationalization to make us feel better. For example, a person who never has enough money may say, "It's better that I don't have a lot of money because I'd just spend it carelessly anyway." A situation that is undesirable or "sour" is made sweet.

Another example might be that for two seasons, your school's basketball team has not won a game. The coach tells the players that he is happy that they are at least getting a chance to play, since the important thing is not winning but playing the game. The unhappy situation has been given a "sweeter" face or outlook by reminding the players that they are indeed getting something in return for their efforts.

Compensation

Compensation is a defence mechanism used when we want to "make up" for something that we feel is lacking in ourselves or our lives, whether real or imaginary. We want to offset the feeling of loss or discomfort created by this lack, so we find something, a behaviour or activity, in which to excel.

Compensation is a widely-used defence mechanism. A person who is short, for instance, may feel inadequate and may choose to become an excellent skier. Skiing achievements make that person feel accomplished, which balances the discomfort of not being tall. The anxiety is reduced and is replaced by a sense of satisfaction or well-being.

Sometimes a person may try to compensate through the achievements of others. Parents who may have once wanted to become a doctor or to compete in the Olympics, for instance, may encourage their children to do so. The parents receive pleasure from the accomplishments of the children. The parents' sense of well-being is enhanced.

Displacement

Displacement is a defence mechanism through which a feeling or emotional response is deflected from the original and true source of the response and transferred to a substitute target. A person cannot direct the feeling or response to the original source because of a possible threat. Therefore, to alleviate the anxiety, the person finds another target.

How many times after a bad day at school have you come home and expressed your feelings of frustration or anger by kicking a door or yelling at someone for no particular reason? The source of your feelings may be someone in authority to whom you feel you cannot safely direct your responses. You choose a safer target, something or someone

who cannot hurt you in return. You displace your feelings from one, potentially threatening, source to another safer source.

Another more common term for displacement is the **scapegoat reaction**. When we displace our feelings, we are using the substitute targets as scapegoats. A parent may be angry with an employer. To avoid a possible argument or more emotional distress by dealing directly with the employer, the parent finds a scapegoat to receive the anger, a son or daughter or maybe a stranger. Once the anger is expressed, the feelings go away and the anxiety caused by them disappears.

Sometimes an entire group of people may be used as a scapegoat for a problem for which they are not responsible. In Canada, in periods of high unemployment, immigrants are sometimes blamed for taking away jobs, and making it difficult for others to find employment. The real reason for the unemployment may be a lack of people with the required skills, a national economic crisis, or the business failure of a particular industry.

Repression

Repression is another defence mechanism which is used to relieve anxiety by forcing unacceptable thoughts or reactions into the unconscious mind. The first three defence mechanisms we discussed change the form of a desire or thought. Repression is a defence mechanism that prevents the expression of a desire. Ideas, impulses and feelings which are disturbing, dangerous, or painful are "forgotten" by the conscious mind and memory. A person who was mistreated in childhood may not be able to cope with the feelings created by those memories. So the mind represses the memories, buries them in the unconscious. The conscious mind is then free of the anxiety.

Repressed material cannot usually be recalled voluntarily nor is the defence used consciously. However, it does not remain dormant. It may present itself at any time and cause abnormal behaviour or emotional disorder in its attempt to reach the conscious mind again.

Repression should not be confused with **suppression** which is the conscious and deliberate attempt to push an unacceptable thought aside. A person who must deal with the impending death of a family member may deliberately push that thought aside or ignore it, until better able to cope with it. The person does not deny the truth, if pushed to admit it; the truth is simply set aside for a time.

ROLE OF DEFENCE MECHANISMS

A defence mechanism is considered to be a reasonable emotional response to an anxiety-producing situation because the stress caused by anxiety is lessened and the mind is kept in a healthier state. Because these defences are common, the behaviour that results is considered to be normal in many cases.

> *"...the vital role which defences play in emotional well-being is clear. They can contribute constructively to emotional equanimity [calmness]. One's satisfactions in living may be increased and personal and professional effectiveness and efficiency may be enhanced."*
> **Henry P. Laughlin,** *The Ego and Its Defenses*

If, however, defence mechanisms are used too extensively as tools to cope with anxieties, they begin to affect judgment. The overuse of defence mechanisms prevents a person from learning more effective ways of working out stresses and resolving anxieties. The anxieties that cause the mind to use a defence mechanism begin to dominate the emotions and personality. The behaviour that results becomes abnormal, affecting the person's ability to function effectively.

PROGRESS CHECK

1. Choose one defence mechanism and outline a situation in which you have seen it being used.
2. Differentiate between suppression and repression.
3. How are defence mechanisms both good and bad for our mental health?

14.4 Mental Disorders

Abnormal behaviours have been studied by the medical community and identified as belonging to certain mental disorders. The milder forms of mental disorder are considered to be emotional disorders. The more serious forms of mental disorder are considered to be thought disorders. Statistics Canada reported in 1981 that one of every eight Canadians can expect to be treated for a mental disorder.

EMOTIONAL DISORDERS

An **anxiety disorder** is an emotional disorder where fear and anxiety are present for a long period of time, are of great intensity, and prevent a person from performing certain actions in a normal way. People suffering from this type of disorder have lost their ability to manage their fears.

Anxiety disorders can be caused by several factors and can produce many different behaviours. For example, people with a **phobia** have a fear of an object or situation that causes them to avoid the object or situation in exaggerated ways. A person who is afraid of snakes because they may be poisonous and acts cautiously when confronted with a snake is acting in a normal way. The reason for the fear is based in reality. When a person reacts with fear and terror when seeing a picture of a snake and avoids magazines that may contain such pictures, then the reaction is abnormal because there is no real danger present.

Many different objects or situations may act as the cause of a phobia. Some examples of phobias are **claustrophobia**, fear of closed spaces, **agoraphobia**, fear of public spaces, and **acrophobia**, fear of heights. The abnormal behaviour that results from a phobia varies from one to another. The claustrophobic person, when confronted by a closed space may exhibit dizziness, sweating, trembling, rapid heartbeat, and laboured breathing. The agoraphobic may show these same symptoms at merely the thought of being in an open space, or may continually check that doors are locked and windows are closed. Acrophobic people may lose their sense of balance and orientation when they find themselves high off the ground.

THOUGHT DISORDERS

Organic mental disorders are examples of thought disorders because they interfere with the thinking process. These disorders are characterized by a change in the structure of the brain which triggers the symptoms related to the abnormal behaviour. These changes or abnormalities may be due to a specific degeneration of a section of the brain or to the presence of diseased organisms. Dementia and delirium are both examples of organic mental disorders which are common in 10% to 15% of older people.

Commonly known as senility, **dementia** is characterized by a deterioration of intellectual abilities like remembering, thinking, or reasoning and is accompanied by emotional impairment. This occurs over a gradual period and is irreversible in 8 out of 10 people with the condition. People with dementia may forget that they have children, be unable to make decisions because they cannot understand situations, may lose the ability to interact with others, and may not be able to control impulses like continuous swearing or yelling.

Alzheimer's disease is an example of dementia where electro-chemical activity in the brain has changed and certain cells waste away. In addition to suffering from impaired memory and judgment, and emotional instability, the person with Alzheimer's must endure confusion, delusions, or depression.

One of the most common forms of organic mental disorders in older people, **delirium** can be thought of as confused consciousness. A person's attention wanders, speech is slurred, or disorientation may occur off and on over a brief period of time. Sleep disturbances are common and the person may have difficulty concentrating.

Unlike dementia, delirium usually develops quickly and may be caused by physical illness due to infection, organ failure, nutritional imbalances, or drug intoxication. Delirium is reversible if the cause is quickly identified and treated. If not, the damaged brain structure causing the condition may be permanent and death may result.

PROGRESS CHECK

1. Phobias are an anxiety disorder. Describe the behaviour that can be associated with two phobias.
2. What is an example of an organic mental disorder? What behaviour can be associated with it?
3. How does delirium differ from dementia?

WHY ARE THERE MENTAL DISORDERS?

There are many reasons for mental disorders. There are also many theories about those reasons. Psychologists and psychiatrists are always working to understand better why some people behave abnormally.

A mental disorder may be traced to an inherited weakness; a genetic defect may be passed on. Personality characteristics may force some people to reject certain events, and to choose a new behaviour to protect that rejection. Brain abnormalities have been identified as reasons for such disorders as Alzheimer's disease. Biochemical factors may also play a part. For example, an excess of specific brain chemicals that transmit messages through the nervous system is thought to cause schizophrenia.

One theory of the cause of mental disorder does not exclude another. For instance, an anxiety disorder in one person may be explained by examining certain events in that person's life that have caused the anxieties mentioned. But, that person may have also inherited a physiological defect that makes certain parts of the nervous system vulnerable to anxieties. Psychologists and psychiatrists assess all contributing factors that may cause a mental disorder. In this way, no one theory dominates in the diagnosis of disorder. A more thorough and accurate assessment of a person's mental health can be made.

TREATMENT OF MENTAL DISORDERS

The abnormal behaviour common to today's mental disorders has likely been with us since the beginning of recorded time. Mention of abnormal behaviours is made in many religious texts as well as the recorded history of the Greeks and Romans. During ancient times, a person with abnormal behaviour was assumed to be influenced by the supernatural in the form of demons or gods.

European society rejected the mentally disturbed who were believed, at times, to be sorcerers or witches. Their care was left to religious organizations who kept them in monasteries and hospitals. The first hospital for the insane is believed to have been built in Valencia, Spain in 1409.

During the 18th century, Franz Mesmer believed that mental illness was caused by the build-up of magnetic fluids in the body. He claimed that these fluids could be removed by using certain special magnetic powers which he possessed. Franz Joseph Gall claimed to have discovered twenty-seven organs within the brain. Each of these was responsible for a particular mental function. He also believed that the overall shape of the skull was affected by these organs. Bumps and other features of the skull, according to Gall, could be measured in order to judge a person's personality. This particular practice became known as **phrenology**.

By the 19th century, the mentally disturbed were treated by using certain mechanical devices. People were fastened into wooden chairs, beds or cages and spun around or dropped into water. All of this was intended to scare the patient in the belief that the mental disturbance would also be scared away. Incoherence, terror, vomiting, unconsciousness and even death were the actual results. In England, a mental hospital called Bedlam, (for the Bethlem Royal Hospital) became known for its wild and brutal treatment of the mentally disturbed. A penny tour of their cells by the general public became a fashionable outing.

Eventually, progressive treatments were developed. In 1798, Philippe Pinel (1745–1826) removed the chains from forty-nine insane patients at the Bicetre Hospital in Paris, and ushered in a new type of treatment based on kindness, understanding, and rehabilitation. Another pioneer in the treatment of mental disturbances, Pierre Janet (1859–1947) described a hierarchy of mental functions. At the bottom of this hierarchy were such conditions as comas. Rational experience, such as problem-solving and conscious activity, were placed at the top. These attempts to classify various levels of mental activity were important for the work that followed from Sigmund Freud (1856–1939) who provided theories on personality development and the unconscious mind that have greatly influenced most forms of modern treatment of the mentally disturbed.

Some Current Treatment Techniques
Today, the treatment of mental disorders takes several forms, all of which are based on a real desire to help discover the basic cause of that disorder, and to bring the patient into a state of mental well-being.

Psychotherapy takes many forms, but is based on uncovering unconscious conflicts that are causing a mental disorder. Such therapy relies on conversation between patient and therapist. These conversations can be between individuals or in groups, such as in family group therapy.

Behaviour therapies are based on the theory that abnormal behaviours are learned behaviours, not related to conflicts in the personality or unconscious mind. Therapy tries to change behaviours, thus eliminating the mental disorder.

One technique used is **positive reinforcement**. A patient with a severe weight loss problem who refuses to eat, for instance, will be given positive reinforcement once a regular eating pattern is established. As weight is gained, rewards are given. Undesirable behaviour is changed, and the desired behaviour is rewarded.

Physical therapies may include the use of drugs for the treatment of mental disorders in which there is an excess of specific chemicals that are causing abnormal behaviour. Drugs may also be used to treat disorders in which brain deterioration has occurred or in which diseased organisms are present. Electroconvulsive shock therapy may also be used to treat disorders which have severe depression as one of the symptoms.

PROGRESS CHECK

1. What are some theories about the cause of mental disorders?
2. How were mentally disturbed people treated in the past?
3. What was Mesmer's theory of mental disorder?
4. What was Janet's theory of mental disorder?
5. Briefly describe some types of current therapy.

OVERVIEW

In this chapter, we have explored certain perspectives on mental health. Good mental health is characterized by the choice of methods used to cope with anxieties. It is also characterized by the degree to which anxiety states are balanced by feelings of control, satisfaction, and self-confidence, for instance.

What does it mean however, when anxieties are not balanced by more positive feelings? How can we learn to deal with anxieties in a positive way so that they do not dominate everyday life?

Concern about one's psychological state is not an unhealthy sign. Thinking about feelings and reactions to certain situations is considered a natural and progressive step towards maturity and learning how to deal with problems effectively. There are times however, when such self-appraisals reveal the need for additional help to resolve or manage personal problems.

Mental health specialists suggest that help or counselling should take place in a trusting relationship, whether with a professional in a counselling centre or with a friend, relative, or adviser. Talking about feelings and concerns with someone we know or who may be familiar with the situation sometimes provides solutions or perspectives that might have been missed. Sometimes such discussions may reveal which problems can be effectively dealt with first while leaving others to a later time.

Other activities may also help to resolve anxieties in an indirect fashion. Staying physically active keeps our body healthy and able to deal with the effects of emotional stress. Taking part in social activities provides needed emotional and social support.

It may also be important to remember that good mental health does not mean the absence of problems or anxieties, but rather the presence of useful and positive methods of coping with the problems and anxieties that are a part of everyone's daily life.

KEY WORDS

Define the following terms, and use each in a sentence that shows its meaning.

statistical model
clinical model
normative model
defence mechanisms
rationalization
compensation
displacement

phobia
claustrophobia
agoraphobia
acrophobia
organic mental disorders
dementia
delirium

scapegoat reaction	phrenology
repression	psychotherapy
suppression	behaviour therapy
anxiety disorders	physical therapy

KEY PERSONALITIES

Give at least one reason for learning more about each of the following.

Henry P. Laughlin   Philippe Pinel
Franz Mesmer   Pierre Janet
Franz Joseph Gall   Sigmund Freud

DEVELOPING YOUR SKILLS

ORGANIZE AND FOCUS

1. (a) Create an organizer that summarizes the types of defence mechanisms used by people to cope with anxiety.
   (b) What questions can we ask about the effects that defence mechanisms have on human thinking and behaviour?
2. (a) Develop an organizer that lists the types of mental disorders and symptoms given in this chapter. In a separate column list the methods used to treat these disorders.
   (b) What questions should we ask about mental disorders and their treatments?

LOCATE AND RECORD

1. Divide the class into pairs and write down the types and examples of fears, anxieties, and doubts that have been experienced. In another column comment on whether they were temporary or longer lasting. In a final column list some of the other feelings and attitudes that help us to overcome and deal with our fears, anxieties, and doubts.
2. In groups have a brainstorming session and list examples of abnormal behaviour that could be included under the headings of Statistical, Clinical, and Normative Models of abnormal behaviours. Share your conclusions with the class and discuss whether these examples fit the model in which they have been included.

EVALUATE AND ASSESS

Discuss why most people adopt socially acceptable behaviours. Analyze what could be the consequences of not doing so.

SYNTHESIZE AND CONCLUDE

1. Make a list of situations which create tension in your life. In which of the following activities do you participate to help reduce and overcome tensions?
   - talking out the problem
   - giving yourself a small holiday
   - becoming less competitive
   - doing something for someone else

   Which ones work best?

2. Analyze people's attitudes toward those who suffer from mental disorders. Are people becoming more knowledgeable and understanding? Are society's attitudes toward those who have mental disorders important?

APPLY

Refer to the organizer that was originally developed to list the types of mental disorders, symptoms, and the methods used to treat them. Refer to the symptoms below and expand upon the symptoms listed in the organizer for each mental disorder.
- silent, listless behaviour
- poor work performance
- being suspicious without cause
- hallucinations
- anxiety over imagined problems
- headaches or insomnia without cause
- sudden mood changes

Once this has been accomplished refer to the chart below listing the types of people who are trained to help those suffering from mental illness. Choose the experts that you think are best suited to treat particular mental disorders.

Some Agencies and People Who Help
- Community mental health centres: provide counselling, treatment, and half-way houses
- Family service agencies: provide practical help and information to families of the mentally ill
- Mental Health Association: provides information on mental health and illness
- School and college counselling services: provide help for students with emotional and psychological problems

- Local crisis centres: provide emergency help 24 hours a day
- Hospitals: may provide emergency or short-term care or, through a mental health clinic, may provide longer term treatment
- Family Physicians: provide emergency care and referrals to appropriate treatment centres or specialists
- Psychologists, psychiatrists, social workers, nurses: provide counselling and therapy.
- Clergy: can provide information and comfort.

COMMUNICATE

1. Write an organizer or short essay, or debate one of the following topics.

- Certain types of abnormal behaviours are healthy in a society because it forces that society to reexamine its beliefs, attitudes, and practices.
- In the long-run some behaviours that are considered abnormal are later adopted by the larger society.
- Anxiety, fear, and doubt are a normal part of the human experience.
- Everyone is abnormal in some way.
- The human mind is too complex to understand.
- Mental patients should not be allowed out into the community until they are cured.
- Mental illness is a family problem; not an individual problem.
- There is nothing to fear except fear itself.

2. Write an essay on a type of mental disorder. Analyze its causes, symptoms, and treatments. Comment on society's attitudes toward those suffering from this disorder and give your opinion of its attitudes.
3. Often you hear someone say that a certain person is *neurotic*. Less often you probably have heard the term *psychotic* used to describe a person's behaviour. Although these terms are not often used today by the medical profession, you might be interested in knowing their definitions. People use the terms far more loosely than should be done. By referring to a reference, write a short explanation of the differences between *neurosis* and *psychosis*. Compare your findings with other class members' findings.

CHAPTER 15

Being a Part of Society

Why do we conform to the norms of our society? What happens if we disregard those norms?

INTRODUCTION Imagine yourself in the following situations. You attend a party with three of your friends. All three start to drink wine from a bottle that one friend has pulled from a coat. You have promised your parents you will not drink alcohol until you are 19 years old. Your friends each take a drink and offer you one. What will you do? A new teacher has come to your school. Her teaching style is different from your other teachers. You are not sure whether she is a good teacher or not. Some of your friends begin discussing her during lunch. They all say that she is not a good teacher. What will your opinion of the teacher be now? You are walking on a crowded street and you see a person beating up another person who is on the ground. The other people on the street hurry by without stopping. It is obvious that the person on the ground is in great pain. What will you do — walk away and assume someone else in the crowd will help — or try to help in some way?

If your actions in these situations were based on what others around you were doing or saying, you were conforming. To conform means to behave in a way that agrees with the behaviour of the majority of society. The pressure to conform can come from several groups. Parents, teachers, peers can all have an influence on our behaviour.

In this chapter, we will examine several of these influences, and why we often conform to their patterns of behaviour. We will also examine some of the classic experiments conducted in the field of social psychology that investigate how conformity works.

15.1 *Social Influence*

In every society, there are certain characteristics of that society that must be agreed upon by the majority of the people in order to maintain harmony. How one dresses, speaks, or behaves in certain situations are all ways in which a society — or certain groups in that society — determine what is acceptable. These shared beliefs or behaviours form what are called "cultural norms". Cultural norms are the set of beliefs, behaviours, or attitudes that describe a society and how it operates. Cultural norms are maintained through the social pressure that certain individuals (or large groups of individuals) bring to bear on other individuals in a society.

In the party situation mentioned in the introduction, your three friends tried to influence you to conform to their behaviour. If you decided to drink the wine, thus going against your own belief or promise, you were pressured by social influence, and conformed to the wishes of your friends.

15.2 *Conformity*

According to social psychologist Elliot Aronson, **conformity** may be defined "... as a change in a person's behaviour or opinions as a result of real or imagined pressure from another person or a group of people." We act or behave in agreement with the actions or behaviours of others in our society.

WHY DO WE CONFORM?
Conformity is a social force that causes people to do things in a similar way. A degree of conformity is necessary for a society to survive because it establishes order and structure. People need to know that they can expect certain patterns of behaviour from each other. A society in which people always acted according to their own feelings and whims would result in chaos.

For this reason, one of the basic tasks of any society, no matter how small or large, is to establish a set of rules that determines what is acceptable behaviour. Those who do not conform to these rules or cultural norms are punished in some way that indicates disapproval or rejection, while those who abide by the norms are rewarded with acceptance. The rules may be the laws that everyone is expected to obey, or they may be folkways, such as the rules of dress for certain occasions.

Many psychologists believe that people conform because they have a basic need to belong. Humans are social animals and have a strong need to fit in. Eric Fromm describes this need in his book *Escape From Freedom*. Fromm believes that we must feel that we belong to something in order to give our lives meaning and direction. Without that sense of belonging, we feel insignificant. We must conform to feel whole.

Conformity occurs at every stage of life from infancy through adulthood. Group pressure or social influence compels us to act in certain ways, to wear certain clothes, or to hold certain beliefs or attitudes.

HOW DO WE CONFORM?

The majority of people in any society conform in one way or another. Today, many people jog and keep fit not only to feel more healthy, but because it is the acceptable and recommended thing to do. We want to fit in, so we do what the

majority is doing. Fashion is one of the principal ways in which social conformity takes place. Clothing designers in such places as Paris, London, and New York create certain looks or styles which are then copied around the world. These styles are popularized and become the accepted look for that season. If we want to fit in, and be accepted by our peers, we conform to the current fashion styles.

In some cultures, conformity is maintained by physical appearance. In China, during the period before the revolution of 1949, small, delicate feet were admired in women. It was customary to bind the feet of young girls to restrict their growth. This forced conformity in physical appearance ensured a higher marriage price for the woman. The cost of this social conformity to the woman however, was pain, deformity, and dislocation of the foot bones.

In some cultures, the natural shape of children's heads has been altered to improve the aesthetic qualities of the skull. The head may be elongated, broadened, or flattened by using headboards of various shapes. As recently as the 19th century, such head deformity took place in France. Parents conformed to this practice because elongated, flattened, or broadened heads were considered beautiful. Another aim of this practice was to "shape the brain" in the belief that children would develop noble character traits like bravery, sincerity, and intelligence.

PROGRESS CHECK

1. Why do all societies insist on a certain degree of conformity?
2. Why do individuals conform?
3. In what ways do we conform?
4. How have different cultures established social conformity through physical appearance?

15.3 Classic Conformity Experiments

Social psychology is a field of social science that studies the ways in which individuals and groups interact with, and influence, one another. In attempting to analyze how social influence works in establishing cultural norms in a society, it is sometimes easier to study social influence on a smaller scale. A study can be done with a few individuals in an experimental setting where the sources of influence can be controlled. The following experiments represent some of the classic attempts to analyze conformity and the process of social influence.

SHERIF AND THE AUTO-KINETIC PHENOMENON

In 1935, social psychologist Muzafer Sherif wanted to investigate the effects of social influence on individual behaviour. He used the **auto-kinetic phenomenon** to test his subjects (auto means by oneself and kinetic means of, or due to, motion.) This phenomenon has to do with the effect of darkness on vision and the perceived movement of a stationary point of light.

A person is seated in a completely darkened room, then a pinpoint of light is projected in front. When people first see the pinpoint of light, they usually report seeing the light move, even though it is stationary. When the light is turned off, then on again, people report seeing it move again, but in a different direction and for a different distance than the first time.

Sherif showed a person a pinpoint of light several times. After turning the light off and on a few times, he asked the person to tell him how far the light had moved. He then placed two or three people together in the same room and noted that they began to influence each other's judgment. For example, if one person had established that the light moved 9 cm while a second person said 3 cm, the two were likely to agree at 6 cm when asked again how far the light had moved.

Sherif's experiment shows that people look to others to help them construct reality. The people in the experiment were uncertain about the movement of light so they looked to others to help them form a judgment.

SOCIAL-COMPARISON THEORY

Social psychologist, Leon Festinger, noted that people are generally confident in their judgment of reality as long as others agree with them. If there is a difference in opinion however, the natural tendency is to assume that one of the opinions is wrong. The social-comparison theory suggests that when a difference in opinion exists, we may try to reduce the difference by changing our opinion to conform with the differing one. We may also try to convince the other person that he or she is wrong.

In the "new teacher" situation in the introduction, you had no set opinion on the new teacher. When your friends decided that the new teacher was not a good teacher, you had to decide whether to agree, disagree, or stay undecided. If you decided to agree with your friends, you were influenced by what Festinger calls social-comparison. You depended on others to provide your information, and decided to reduce the difference between your opinion and theirs by agreeing with theirs. You conformed to the majority opinion.

The Asch Experiment

Social psychologist, Solomon Asch, conducted a series of experiments in 1956 to discover the ways in which people might maintain their independent opinion when faced with disagreement. His results were surprising. The following experiment is typical of the kind that Asch conducted.

Asch assembled a group of eight college students. Each student was shown a series of cards on which four lines were drawn. Two of the four lines were the same length. Each student was then asked to indicate which two lines matched. Seven of the eight students in the group had been "planted" by Asch. They gave wrong answers. The eighth person, the actual subject being tested, had to decide whether his or her perception was correct and therefore to stay with that opinion or to agree with what the majority had said about the length of the lines.

About 33% of the time, the unsuspecting eighth person yielded to the opinion of the majority. At one point, when the difference in the line lengths was 18 cm, some of the students being tested still yielded to the majority opinion. Why did some of the students not believe their own eyes? During interviews conducted after the experiment, they gave three types of reasons.

Some believed the majority was wrong but did not want to appear different by maintaining their opinion, even though they believed their answer was correct. Their need for social approval was greater than their need to keep to their opinion. A second group did not trust their own judgment when faced with unanimous disagreement. They were not sure of themselves, although they thought they knew the right answer. They began to doubt their own information or perception. They concluded that they must be wrong since logic suggested that the majority could not be wrong. A third group said they saw the lines the way the majority claimed to see them. These students actually convinced themselves that they saw something different than what was presented to them. They changed their perception to fit the one described by the majority.

"In their quest for knowledge, experimental social psychologists occasionally subject people to some fairly intense experiences . . . These procedures may raise serious ethical questions. For a social psychologist, the ethical issue is not a one sided affair. In a real sense he is obligated to use his skills as a researcher to advance our knowledge and understanding of human behaviour for the ultimate aim of human betterment."

Elliot Aronson, *The Social Animal*

THE LATANÉ EXPERIMENTS

Bibb Latané is a social psychologist interested in examining the conditions under which sources of influence have the greatest impact. He devised experiments, during the 1960's and 1970's, to test the theory that the impact of the source of influence decreases when the number of people involved increases. Latané and another social psychologist, Judith Rodin, staged the following experiment involving "a lady in distress".

A female experimenter asked college students to fill out a questionnaire. While the students were working, the experimenter went into an adjoining room. She said that she would return when the questionnaire was complete. Shortly after she left, she turned on a recording. There was a loud crash that sounded as though she had fallen to the floor while climbing up on a chair. An anguished statement followed: "Oh, my foot . . . I, I can't move it. Oh . . . my ankle . . . " These cries continued for a short time and then stopped.

Of the students who were alone in the other room, 70% left the room to offer help. However, when students were working in pairs, only 20% offered to help. The conclusion reached by the experimenters was that the presence of another bystander hindered action. Each of the students who was unhelpful was interviewed after the incident. All said they did not feel the accident was serious, partially because the other student in the room had not responded to the cry for help.

In the "crowded street" situation in the introduction, a decision to walk away from the emergency, to conform to the behaviour of others, is an example of this phenomenon. Because there were a number of people walking away from the beating, the influence to help the victim was lessened. When there are more witnesses in an emergency situation, there is less pressure on each of them to act, and it is less likely that the victim will receive help from any of them.

Progress Check

1. Briefly describe each of the three experiments discussed.
2. What do these experiments tell us about human conformity?
3. Can experiments performed with a few individuals describe accurately society as a whole? Why or why not?

15.4 Non-Conformity

People who do not behave like the majority of a society are called **non-conformists**. An individual can be a non-conformist or an entire group of people in a society can be non-conformist. When large groups of people behave in a different fashion than the majority, they are sometimes identified as a sub-culture or as a counter-culture.

A **sub-culture** is a group or a part of society whose members share beliefs, norms, and values that are different from the majority. A sub-culture has distinctive features that set it apart from the majority, while keeping some features of the larger culture. The term sub-culture does not suggest an inferior culture. It is rather a separate section of a culture. In our society, teenagers form a sub-culture. They recognize the norms and values of the wider Canadian culture while keeping their own distinctive interests, behaviour, and values.

A **counter-culture** is a group or a part of society whose members also share certain beliefs, norms, and values that are different from the majority but their behaviour and beliefs are offered as an alternative, another possibility. Where a sub-culture is considered to be a desirable separate part of the culture, a counter-culture is frequently considered an undesirable part and its members are frequently shunned or ostracized by the majority.

In the 1960's and '70's, many people rejected the norms of North American society. They resented the war in Viet Nam, and they thought that society generally was too repressive and conservative. These "hippies" became a counter-culture.

Males let their hair grow long as a protest to the traditional short haircuts of established society. Females put flowers in their hair to symbolize their desire for peace, for a return to a more rural time. The hippies lived in communes, sharing food, accomodation, and sex partners. The leaders of this counter-culture wanted to prove that, by living co-operatively, society could be improved. Barriers between people could be broken down.

Some members of the hippies counter-culture began to take illegal drugs in a further attempt to break away from what they saw as a restrictive society.

The larger society found it very different to accept the behaviours of the hippies who seemed to be ignoring and breaking many of the accepted cultural norms. But society, and its norms, change with time. Political issues, such as the Viet Nam war, began to be criticized by a larger number of people. Peace became the goal of most people. Some of the attitudes of the counter-culture became part of the larger culture.

Progress Check

1. What is a sub-culture?
2. What is a counter-culture?
3. Why was the "hippie" movement considered a counter-culture?

CASE STUDY

The Hutterites

The Hutterites are an example of a sub-culture living in harmony within the larger North American culture. There are at least 200 Hutterite farming colonies in the Canadian provinces of Alberta, Saskatchewan, and Manitoba and in the American states of Montana, Minnesota, Washington, and the Dakotas. Many are descendants of settlers who fled religious persecution in Europe about 200 years ago. Hutterites believe that they can best serve God by living in a communal society where all rewards and all work are shared equally. They observe strict religious beliefs and practices and are devoted pacifists.

The society of the Hutterites operates on the principle that each person receives according to her or his need and gives according to her or his ability. Each person in the colony is assigned a specific task for each week. All work in the colony is considered equally important; no person has more prestige or status because of the type of work he or she does.

Hutterite clothing styles have not changed in over a century. Women and young girls wear simple long dresses and cover their heads with scarves. At the age of 15, Hutterite children become "young people". This allows them to join the adults at work and at the dining table. Formal education usually ends at age 16. As in other societies, young Hutterites who break the rules are punished. If they are caught smoking or attending a movie in a neighbouring town, they have to kneel in front of the entire congregation and confess their sins. Later, they will be compelled to sit with the younger children.

Marriage and the family remain strong institutions. This can be seen in the complex set of rules that must be followed in order to marry. If a young man wishes to marry he asks his parents and the preacher for permission to do so. The preacher, in turn, requests permission from the "council of baptized men". Once consent is given, the young man's father will visit the preacher in the colony where the bride lives if she does not reside in the same colony. The father requests the permission of the preacher to hold the marriage. In turn, the preacher in the bride's colony will request permission from that colony's council. Once consent is received, the father of the young man asks the father of the young lady for his permission for his daughter to marry. In the majority of cases, this process leads to matrimony; however, there are cases where one side does not agree and the marriage does not take place.

The Hutterites live a life dramatically different from the rest of society. Yet, as a sub-culture they do not challenge the larger culture to change. They abide by the laws, take part in the economic system, and interact with people in the towns and communities in their area.

1. *Why do the Hutterites stay in colonies?*
2. *How strict are their rules of behaviour?*
3. *How can the Hutterites be non-conformists, and still be part of North American society?*

15.5 Why do some people feel alienated?

People learn ways of belonging to society. People can also reject society's norms of behaviour and form sub-groups with their own accepted norms. In every society, there are also people who want to identify with, or be part of a group, but who have lost their ability to do so. These people are alienated.

THE ALIENATED

Alienation is not a mental illness, but a "state of mind, a cluster of attitudes, beliefs, and feelings in the minds of individuals". The first person to describe the kind of society that creates alienation was Karl Marx. He believed that industrial society was too big, too bureaucratic, and too impersonal. He thought that a group became alienated when society denied its members their individuality.

For example, a farmer who grows his own food on land his family has always farmed, who knows what he will plant and to whom he will sell, and who lives in a community where everyone knows each other, is not likely to become alienated. But if you take that farmer's land from him and force him to take a factory job in a city where he has no social ties, and if in that factory job he has no say in what he produces or where the profit of his labour will go, he is almost certain to feel alienated.

Marx wrote about alienation in terms of groups of people. In the present day, it is as likely to be discussed in terms of individuals. Alienated individuals are likely to be anxious, hostile toward others, and resentful of authority. They are likely to have a low self-esteem and to feel insecure when their environment changes.

People can become alienated when they lose power, or when society changes faster than its members can adjust. If traditional social values, such as shared religion, family ties, and social obligations, are weakened, people can feel alienated.

Six Traits

The American sociologist, Melvin Seemans, noticed six traits often found in the alienated individual.

Meaninglessness: People see their lives as meaningless when they cannot predict the effect their behaviour will have on those around them. They do not know whether they will be praised or blamed for any given action. Factory assembly-line workers who never see their employers and who are expected to do their work with little or no involvement in the process of production, may feel that they, as individuals, have no effect and that their lives are meaningless.

Self-estrangement: People who are not sure of their identity will find it difficult to express themselves. Some immigrants, for instance, who find themselves in a new culture may adopt certain customs and beliefs of the new society. Yet, they may still feel separate from that society because they have not completely adjusted to the loss of their traditional beliefs. Their identity has changed but they may not completely understand it and they may have difficulty expressing their interests. People who do not know themselves, who are confused by their identity, cannot usually express themselves. Many different people in society suffer from this self-estrangement.

Powerlessness: People feel powerless when they feel that they have no authority over their place in society. An example would be Marx's factory workers, who had no control over what they produced or what was done with it, and who had no share in the profits that came from their work.

Cultural estrangement: An alienated person often refuses to accept society's moral or cultural values. The "punkers" and "skinheads" of the 1980's were believed to be culturally estranged from society.

Social isolation: In some ways, this is the opposite of cultural estrangement. It is the feeling that society has rejected you. Poverty-stricken families forced to depend on food and clothing "banks" for their basic needs are examples of people suffering from social isolation. These groups see no improvement in their situation nor any major concern for their problems from the society at large.

Normlessness: This is the belief that you have to break society's rules to get what you want. The person who kidnaps a famous personality to get attention in order to be heard on a problem or an issue is an example of normlessness.

A Condition of Society

Alienation is primarily a social condition. It affects both groups and individuals. Alienated groups see themselves as outside the power structure of society. Alienated people see themselves as lacking something that other people have. It is not a permanent social condition however. In certain situations, if the reason for the alienation can be eliminated, then a general improvement in society occurs.

Modern alienation is a product of the industrial revolution, which was probably the greatest social upheaval in history. It is still going on, and the world has grown increasingly more complex since it began, and more and more difficult for any one person to understand. Society has not yet adjusted to all the changes. One result has been that many people feel lost in their own culture. In the book, *The Messianic Legacy*, the authors wrote, "One of the most basic needs is the need for meaning, the need to find some purpose in our lives. Human dignity rests on the assumption that human life is in some way significant. We are more prepared to endure pain, deprivation, anguish, and all matter of ills, if they serve some purpose, than we are to endure the inconsequential. We would rather suffer than be of no importance."

PROGRESS CHECK

1. Define alienation.
2. How did Marx see alienation developing in a society?
3. What are some characteristics of the alienated individual?

15.6 Suicide

Sometimes an individual who feels alienated from society attempts suicide. Often a suicide attempt is a cry for help. It is sometimes the only method of expression that people feel they have in the face of emotional and psychological distress. Although there may be many ways of coping with the distress or resolving the problems that cause the distress, the suicidal person is frequently unable to see these solutions.

The suicide rate in Canada in 1985 was 15.1 people per 100 000. This is higher than the rate in the United States, and a little higher than that of most other industrial nations. The lowest rates in this country are in the Maritimes. Newfoundland has the lowest of any province (6.23). The province with the highest rate is Quebec (18.5). The suicide rate is increasing in every region of Canada.

In 1985, nation-wide, men were almost four times more likely than women to commit suicide. Among people aged 20 to 24, the rate among men is 7.3 times higher than for women.

Age and sex are both important factors. People between 15 and 24 are still the least likely group to commit suicide but the rate at which they do so is increasing more quickly than for any other group; it is almost 3½ times higher than it was in 1960. There is some question whether the statistics are accurate; the increased number may be due to the fact that suicides are being reported more than they used to be. However, suicide is the second leading cause of death, after traffic accidents, in this age group.

Divorced people have a higher suicide rate than single people. Married people have a lower rate than either. Childless people have a higher rate than people who have children.

Native people have the highest suicide rate in the country. Immigrants appear to follow the suicide rates of their country of birth. The rate in a given ethnic community seems to mirror the rate in that community's home country.

THE SOCIAL AND PSYCHIATRIC CAUSES OF SUICIDE

Emile Durkheim was the first social scientist to study why people commit suicide. He said that there were four kinds of suicide, each the result of different social forces.

Egotistic suicide: In closely knit groups, suicide rates are likely to be low. An individual is more likely to commit suicide if he or she feels a loss of identity from being cut off from society by some loss: of a spouse, of a job, of the family farm, of anything that may define that person's role.

Altruistic suicide: In this type, social ties are too strong. The person feels obliged to commit suicide for the common good. It is rarely found in modern industrial societies. Japanese warriors who committed hara-kiri (stabbing with ceremonial dagger) and soldiers in wartime who jump on grenades to save their friends are some examples of this type.

Anomic suicide: In this type, social obligations or ties are lost, and the individual loses a sense of belonging. Norms and values may be lost and the person feels no connection to the society at large. The homeless street person who commits suicide, might be considered an example of anomic suicide.

Fatalistic suicide: This type is rare. It is the opposite of anomic suicide in that it is caused by the power of a society to determine, in some cases, the fate of an individual. A prisoner who kills himself after being sentenced to a life term is one possible example.

Social explanations of suicide do not explain why specific people kill themselves while others in exactly the same situation do not. There are some common factors found in people who commit suicide. The presence of these factors in a person, however, does not predict suicide. Physical illness is especially a factor among the elderly. Many people who killed themselves were suffering from depression. Almost 30% of suicides in Canada in 1985 were alcohol-related; alcoholics tend to suffer from depression. People who commit suicide are more likely to be taking prescription sedatives or tranquillizers than people who do not.

It is thought that one or more of the following may lie behind much of suicidal behaviour: guilt, emotional pain, feelings of helplessness, the wish to punish oneself, the hope that the life after the present one will be better, or escape from an intolerable situation.

As well, people who commit suicide are likely to have poor coping skills. They are likely to have suffered a recent traumatic event, and to feel thay have no one to whom to talk.

WHO IS AT RISK?

A study by Canada's Department of Health and Welfare reported that certain groups are at a higher risk of suicide than others. Among them are depressed people, alcoholics, young people, the elderly, and native people. The reasons are different for each group.

Depression: As many as 90% of people who committed suicide had a mental disorder at the time of their death. As well, 90% of elderly people who committed suicide suffered from depression. Depressed people often feel friendless and powerless, and have a low self-esteem.

Alcoholics: Alcoholics have a far higher divorce rate than non-alcoholics. They are more likely to have problems on the job, with the law, and with society in general. Low self-esteem is also common in this group, as are feelings of helplessness and isolation. All these difficulties are made worse by the sufferer's inability to break an addiction. Alcoholics may use alcohol to "find the courage" to commit suicide. Alcoholism may also be a form of suicide — it is profoundly self-destructive.

Young people: Young people are still the least likely group to commit suicide. The concern is that the suicide rate in this group may be increasing more quickly than in any other, especially among the males. Many of the contributing factors in this group are common to all high-risk groups: depression, low self-esteem, poor social skills. As well, young people who commit suicide tend to have unhappy home lives and problems at school. They are likely to drink more than others their age, to have difficulty accepting authority from family, school, and the law. Suicides in this group are likely to have felt alienated from their family and friends, to have felt rejected by their peers, to have seen themselves as failures, and to have blamed themselves for things they did not control.

The elderly: Elderly people who commit suicide tend to have physical health problems and to suffer from depression. The two are often connected — people who feel ill tend to feel depressed, and depressed people tend to feel ill. Elderly suicides are likely to have gone through a major and recent role change, for example, retirement, the loss of a spouse, the selling of the family home. Any of these would make them vulnerable to feelings of isolation and loneliness.

Native people: Native people in Canada commit suicide twice as often as other Canadians. Of those who did, most had easy access to a gun. They were likely to drink heavily and to show strong signs of alienation and depression. They tended to be single and poor, and to come from unstable families in which other members had committed suicide.

PREVENTING SUICIDE

The suicide rate can be lowered by changing the social conditions that foster it, and by providing help for people at risk. A federal report, *Suicide in Canada*, published in 1985, contained a number of recommendations. Among them were the following.

(1) The problem of suicide must be studied more fully. For this to be possible, several things must happen: cases of suicide must be evaluated more carefully; coroners and other health care workers must be encouraged to report it. (At present many professionals are reluctant to classify a death as suicide because of its effect on the surviving family and the legal and religious consequences.)

(2) The media should be encouraged to report suicides more responsibly, to prevent the occurence of imitative or "copy-cat" suicides.

(3) There should be more public-education programs aimed at reducing the imagined disgrace attached to seeking help

for depression, informing the public of the warning signs of suicide, and showing people how to cope with stressful situations.

(4) The government should enact stronger gun-control laws. At present, 30% of suicides are committed with fire-arms.

(5) Dangerous but legal drugs — tranquillizers, sedatives, pain-killers — should be more closely controlled; 15% of people who committed suicide used drugs to do it.

(6) Teachers, police, doctors, and clergy should be better trained to recognize and deal with suicidal people.

(7) There should be more, and better, treatment programs for alcoholics and their families.

(8) There should be mental health workers stationed on reserves.

HOW TO HELP
A number of programs to help people in crisis are already available. Every large city has at least one hotline for people in distress, and most smaller communities have a place where people can go to ask for professional help.

Many people who killed themselves showed signs that they were thinking of it. Many withdrew from activities they used to enjoy and showed signs of depression. They often warned people of their intentions in an oblique way, with statements such as, "I won't be around much longer". Many gave away their possessions, lost interest in their work, and neglected their physical appearance. Often they started to drink heavily or abuse drugs.

If you are worried that someone you know is thinking of committing suicide, the most important thing you can do is tell someone about it — your family, your teachers, anyone you think would be able to help.

If you think the person is in immediate danger, stay with him or her. Encourage the person to talk to you, and listen intelligently. Urge them to seek help and try to show them that they are wanted and that people care.

PROGRESS CHECK

1. What are Durkheim's four types of suicide?
2. Name three traits that are often found in people who commit suicide.
3. In your community, where can people go or to whom can they talk, if they are thinking of committing suicide?

OVERVIEW

In discussing the socialization process and various cultures we have learned that we must conform to certain rules of behaviour. Any culture needs some conformity to function in an orderly and democratic fashion.

We have also learned that we have an individual need to conform, to belong to a group. Experiments done by social psychologists reinforce that we sometimes will go to great lengths in order to conform to what we think is the majority opinion.

Yet, non-conformity does happen. Sometimes a group can determine its own rules of behaviour and still co-exist with a larger society — such as the Hutterites. Sometimes, a non-conformist group can point out a new direction for society's norms — such as the hippie movement.

However, sometimes non-conformity can be a symptom of alienation from society. If the problems of aging are not considered by a society, the elderly can feel alienated from that society. If the concerns of the native population are not considered, some native people can feel alienated. The same is true for adolescents.

Severe alienation can lead to suicide. It is important that our society responds to the concerns of the individuals who make up our society. Not everyone needs to be totally a part, totally conforming to the norms of our society. But, no one should feel outside of our society, an alienated individual.

KEY WORDS

Define the following terms, and use each in a sentence that shows its meaning.

conformity
auto-kinetic phenomenon
non-conformist

sub-culture
counter-culture
alienation

BEING A PART OF SOCIETY 433

KEY PERSONALITIES

Give at least one reason for learning more about each of the following.

Elliot Aronson
Eric Fromm
Muzafer Sherif
Leon Festinger
Solomon Asch
Bibb Latané

Karl Marx
Melvin Seemans
Emile Durkheim

DEVELOPING YOUR SKILLS

FOCUS AND ORGANIZE

1. What questions can we ask about conformity and the reasons why people conform?
2. What questions can we ask about alienation and why it occurs only among some people?
3. (a) Design an organizer that lists the ways in which suicide can be prevented. In another column, list the symptoms associated with people who have suicidal tendencies.
   (b) What questions can we ask about the causes, symptoms, and help available to suicidal people?

LOCATE AND RECORD

1. Examine Melvin Seeman's list of traits commonly found in the alienated. Have you ever felt alienated? If so, what were the reasons — how did you cope? Write down your thoughts with some of your classmates. Have they experienced similar feelings and situations?
2. Write down your attitudes toward those people who are obviously alienated from the rest of Canadian society. Are your attitudes similar to the rest of society? Do these attitudes help to lessen the negative feelings that the alienated are experiencing — or do these attitudes make their situation worse? Share your findings and conclusions with other class members.
3. Create groups and have each group research one of the following counter-cultures: criminals, drug users, gangs, terrorists. Outline their beliefs, attitudes, and behaviours and compare them to those of Canadian society. Why might their behaviours be considered undesirable?

4. In groups, find information about different cults. Do research in the library on their belief system, methods of recruitment, their lifestyles, and purposes. Determine the types of people who are most vulnerable to being indoctrinated into a cult. Invite cult deprogrammers into your class to speak and answer questions about the nature of their work. Make a video tape of their presentation for future reference. Are cults counter-cultures or sub-cultures?

EVALUATE AND ASSESS

1. In groups, assess the truthfulness of the following quotation. Give examples of how sub-cultures have influenced Canadian culture and how the opposite has occurred. What kind of society would Canada be today without the influence and contributions of sub-cultures? Would it be richer or poorer?

    "The relationship between the overall culture of a society and its sub-cultures is not static. There is a continuous flow of influence from culture to sub-culture and vice versa. For example, the jargon of the jazz musician eventually found its way to the general public, just as the jargon of the rock musicians has become popular among the young in the more recent past".
    John A. Perry and Erna K. Perry, *Introduction to Social Science*

2. Read over the case study on the Hutterites and, on a sheet of paper, list the types of values, norms, attitudes, and traditions that are part of their subculture. In a separate column, list your own values, norms, attitudes, and traditions. In another column, give your opinion of the Hutterite customs. In a final column, analyze your opinions to see if they are biased.

SYNTHESIZE AND CONCLUDE

1. Write down the reasons which explain why members of the following groups commit suicide: people in Quebec, men, divorced people, the elderly, alcoholics, native people, young people, depressed people. Is depression a common link among these groups? Explain.
2. Create a chart which lists the characteristics that are associated with the "punk movement". In another column,

list the reasons and causes for the "punk movement". Comment on whether you think it would be more accurate to call the "punk movement" a sub-culture, rather than a counter-culture. Do you believe that it is a fad or a permanent popular type of behaviour among some young people?

APPLY

1. Divide the class into smaller units and list examples of groups of people who have been alienated from the rest of society. Use Melvin Seeman's list of traits and apply them to the groups listed. Which characteristics appear to best describe each alienated group? Share your group's conclusions with the rest of the class.
2. In groups discuss what your reactions would have been in the experiments performed by M. Sherif, S. Asch, and B. Latané. Write down how your reactions would be similar or different. On the basis of your conclusions, in what respects are you a conformist? Share your conclusions with the class.

COMMUNICATE

1. Write an essay on "Conformity" by answering the following questions:
   (a) In which ways do people conform?
   (b) What influences our like or dislike of someone or something?
   (c) Do you tend to change your opinions more often on important or non-important matters? List examples. What influences our opinions?
   (d) Can you think of examples when you did not agree with the opinions of the majority? Did you do something about it and if so, what were the results? Would you do it again?
   (e) Were there ever situations when you yielded to the pressures of a peer group or someone else even though you knew it to be wrong? Was it a learning experience? Would you do it again?
   (f) How would you describe your need to belong — is it overwhelming, average, or non-existent? Have there been times when you experienced all three feelings in different situations? Give examples. Would you label yourself a conformist or a non-conformist?
   (g) What are your conclusions about other people and yourself on the subject of "conformity"?

2. In this unit, you have learned a great deal about human behaviour and human societies. Take this opportunity to explore the fact that aggressive behaviour occurs in most human societies. Aggression refers to the several ways in which one person may aim to hurt another person physically or emotionally. Several experts have put forward theories about why people are aggressive. Research one of the following, presenting your findings to your classmates:
   (a) Determine the basic differences between the Instinctive Theory of Aggression and the Social Learning Theory of Aggression. Which experts promote one or the other of these theories? (An introductory university level Sociology textbook would be an excellent reference.)
   (b) Eric Fromm, in his book, *For the Love of Life*, published in 1986, describes a group of people who do not behave aggressively toward one another. Investigate Fromm's ideas on aggression and violence, noting particularly his views on why non-aggressive societies can exist.
   (c) Refer to Konrad Lorenz' book, *On Aggression*, published in 1974, to present a report on his theory of aggressive behaviour. A reference which summarizes his ideas, such as a university level Sociology or Anthropology textbook is another source you might use for your research.

CHAPTER

16

Current Social Issues

What are some of the problems facing society today? What can we do to help?

INTRODUCTION Every generation has important social issues with which they must deal. They may be new issues which previous generations did not have to face, for example AIDS. They may be old issues that were ignored by society in the past, such as wife and child abuse. They may be issues that in the past were not so important as they are currently, such as drugs and aging. Sometimes answers can be found to the issues, sometimes not. Sometimes it take several generations working on the same problem to find a solution.

In this chapter we will deal with four issues that you may be confronting now, or that you may confront in your lifetime — aggression, AIDS, aging, and drugs.

16.1 *Aggression*

Aggression is a word that is used for a variety of acts in which hostility (antagonism, unfriendliness), attack, and often violence are involved. Social scientists tend to define aggression as any behaviour aimed at hurting another person physically or emotionally. The Canadian legal system defines aggression as any intentional use of force against another person, without his or her consent. An attempt to use force or a threat to use force is also an assault.

Every day in the newspapers, on the radio, and on television we read and hear about acts of aggression. Sometimes it is the aggression of one country warring against another, or one gang fighting another. It may be a fight that occurs during a hockey game. Sometimes the aggression takes the form of family violence in which a family member is physically injured, or emotionally battered.

Anthropologist Ashley Montagu believes that aggression is the result of our over-crowded, highly competitive, threatening world. We learn aggression through our social and cultural environment. Another anthropologist, Richard Leakey, agrees that "humans are not innately disposed powerfully either to aggression or to peace. It is culture that largely weaves the pattern in human societies."

VIOLENCE IN THE HOME

There are different reasons why people act aggressively: out of fear or frustration, out of a desire to cause fear or frustration in others, or out of a desire to push forward their own ideas or interests. Most people can deal with the frustration they encounter in day-to-day living in non-violent ways, for example, by channeling it into sports. Some people, though, tend to strike back at a less powerful person when they experience frustration. Aggression gives them power, satisfaction, and often recognition.

Wife Assault

In the case of wife assault, the husband wishes to control his wife's behaviour by causing her to fear him. He may use physical assault or psychological abuse. Physical assault may include slapping, punching, kicking, shoving, choking, sexual activity without consent, throwing things, inflicting burns, beating with objects, and using weapons such as guns, knives, or axes. Psychological assault can include making threats of various sorts, verbally attacking her, controlling her activities, depriving her of sleep, food, or money, or isolating her from her family and friends.

Until recently, wife assault was not thought to be a widespread problem. Since wife assault usually takes place within the home, few people knew about it, and few people felt that it should concern those outside the family. However, the women's movement has brought cases of assault into the open, and demanded that a solution be found. Since the

1970's shelters for abused women have been created in many communities. They are frequently so full that women and children must be referred elsewhere. These shelters offer a safer environment in which the woman can recover and think about her situation.

Most of the newly created shelters try to provide counselling for women and their children. Some have treatment programs for abusive husbands that will help them take responsibility for their actions and teach them other ways of expressing emotions and dealing with problems. But these services are still poorly funded and in many areas are nonexistent.

Now that women victims of violence feel that there is some help available, and are seeking it, some startling statistics are emerging.
- 10% of women in Canada are subjected to wife assault. (This estimate is considered by most experts to be low, since most incidents of wife assault go unreported.)
- Women seeking help have usually been beaten several times in the past before they contact the police. Prior to this they have kept their abuse a secret out of fear and shame.
- 20% of all the homicides in Canada are women killed by their spouses.
- Boys exposed to family violence as they grow up are far more likely to become wife and child abusers than boys

from non-violent families. Girls exposed to family violence frequently become abused wives.

It is going to be difficult to break the cycle of family violence. However, the first steps have been made. Society is now aware that it is a serious problem affecting a substantial proportion of the population. Public awareness and the creation of shelters have encouraged many women to end their silence and to seek help. Men are finding out that wife assault is a criminal offence.

Much more funding is needed. One area that needs attention is the counselling of children from homes in which there has been violence. If these children are ignored, they may become part of a new generation of abusers and victims, and the problem will be perpetuated.

Child Abuse

The home is supposed to be a place where children know they can find love, caring, warmth, and safety. In many homes, however, children live in fear. The warmth has been replaced with tension; the safety replaced with violence and threats.

In the past few years people have been shocked by the rising number of cases of reported child abuse. In over 75% of these cases, one or both parents were the abusers. Table 16.1 below shows the number of recorded cases of child abuse for the Children's Hospital of Eastern Ontario over a seven-year period.

Part of the reason for the rise in reported cases is probably the increase in awareness of child abuse. New Ontario legislation requires everyone to report cases of suspected child abuse. Professionals who come into contact with children, such as doctors, dentists, teachers, social workers, clergy, and child care workers can be fined $1000 for failing to report suspected child abuse.

What kind of parents abuse their children? Barbara Star of the School of Social Work at the University of Southern California developed a personality profile of the typical child abuser. She found that they often expressed feelings of isolation, low self-esteem, depression, hopelessness, and powerlessness. They were often unprepared, either from lack of education or lack of emotional maturity, to take the role of parent. Many had suffered abuse and neglect as youngsters, and many had lacked appropriate models from whom they

TABLE 16.1

1980	189
1981	213
1982	219
1983	304
1984	457
1985	650
1986	625

might learn parenting skills. Most of the child abusers in Star's study were in their late teens to mid-twenties, lacked high school graduation certificates, and were in the lower income bracket.

In many communities it is now possible to learn parenting skills through training programs. As Dr. Thomas Gordon, creator of Parent Effectiveness Training points out, "Millions of new mothers and fathers take on a job each year that ranks among the most difficult anyone can have — taking an infant ... assuming full responsibility for her/his physical and psychological health, and raising her/him so she/he will become a productive, cooperative, and contributing citizen. What more difficult and demanding job is there? Yet, how many parents are trained for it?"

Progress Check

1. What is the Canadian legal definition of aggression?
2. Why do some people act aggressively?
3. Why is society now recognizing that wife assault is a significant social problem?
4. What are some of the reasons that parents abuse their children?

16.2 *Aging*

Canadians are living longer. In 1900 the average lifespan of a Canadian was 45, by 1930 it was 60, and by 1987 it was 75. Experts believe that the upper limit for the average lifespan may some day be 115. Part of this increase in life expectancy is attributed to the control of diseases such as tuberculosis, polio, cholera, and smallpox. Through medical discoveries such as antibiotics and vaccines and through better health care the **mortality rate** or death rate from illness and accidents has been reduced.

The longer life expectancy has meant an increase in **senior citizens**, people 65 years and over. At present, approximately 11% of Canada's population consists of senior citizens; by the year 2031 that figure is expected to rise to 25%.

As the population of seniors grows, more attention is being paid to the quality of their lives and the problems they face. Some experience loneliness and depression, often over the death of spouses, relatives, and close friends; others may be in poor health or have loss of sight or hearing. Some feel frustrated by their lack of independence. Many have severely limited finances. According to the National Council of Welfare, in 1985 one-fifth of all elderly Canadians were considered poor, that is, they had to spend more than 60% of their income on food, clothing, and shelter. At least two-thirds of poor seniors are women.

Another problem with which seniors must deal is the stereotyped image of elderly people as slow, old-fashioned, narrow minded, forgetful, feebleminded, or ill. This stereotype ignores the fact that elderly people are individuals who have feelings, needs, and abilities just the same as younger people.

In many countries, China for example, it is expected that the elderly will live with their children and be cared for by their children and grandchildren. In Canada, however, 9.5% of the elderly live in institutions. Gordon Winocur, a neuropsychologist at Trent University, reports that elderly people who are institutionalized often lose their mental abilities and memories. Everything is done for them physically, but very often little is done for their minds. They are not challenged or stimulated.

It is estimated that one-quarter of those seniors now living in institutions could be living at home if certain services were made available to them. Meals-on-Wheels, a service that delivers prepared meals is one such service. It has been found that those seniors who remain in their own homes are healthier, happier, and more confident. It has also been found that it is cheaper to provide community support services than institutional care.

The Ontario government is experimenting with "granny flats" to try to meet some of the housing needs of seniors. Granny flats, which rent for $300 to $386 per month, are portable houses that can be installed near existing family homes. Senior citizens can be close to their families, while keeping some independence.

Aging can be a positive process. Many people look forward to retirement as a time for taking up new interests, travelling, and socializing with friends. Some take up volunteer work, start a new business, or return to school. Seniors on university campuses are becoming more

common. In 1985 the National Advisory Council on Aging reported that 37% of Canada's elderly regularly took part in sports or exercised. Regular exercise is important to the elderly for it results in better muscle tone and improved heart and breathing abilities.

PROGRESS CHECK

1. Why is the senior citizen population increasing?
2. What are some ways in which older people can keep some of their independence?
3. Give some of the positive aspects to aging.

CAREER PROFILE
GORDON SINCLAIR

Gordon Sinclair came from a poor family in Toronto, Ontario but through his talents, abilities, and efforts he became a self-made millionaire and a Canadian celebrity. He had only a Grade 8 education, yet he became a well-known and respected newspaper journalist and an author of nine best-selling books.

Mr. Sinclair was well known on radio and television for his bluntness and his often controversial opinions. On three separate occasions his nose was broken by people who had been offended by his opinions. He had a very successful radio career; at the age of 83, the year before he died, he was doing 14 shows a week. In 1957 he became one of the original hosts of Canada's longest running non-news television show, *Front Page Challenge*. He was still doing the program in 1984, the year he died.

Mr. Sinclair was proud of his accomplishments and honours, including Canada's highest award, the Order of Canada. In his own words, "If I've had success, and I guess I have, it's just because I don't copy anybody. I'm what I am, good or bad, not what people want me to be ... I had to live with myself and growing old was part of it. You've got to be yourself; to live with yourself, and you're never going to get rid of yourself. Sure you think about death at my age. But with me it's a matter of curiosity ... Down deep in the human spirit there is something eternal and you and I are part of it."

16.3 Aids

AIDS (acquired immune deficiency syndrome) is an infectious disease caused by a virus. The virus, called HTLV-3, attacks and destroys the body's T-helper cells. These cells are important in our immune system, the system that enables us to fight off disease. AIDS causes the immune system to break down, and the body cannot resist diseases that normally do not present a problem.

AIDS is one of the biggest challenges young people must face today. No cure has been found yet for it; the only method available to combat the spread of AIDS is public education and prevention.

According to the World Health Organization, more than 100 000 people around the world have died from AIDS, and somewhere between 5 million and 10 million more people are believed to be infected. It is not yet known if all the people who are infected with the AIDS virus will develop the disease. Some researchers feel that they will.

AIDS is spread through sexual contact, through intravenous drug use where needles are shared, and from an infected mother to her newborn. Abstaining from sex, having a monogamous sexual relationship with a partner who has not been exposed to the AIDS virus, or the use of latex condoms are three choices that can be made to avoid contracting AIDS. Not using drugs that are injected, not sharing a needle, or sterilizing needles before use are three methods for drug users to avoid AIDS. As far as it is known, AIDS cannot be spread through casual contact.

In the past, some people contracted AIDS through blood transfusions. All blood donors are now screened for the virus. As well, all blood supplies used in hospitals are routinely tested.

At the moment, there is no known treatment for AIDS. There are about a dozen drugs currently being tested that show some laboratory signs of controlling the AIDS virus. Recently, researchers have been successful in inoculating laboratory animals with the disease. This will enable them to study the natural life cycle of the disease and to test the effects of new drugs and vaccines.

Dr. Elisabeth Kübler-Ross, a psychiatrist and renowned authority on the subject of death and dying, has written a book entitled *AIDS The Ultimate Challenge*. In this book she examines the ways in which AIDS patients should be treated by the medical profession and by their families and friends. AIDS patients have often experienced public and professional fear and discrimination. However, there is a growing support movement consisting of AIDS patients themselves, their families and friends, volunteers, and health care professionals. These people are committed to providing patients with the best care possible and with the understanding, love, and compassion they need. Many terminally ill patients, given a choice, prefer to die in the familiar setting of their homes rather than in institutions. With the help of trained support workers and volunteers, many AIDS patients can remain in their homes for much of their illness. As Kübler-Ross notes "It is an unpredictable disease that can swing back and forth between 'close to death' and practically well." Palliative care (relieving illness without curing it) can be provided by such places as Casey House, that opened in March, 1988 in downtown Toronto.

Casey House, Toronto

What can be done in the future to combat the spread of AIDS? Kübler-Ross points to the role of the parent in guiding young people. "What we are learning now ... is to ...keep the doors of understanding, mutual respect, and unconditional love open and to educate ... children about issues we are too bashful to talk about until it is too late. What this epidemic teaches us is to become honest again, to talk and listen to each other, to accept and love each other more."

Progress Check

1. How is AIDS spread?
2. How can you avoid contracting AIDS?
3. Why should AIDS patients receive palliative care?
4. What does Dr. Kübler-Ross think can be learned from this disease epidemic?

16.4 Drugs

Drugs are an accepted part of our everyday life in North America. You probably use drugs without realizing it.

What is a drug? A **drug** is a substance other than a food, that when taken, changes the way the body or mind functions. Some act as stimulants, some as sedatives. Some produce a tranquillizing effect, some relieve pain, while others produce hallucinations.

From the above definition of a drug, you will see that many common substances you come in contact with during your day are drugs. Coffee, tea, chocolate, cola drinks, and some medications contain a drug called caffeine. Tobacco contains a drug called nicotine. Wine, beer, and liquor contain a drug called alcohol. Cough medicines and nose drops contain drugs. Toothpaste often contains a drug — fluoride. When you have an illness caused by a bacterial infection you may be given an antibiotic such as penicillin.

Some drugs are produced from plants, and some are manufactured from chemicals in a laboratory. Many drugs that have a medical use can be bought legally at drug stores with a doctor's prescription (prescription drugs), or without a prescription (over-the-counter drugs). Some drugs that do not have a medical use are manufactured in laboratories for the illicit drug trade, and are sold illegally by individuals.

Recreational Drug Use in Canada

The Legal Drugs

Most cultures and societies approve of the use of certain recreational drugs because they are not thought to be harmful. Caffeine, nicotine, and alcohol are the most common. They are sometimes called invisible drugs because many people do not think of them as being drugs.

Often, however, there are some restrictions in the use of the invisible drugs. In Canada, the possession and use of tobacco products and alcohol are prohibited to anyone underage, and the use of tobacco products in many public places across Canada is banned.

In Canada caffeine and alcohol are considered by most people to be acceptable drugs, and they are consumed regularly by many adults. Nicotine used to be considered an acceptable drug, however, public opinion has changed recently. The number of people who smoke is declining as people have become aware of the hazards involved. Medical studies have proven that smoking is dangerous to the health, not only of smokers, but also of those who are close enough to inhale the second-hand smoke. Most Canadians who use socially approved drugs use them in moderation, but some people overdo them. A couple of cups of coffee may make a person feel more alert. However, consistently drinking four cups of coffee a day can lead to a physical dependency. Stopping may result in headaches, irritability, and fatigue. The negative effects which result from going off a drug are called **withdrawal symptoms**.

Drinking alcoholic beverages is considered socially acceptable by most people, as long as it is done in moderation. Many people will drink a glass of wine or beer with friends — often they find the alcohol makes them more relaxed and outgoing. You have probably at some time or other witnessed the results of overindulgence in alcohol. A person who has consumed too much alcohol may become sloppy, obnoxious, sick, and out of control. A drunk person is not a pretty sight. Someone who drives a car in this condition is a potential killer.

The Illegal Drugs

There are also drugs available in Canada that are illegal to sell or use. Some of the most common are marijuana, hashish, LSD, heroin, and cocaine.

Cannibis: Marijuana and hashish come from a plant called Cannibis sativa, commonly known as hemp. Some of the effects of smoking or ingesting cannibis are an impairment of concentration, distortion of time and space, and an impairment of balance and coordination. The smoke from cannibis contains higher levels of cancer-producing substances than does tobacco smoke. One study showed that smoking five marijuana cigarettes a week is more harmful than smoking six packs of cigarettes in the same time period. Another recent study has shown that the regular use of marijuana can impair the body's immune system.

LSD: LSD (lysergic acid diethylamide) is produced in laboratories for the illegal drug trade. It is classed as a **hallucinogen**, a drug that causes a person to sense things that are not actually present. Users often experience intensified colours and sounds, distortions of vision and time, and extreme swings in mood. Some of the differences in users' reactions may be due to the drug's strength and purity. LSD is sometimes mixed with cheaper drugs such as PCP, called "angel dust". PCP is a very dangerous drug, an overdose of which can cause convulsions, coma, and death.

Opiates: Drugs that are derived from opium are called **opiates**. Opium was developed from poppies by ancient Egyptians and used recreationally as well as to treat illness. During the nineteenth century, morphine was developed from the active ingredient of opium. It was used to ease the pain of wounded soldiers, but morphine addiction soon became too serious a problem to continue the use of the drug. In 1898 a new drug, heroin, was developed from morphine. Heroin is one of several drugs classified as a **narcotic analgesic** because it numbs (narke = numbness) and relieves pain (an = not; algeo = feel pain).

Heroin is an effective pain killer and can be prescribed for terminally ill patients. Regular heroin users become physically dependent on the drug. When denied the drug they suffer severe withdrawal symptoms.

Cocaine: Like nicotine and caffeine, cocaine is classified as a stimulant, because it produces a temporary increase in energy in the user. Cocaine is a white powder prepared from the leaves of the coca bush of South America. It is usually inhaled through the nose. Regular users of cocaine may experience hallucinations, insomnia, and impotence.

CAREER PROFILE
STEVE COLLINS

In 1980 Steve Collins stunned international ski jumping enthusiasts. Only 15 years old and a mere 48 kg, he won the 90 m World Cup ski jumping competition in Lahti, Finland and became an instant sports hero. Reporters from Europe and North America pursued him for interviews and photographs. Many newspaper and magazine articles were written about this amazing young Ojibwa from a reserve near Lake Superior. Europeans especially were fascinated by his Ojibway heritage, and reporters called him "Big Chief of the Hill".

Steve had begun ski jumping at the age of 10, and within a few years he was jumping off the 100 m runway at Big Thunder in Thunder Bay, Ontario. He reached speeds of 100 km/h and in 1980 set a jump record of 128.5 m at Thunder Bay, Ontario. His coaches regarded him as a gifted athlete with a promising future.

Steve had never liked school, and dropped out after Grade 10 to devote himself full time to jumping. However, after he won the junior title in Sweden in March 1980 and the 90 m World Cup event in Finland a week later, he began to experience serious problems. Being famous at the age of 15 was difficult for this shy, private person. All the attention and admiration, the interviews and social functions, the world class competitions, and the months of travel away from home were not easy to handle. He was also under enormous pressure to repeat his earlier World Cup winning performance.

In January of 1981 he did win the 90 m Canadian ski jumping championship, but after this victory his performance began to slip. As he grew taller and gained in mass it was necessary for him to adapt his techniques to his maturing physique. Despite efforts to improve, victories eluded him. He turned increasingly to alcohol and marijuana to cope with the pressures and the disappointments.

In 1984, many Canadians were shocked to hear that Steve had entered a drug rehabilitation program in St. Paul, Minnesota, and that he had been removed from the Canadian ski jumping team. He spent 27 days at the rehabilitation centre during which time he attended daily lectures on dealing with and overcoming alcoholism.

Steve has been very open about his struggle with drugs. He now reports that he has been successful in dealing with his drug problem, and that for the first time in years he feels good about himself. Twice a week he attends AA meetings. He admits though that the desire to drink is with him every day, and that it is a daily struggle to resist taking a drink.

Since his rehabilitation program, Steve's ski jumping results have improved, and he now feels more relaxed and confident about the future. On February 2, 1986 he won the 90m Canadian ski jumping championship — his first win since 1981. He is determined to win future titles.

Steve's determination, courage, and honesty in dealing with his personal problems and his drug dependence make him an excellent role model for young people who have drug problems.

Drug Abuse and Addiction

Drug abuse can be defined as the use of a drug for a period of time (at least a month) in quantities sufficient to intoxicate the person throughout the day. A **drug addict** is a heavy user of a particular drug who requires the drug in order to function normally.

In Canada the most prevalent type of drug addiction is alcoholism. Alcoholics outnumber opiate addicts by about twenty to one.

When people start taking drugs they do not expect to become addicted. However, when addictive drugs are used over a long period of time, the user develops a **tolerance** for them. Larger and larger doses of the drug are needed to produce the same effects. Eventually the user discovers that it is difficult to cut back or stop taking the drug.

SOCIETY AND DRUG ABUSE

Different parts of society have taken differing stands on how to deal with drug-related problems. Some groups feel that drug education is the best way to discourage people from experimenting with drugs. Once the negative effects and the hazards of the different drugs are known, people will sensibly avoid taking them. By setting up compulsory drug education programs in schools, they believe students will avoid or will exercise strict control over drug taking.

There are some disadvantages to an educational approach. Many young people dismiss warnings about the negative effects of drugs as being exaggerated or emotional as opposed to being factual. It is true that some campaigns against drug use have taken an emotional rather than a factual approach. Some young people choose to rebel against authority as a matter of principle.

The exercising of parental control is another approach. Parents are encouraged to supervise with love, but to set limits and act firmly if the rules are broken. They are urged to make it their business to know their children's associates and whereabouts. Parents are advised to know the symptoms of drug abuse and misuse — skipping classes, lower marks, a lack of concentration, red and glassy eyes, an inability to stay awake, and secret telephone calls — these are possible signs. Should they discover that their children have a drug problem, parents should be aware of the community services that can help. Parents are also important as role models for their children. If they are seen to deal effectively with stress and problems, they will be setting an example their children may wish to follow.

Another approach to dealing with the drug problem is to impose higher fines and longer jail sentences. Potential users would then have to weigh the risks and decide whether it was worth it to take drugs. Some of the groups who support increasing penalties also advocate greater police powers in searching for illegal drugs.

Many people oppose jail sentences for drug users. They feel that a prison sub-culture would not help drug users and would not rehabilitate addicts. Instead, drug offenders might learn to use still more addictive drugs and to support their drug habit by illegal means.

In some countries, the government has stepped in to exercise stricter controls. Sweden, for example, rationed alcohol between 1917 and 1955. Some Canadians recommend government controls that include rationing of alcohol, shorter drinking hours, fewer establishments licensed to serve and sell liquor, a higher drinking age, a complete ban on alcohol advertising and on showing drug use on television programs, and government-financed campaigns designed to greatly reduce drug use.

Lately there has been a great deal of controversy over drug testing in the workplace and in sports. Some people argue that drug testing should be widely used by employers on job applicants as well as on people already employed. Applicants found to be using drugs could be refused employment, while employees who use drugs could be offered therapy and treatment to overcome drug problems. It is felt that ridding the workplace of drugs will make it a safer environment for employees, and a more efficient place to carry on business.

Some companies, like CNR, currently require applicants for jobs such as engineers and brake persons to undergo urine drug tests. In positions such as these, where many lives are at stake, it is necessary to ensure that employees are not impaired while carrying out their duties. Applicants who refuse to take the tests are not hired. Canada's top athletes regularly undergo drug tests to ensure that they are not using drugs to increase their strength and stamina during competitions.

It is clear that the Canadian people and the lawmakers have many important decisions and policies still to be worked out with regard to drugs. Should drug abuse and addiction be regarded as a treatable disease, or should it be regarded as a social problem? If it is a disease, education, medical treatment, counselling and therapy are the most obvious approach. If it is a social problem, are stricter controls and penalties the answer? How far should the government be prepared to go to protect the public from itself? Are human rights in jeopardy?

Progress Check

1. List some drugs you have contact with most days.
2. Why are some substances called "invisible drugs"?
3. There are many different proposals to limit the abuse of drugs. Write about three of the proposals. Do you think any of them will be helpful?
4. Do you believe in mandatory drug tests in the workplace?

CASE STUDY

The Addiction Research Foundation

The Toronto-based Addiction Research Foundation (ARF) is technically a branch of the Ontario Ministry of Health, set up to serve the people of Ontario. However, it has an international reputation, having been designated as a collaborating centre for the World Health Organization. Important researchers from all over the world work at the ARF, carrying on studies into the prevention and cure of substance abuse. Founded in 1949, the ARF today has four major roles: research, treatment, prevention, and information services.

Research: *In recent years, the ARF has looked into ways of treating alcoholics by immunization. The method, as yet unperfected, would involve the injection into alcoholic patients of antibodies generating an allergic reaction to alcohol. In 1985-86, the ARF developed the 'skinalyzer', an instrument that measures the alcohol intake of unconscious patients.*

Treatment: *The ARF's Youth Clinic was opened in 1986. The treatment process for young drug abusers attempts to involve the patient's family, a step that has shown improved results in the cure of drug abusers. The ARF has collaborated with United Nations agencies in setting up drug-treatment programs in countries such as Thailand and Nigeria.*

Prevention: The ARF has produced video-teaching packages on alcohol and drug abuse, distributed to schools across Ontario. The 'server-intervention' program trains bartenders and other licensed-premises personnel to take more responsibility for their patrons' consumption. The Campus Alcohol Policies and Education program (CAPE) encourages careful use of alcohol among university students.

Information Services: Dial-a-fact *and* The Journal *are the key elements of the ARF's public-awareness program. Available cost-free throughout Ontario,* Dial-a-fact *provides ready access to pre-taped telephone information on facts about and treatment for problems related to substance abuse.* The Journal, *the ARF's monthly newspaper, has a circulation of well over 20 000 and correspondents throughout Canada and the world. Aimed largely but not exclusively at health-care professionals, it covers the entire spectrum of issues related to alcohol and drugs.*

1. What kinds of services does the ARF currently provide?
2. Suggest some other kinds of services which an organization like the ARF might offer.

OVERVIEW

Young people today live in a world that is full of choices and information that older people did not have when they were young. Never before have people been so aware of problems and issues. Individuals and the larger society are trying to deal with, reduce, and eliminate situations that are harmful to others. Issues like drunk driving are of major concern to all Canadians — especially young people because it is a major cause of injury and death in this age group. Police RIDE programs have been created to find the people who drink and drive. More important they are warnings to those who are tempted to break the law. These programs were brought into effect a few years ago because of society's growing awareness that this issue had to be addressed.

Public education is necessary to keep everyone aware and informed about issues. Ignorance can result in wrong decisions which can be harmful to others. A major challenge to everyone is to keep up with and to be informed about issues that can affect the lives of Canadians. This is an ongoing challenge in a world that is becoming more complex and is always changing.

456 HUMAN SOCIETY

KEY WORDS   Define each of the following terms, and use each in a sentence that shows its meaning.

                               mortality rate   opiates
                               senior citizens   narcotic analgesic
                               drug   drug addict
                               withdrawal symptoms   tolerance
                               hallucinogen

KEY
PERSONALITIES   Give at least one reason for learning more about the following.

                               Ashley Montagu   Thomas Gordon
                               Richard Leakey   Gordon Winocur
                               Barbara Star   Elisabeth Kübler-Ross

DEVELOPING
YOUR SKILLS

FOCUS

The focus of this chapter is on some specific issues that are considered to be important in Canadian society. Divide the class into groups and have each group brainstorm a list of possible questions that can be asked on one of the following topics: wife abuse, child abuse, drugs, aging, AIDS.

At the end of the brainstorming session, each group can share its questions with the rest of the class.

ORGANIZE

1. In an organizer list examples of wife abuse. In another column list the ways in which society is trying to deal with wife abuse.
2. Develop an organizer that lists the following drugs and their effects on users. In a separate column explain the public's attitude toward the drug: caffeine, alcohol, nicotine, cannibis, LSD, opiates, cocaine.

LOCATE AND RECORD

1. Write down some of the problems experienced by elderly people in Canadian society. Conduct interviews with elderly persons and ask them about their experiences as senior citizens. Ask them if they have experienced any of

the problems that you recorded earlier. Ask them for suggestions about how their conditions could be improved. Record your research on paper and share your results with other class members.
2. Have people who work in shelters for victims of wife abuse come into the class to discuss their experiences. What services do these shelters provide and what additional services could be provided?
3. Invite police officers into the class to speak on drunk driving. Why is it important to educate all Canadians about this topic? What is being done to reduce drunk driving in Canada? Can more be done?
4. The class can be divided into groups to discuss the topic of human aggression. List examples of human aggression that group members have experienced. For homework listen to the radio news and watch the news on television. List the incidents of aggression that are reported. Read the newspaper and cut out articles that contain examples of human aggression. Share your research with group members. Beside each example or report of human aggression write down whether this aggressive action was learned, innate, or a combination of both types of behaviour.

EVALUATE AND ASSESS

What are the most important reasons for drug use by young people? Do you know people who take drugs? If so, why do they take them? Are their reasons similar to those expressed in this article?

"Sometimes people start taking drugs out of boredom. They are dissatisfied with their lives, but they may lack the initiative or the imagination to create or develop new interests. They often turn to drugs out of curiosity in the hope that drugs will provide temporary excitement, pleasure, or fulfilment.

Other people start taking drugs because their friends are. Peer pressure is often a powerful influence. A desire to fit in and to take a risk can sometimes be the beginning of drug dependency.

Young people sometimes find it difficult to accept the values, expectations, and goals of their parents and society. Taking illegal drugs is a form of rebellion against authority. It can be an attempt to demonstrate to themselves and others that they are grown up, and that they can make decisions for themselves.

Modern day living is full of pressures and stresses. For some people, taking drugs helps them to relax and reduces the pressures they feel. Alcohol is used by many people as a means of temporarily forgetting about problems and worries. Other people feel inadequate in social situations, and take alcohol to bolster their self-confidence.

There is some evidence indicating that parents who abuse or are addicted to drugs become a negative role model for their children. Bob Graham is the director of a centre in Ottawa that treats teenagers who are addicted to alcohol and other drugs. He has observed that if one parent abuses alcohol or other types of drugs, there is a 50-50 chance that the children will also abuse drugs. If both parents abuse drugs or are addicted to them, there is an 85% chance that their children will also abuse drugs later in their lives.

Some studies indicate that alcholism runs in families due to genetic and environmental factors. To see if a tendencey toward alcoholism can be genetically transmitted, Dr. Donald Goodwin studied the behaviour of over 5000 male adults living in Denmark who had been adopted at birth. He discovered that the adopted babies born to alcoholic parents were almost four times as likely to become alcoholics as the babies born to non-alcoholic parents.

Other studies support the idea that children of alcoholic parents are 20% more likely to develop a drinking problem than children of non-drinking parents. It is estimated that half of the ten million alcoholics in North America inherited the tendency toward alcohol abuse from their parents."

SYNTHESIZE AND CONCLUDE

1. In an organizer list the types of child abuse. In a separate column write the possible causes for these types of child abuse. Discuss your conclusions with classmates.

    "Types of Child Abuse: In its broadest sense, child abuse and neglect mean the physical or emotional injury, including sexual asault, negligent treatment or maltreatment of a child by the person(s) caring for or responsible for the child's health and welfare.

    Child abuse usually refers to a form of unreasonable physical punishment or attack on a child, and can include verbal assault as well as extreme forms of discipline such as long periods of solitary confinement.

Child neglect usually means interference with a child's maximum development by failure to provide the child with the necessities of life, such as nutritional food, clothing, shelter, medical attention, reasonably sanitary living conditions; and can include leaving a child unattended. Child neglect also includes failure to give the child adequate emotional support and/or love and affection. Child neglect is considered a form of child abuse."

2. Every week local newspapers have updates on the number of reported AIDS cases in Canada, the U.S.A., and the world. Compare these figures from March 11, 1988 with the updated figures. To what extent have the number of casualties changed? When examining the news report look for shifts in the groups that are most affected. If there are changes, give the reasons. Does the news report give information about new drugs or treatments for AIDS victims? Has there been a breakthrough in the prevention of this deadly disease? Report and discuss your conclusions with the class.

Number of cases

	Living	Dead	Total
Canada	742	902	1 644
Ontario	271	377	648
U.S.	24 008	30 715	54 723
World			80 912

APPLY

1. What do the following statistics reveal about the types of child abuse that exist in the world today? What could be done to lessen or eliminate these forms of abuse? Why are children especially vulnerable to abuse?
- 11 million children in Brazil live in the streets
- in underdeveloped countries child beggars are a common sight
- the World Health Organization (WHO) reports that millions of children in Latin America are child prostitutes or are recruited by guerrilla movements to fight for their various causes
- the International Labour Organization estimates that there are 52 million child labourers in the world

- millions of children are threatened throughout the world by wars, malnutrition, and poverty

2. Divide the class into groups and discuss what your reactions and feelings would be if you found out that a family member, friend, or classmate had AIDS. Would you treat that person differently? fairly? with compassion?
3. Analyze this graph by answering the following questions.

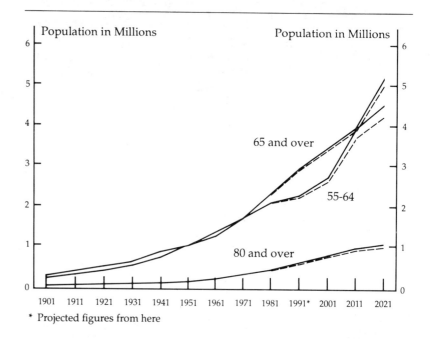

* Projected figures from here

(a) What has happened to the number of people in Canada over 65? over 80?
(b) What predictions does this chart make about the number of elderly in the year 2021? How old will you be?
(c) Why do you think it is important to understand the elderly and their situations?

COMMUNICATE

1. Seminar Activity: Members of the class can choose one of the following issues in Canadian society. Research can be done individually or in groups in the library to find information concerning the causes of the issue, the arguments that are presented by opposing sides, and a

decision by yourself concerning where you stand on the issue. Why is it important for Canadian society to solve this issue? Present your conclusions to the class.

- abortion
- death penalty
- prison reform
- pollution
- euthanasia
- genetic engineering
- prostitution
- censorship
- wiretapping
- the right to strike
- civil liberties
- the nuclear threat
- drunk driving
- the rights of criminals and their victims

2. Write a short essay or debate one of the following topics.
- Governments should be spending as much money to find a cure for AIDS as they spend on national defence.
- Wife abuse has always existed but is now being challenged and opposed because of people's growing awareness that women are equal to men in all areas.
- AIDS victims and AIDS carriers should be isolated from society.
- Everyone should be obliged to take a blood test for AIDS.
- People should not be required to retire when they reach the age of 65.
- People who abuse others should be required to obtain compulsory psychiatric treatment until they are cured.
- Spanking is a form of child abuse and anyone guilty of this offence should be charged with performing a criminal act.
- The LeDain Royal Commission was wrong to recommend in 1972 the decriminalization of the use of marijuana in Canada.
- Drug addicts should be supplied with the drug to which they are addicted by the government if they are in rehabilitation programs to help them with their problems.
- Drug addicts should be treated in the same manner as alcoholics by society.
- People who drink and drive should be charged with attempted murder.

3. This chapter and the others in this textbook have concentrated mainly on society as it is today, and was in the past. What does the future hold? Read one of the following books, written by futurists who have predicted what they think life will be like: John Naisbitt's **Megatrends** or Alvin Toffler's **Future Shock** or **The Third Wave**. Present your findings as a brief written report and be prepared to discuss this futurist's perspective in a class discussion.

Index

Abella, Rosalie, 148, 162
   career profile, 164
Abnormality, definitions of, 395-96
Aboriginal peoples, 186-93. See also Indians; Métis; Native Peoples
Achieved roles, 80
Achieved status, 80
Acrophobia 404
Active avoidance, 72
Activists, 137-38
Addiction, 452
Addiction Research Foundation, 454-55
Adler, Alfred, 108, 109
Adopted children, 41-42
Advertising
   and children, 253-54
   discrimination in, 254
   in media, 253-55
   and quality of life, 254-55
Affinal kin, 286
African National Congress (ANC), 181
Age of enlightenment, 7
Aggression, 437-38
Aging, 441-43
Agoraphobia, 404
AIDS, 445-47
Alcoholics, 429
Algonquin Indians, 5
Alienation, 424
   six traits of, 425-26
   as social condition, 426
Allport, Gordon, 99
Altruistic suicide, 428
Alzheimer's disease, 404, 405
Ancestors, human, 52-56
Androgens, 154

Angel dust (PCP), 449
Anglo-Saxon, 231
Animal kingdom, 32-36
Animals
   classification of, 33
   primates, 33
Anomic suicide, 428
Anthropology, 10-11, 20-24
   cultural, 20-21
   physical, 21-24
Anti-semitism, 136
Anxiety disorder, 403
Apartheid, South Africa, 179-83
Asch experiment, 418-19
Asch, Solomon, 418
Ascribed roles, 79
Ascribed status, 79
Assembly of First Nations, 191
Associations, 287-88
Auto-kinetic phenomenon, 417
Autonomy, 118

Bassett, Isobel, 149
Bedlam, 407
Behavioural sciences, 10-11
Behavioural assessment, personality, 106
Behaviour therapies, 407-408
Bernstein, Basil, 219
Bibby, Reginald, 348, 369
Biculturalism in Canada, 217
Bilingualism in Canada, 217
Bill C54, 262
Bipedal movement, 279
Birenzweig, Esther, 330
Birth order, 108-109
Bishop, Jay, 255
Blissymbolics, 221
Boas, Franz, 20, 213-14

Boat peoples (Vietnamese), 200
Body language, 220
Boll, Eleanor, 110
Books, 249
Bossard, James, 110
Boyd, Monica, 161
Braille, 234
Brown, Georgina, 151

California Psychological Inventory, 105
Canadian Broadcasting Corporation (CBC), 244-45
Canadian Charter of Rights and Freedoms, 204-206
Cannibis, 449
Casgrain, Thérèse, 145
Censorship, 262-63
Child abuse, 440-41
Child-rearing, non-family, 334-35
Children
   adopted, 41-42
   and advertising, 253-54
   and media violence, 255-56
Chimpanzees, 33
Chinese immigrants, 199-201
Cinderella complex, 158
Civilization, 56
Civil servants, 162
Clan, 287
Classical conditioning, 71
Claustrophobia, 404
Clinical model of abnormality, 395-96
Clinical psychology, 18-19
Cocaine, 449
Collins, Steve, 450-51
Compensation, 400
Comte, August, 12

Conditioning, 71
　classical, 71
　operant, 72-73
Conformity
　Asch experiment, 418-19
　auto-kinetic phenomenon, 417
　classic experiments, 416-20
　defined, 414
　Latané experiments, 419-20
　reasons for, 414-15
　social-comparison theory, 417-18
　ways of achieving, 415-16
Conform, need to, 69-70
Consanguineal kin, 286
Conscious self, 19
Constitution Act, 1982, 186, 191
Counter cultures, 218-19, 420-21
Coureurs de bois, 189
Courtly love, 345
Courtship
　current practices, 347-48
　privatization, 346-48
Creation, Old Testament theory of, 5-6
Cro-Magnon man, 24, 54, 55-56
Cultural anthropology, 20-21
Cultural estrangement, 425
Cultural-relativism, 274
Cultural sovereignty, 263
Culture, 2
　analyzing, 285-88
　anthropologists' study of, 282-84
　classifying, 283-84
　defined, 4, 47-48, 274-78
　development of, 278-81
　early human, 52-56
　economic exchange systems of, 290-91
　economic structure of, 289-91
　enculturation, 275
　environment of, 288-89
　ethnography, 282-83
　ethnology, 283
　field work, 282
　food, 277-78, 289-90
　human body, structure of, 279
　human brain, 280
　land ownership, 290
　language and, 280

　material, 284
　non-material, 284
　program for survival, 275-76
　religious customs of, 295-96
　reproduction, 278
　social control, 295-96
　social interaction, 288
　social structure, 285-88
　socialization in, 48-50
　text for living, 275
　way of life, 276-77
Curriculum, 370-71

Dafal, 304, 306
Darwin, Charles, 22-23
Daycare, types of, 334
Defence mechanisms, 111-12, 397-403
　compensation, 400
　defined, 398
　displacement, 440-401
　rationalization, 398
　repression, 401-402
　role of, 402-403
　saving face, 398-99
　scapegoat reaction, 401
　sour grapes, 399
　suppression, 402
　sweet lemon, 399-400
　types of, 398-402
Delirium, 404-405
Dementia, 404
Dene, 191
Department of Labour stereotype study (1985), 83
Depression, 429
Descent group, 286
Dialects, 216
Disciplines, 2
Discrimination, see also Prejudice
　aboriginal peoples, against, 189-90
　in advertising, 254
　Chinese immigrants, against, 200
　defined, 137
　Irish immigrants, against, 197
　Japanese immigrants, against, 201
　laws against, 150
　Ukrainians, against, 199

　United Nations convention to eliminate, 162
　women, against, 146-50
Diseases, inherited, 38
Displacement, 400-401
Divorce, 353-55
　children, effect on, 357-58
　men, effect on, 362-63
　teenagers and, 358-60
　women, effect on, 360-62
Dominant group, 137
Dreikurs, Rudolf, 108, 109
Drug abuse, 452-54
Drug addict, 452
Drugs, 447-54
Durkheim, Emile, 12, 13, 427
Dyslexia, 228
Dyson, Margaret, 357

Ecology, 288
Economic exchange, systems of, 290-91
Ectomorph, 102
Education
　in Canada, history of, 372-74
　compulsory, 372, 374
　enrolment in schools, 374
　government reports on, 377
　and jobs, 370-71
　and personal fulfillment, 372
　purposes of, 369-72
　reform in 1980s, 379-82
　reform in 1970s, 377-78
　skills, lack of, 378
　and social order, 371
　and status, 371
　system, alternatives to, 376-77
　system, expansion of, 374
　and transfer of culture, 371
Educational psychology, 17-18
Edwards, Henrietta Louise, 143, 144
Ego, 19
Egocentric, 68
Egotistic suicide, 427
Electra complex, 116
Elkin, Frederick, 324
Emotional disorders, 403-404
Empirical data, 8
Enculturation, 275
Endomorph, 102

English language, development of, 231-32
Environment, 40
    defined, 4
    genetic adaptations to, 30-40
    heredity and, 37-38
    influence of, 41-42
Erikson, Erik, 117, 124
Erikson's theory of personality development, 117-23
Estrogens, 154
Ethnic groups, 33
Ethnic prejudice, 177
Ethnocentrism, 177, 275
Ethnography, 282-83
Ethnology, 283
Evolution, theory of, 22
Evolve, 39
Extended family, 286
    in nineteenth century, 327-28
Extrovert, 101
Eysenck, Hans, 102

Families, 323
Families in Transition (FIT), 356
Family
    defining, 322-23
    elements of, 324
    functions of, 324-26
    husbands and wives, changing roles of, 333-34, 349
    importance of, to socialization, 66
    and personality, 107-110
    polyandrous, 323
    polygamous, 323
    polygynous, 323
    post-modern, 331-35
    reconstituted, 323
    single-parent, 323
    size, 110
    socialization and, 332
Family law, 342-43
Famous Five, 143
Fatalistic suicide, 428
Feminists, 136
Feral children, 48-49
Fergusson, Muriel, 145
Festinger, Leon, 417
Fetus, 154
Field work, 282

Films, 246
Fire, discovery of, 54
Folkways, 68, 284
Food
    cultural differences in, 277-78
    supply, 289-90
Fossil discoveries, 23-24
Fox, Terry, 260-62
Freud, Sigmund, 18-19, 407
Freud's psychosexual theory of personality development, 111-16
    ego, 111
    Electra complex, 114
    fifth (genital) stage, 114-15
    first (oral) stage, 112-13
    fourth (latency) stage, 114
    id, 111
    Oedipus complex, 114
    second (anal) stage, 113
    superego, 111
    third (phallic) stage, 113-14
Functionally illiterate, 227-28
Futures, A Choice of, 377

Galdikas, Biruté, 292-94
Galen, 100
Games stage, role learning, 78-79
Gatekeeping, 163
Generativity, 121
Gender, influence on personality, 109-110
Gender expectation, 83
Gender roles, 82-86
    biology v. socialization, 154-59
    stereotypes, 83
Gene, 37
Genetics, defined, 32
Glaz, Avis, 84
Global society, individual in, 3-4
Global village, 252
Granny flats, 442
Gravity, Law of, 6
Great Hare, The, 5
Gutenberg, Johann, 248

Habitat, 40
Hallucinogen, 449
Hamilton, Richard, 139
Harlow, Harry F., 67
Harlow's monkey studies, 67

Harris, Louis, 353
Hashish, 449
Head tax, Chinese immigrants, 200
Heredity, 37-38
    defined, 32
Heroin, 449
Hewitt, Foster, 244
Hierarchy of needs, 44-45
Hippies, 421-22
Hiring equity, 147-48
Hite, Shere, 350
Hitler, Adolf, 136
Holocaust, 34-35
Hominids, 52
    development of, 279-81
    evolution of, 53-56
Homo erectus, 54
Homogamy, 352
Homo sapiens, 33
    hunters to farmers, 56-57
Hormones, 154
Horney, Karen, 110
Human body, structure of, 279
Human brain, 280
Human groupings, 34-36
Human needs, 44-46
Husbands, changing role of, 333-334, 349
Hutterites, the, 422-24
Hypothesis, 8-9

Id, 18
Illich, Ivan, 375
Illiteracy, 227
    causes of, 228-29
Immigration policies, Canadian, 194
Immortality, 5
Immutable species, 22
Incest taboo, 325
Indian Act, 1876, 190
Indians, see also Aboriginal peoples; Métis; Native Peoples
    current living situation, 190-92
    nomadic tradition, 188
    problems resulting from European immigration, 187-89, 190
    status and non-status, 186-87, 190

Individual in society, 2-6
Industrial Revolution, 7-8
Inferiority complex, 109
Instinct, 37
Instinctive behaviour, 37-38
Interviews, personality assessment, 105
Introvert, 101
Inuit, 5
Inuit Committee on National Issues, 191
Irish immigrants, 196-97
Isolates, 49-50
Itard, Jean-Marc-Gaspard, 51

Jackson, Michael, 247
Janet, Pierre, 407
Japanese
    immigrants, 201-204
    WWII incarceration, 203
Jargon, 218
Jung, Carl, 101
Jung's theory of personality development, 116-17
    persona, 116
    shadow, 116

Kay, Teri, 357, 361
Keller, Helen, 233-35
Kibbutzim, 334-35
Kindred, 287
King, William Lyon Mackenzie, 203
Kinship groups, 285-87
    affinal, 286
    clan, 287
    consanguineal, 286
    descent group, 286
    extended family, 286
    kindred, 287
    nuclear family, 286
    tribe, 287
Koko the gorilla, 222-24
Kübler-Ross, Dr. Elisabeth, 445-46
Ku Klux Klan, 177

Lambert, Wallace, 373
Land ownership, 290
Language
    Blissymbolics, 221
    body, 220
    communication, for, 225
    and culture, 236-37
    and culture development, 280-81
    defined, 214-15
    dialects, 216
    earliest uses of, 280-81
    English, development of, 231-32
    illiteracy, 227
    international, 237
    learning, 229-30
    listening, 225-27
    moral meaning of, 217
    sexist, see Sexist language
    sign, 222
    and social class, 219
    and social identification, 217-19
    and society, 216
Latané, Bibb, 419
Latané experiments, 419-20
Laws, 69, 284
Leakey, Louis and Mary, 24, 53
Leakey, Richard, 438
Learning
    active avoidance, 72
    and conditioning, 71-73
    how to fit in, 73
    negative reinforcers, 73
    omission training, 72-73
    passive avoidance, 72
    Piaget's theory of, 74-77
    positive reinforcer, 73
    reward training, 72
Legends, defined, 5
Levin, Martin, 358
Life chances, 79
Listening, 225-27
    three levels of, 226
Literacy, 227
    developing, 229-31
    media, 249, 252
Living and Learning (1964), 377
Looking-glass self, 125
Love, 345
    courtly, 345
    importance to relationship, 351
    romantic, 345-46
LSD, 449

McClung, Nellie, 143
McCurdy, David W., 282
McGibbon, Pauline, 145-46
McKinney, Louise, 143, 144
McLuhan, Marshall, 250-51
MacPhail, Agnes, 145
Magazines, 248-49
Mandella, Nelson, 181
Mandella, Winnie, 181
Manipulate, 243
Marijuana, 449
Market exchange, 290
Marriage
    arranged, 343-44
    customs in Canada, developing, 344-45
    decline of, 349
    defining, 342-43
    love and, 345-48
Marriage contract, 342
Marx, Karl, 13, 424
Masai tribe, 308-12
    clothing and ornaments, 309-10
    customs and social structure, 310-11
    food gathering, 308-309
    hunting, 310
    religion, 312
Maslow, Abraham, 44, 46
Mass media
    advertising in, 253-55
    books, 249
    defined, 243
    films, 246
    magazines, 248-49
    newspapers, 248
    radio, 244-45
    rock videos, 247
    television, 246-47
    violence in, 255-59
Material culture, 284
Maternity leave, 161
Matriarchy, 139
Mead, George Herbert, 78
Mead, Margaret, 20-21
Meals-on-Wheels, 442
Meaninglessness, 425
Media literacy, 249-52
Medicine-men, 4
Men, effects of divorce on, 362-63
Mental disorders
    anxiety, 403

delirium, 404-05
dementia, 404
emotional, 403-404
organic, 404
phobia, 403
reasons for, 405
thought, 404-405
treatment of, 406-408
defining, 396-97
Mesomorphs, 102
Métis, defined, 187
Métis National Council, 191
Minnesota Multiphasic Personality Inventory, 104, 106
Minority groups, 134
attitudes towards, 136-37
ethnic, in Canada, 196-204
Missionaries to Canada, 188, 189
Monkey babies, study of, 66
Monkey studies, Harlow's, 67
Monotheistic, 312
Montagu, Ashley, 33, 438
Mores, 284
defined, 68
Mortality rate, 441
Motivation, defined, 15
Multiculturalism, 193-96
Multicultural society, 195
prejudice in, 195-96
Murdoch, Irene, 342-43
Murphy, Emily, 143
Mutations, 39

Narcotic analgesic, 449
National Council of Women, 141
Native Council of Canada, 191
Native people, see also Aboriginal People; Indians; Métis
and suicide, 430
Natural selection, 23
Natural sciences defined, 8
Neanderthals, 54-55
Needs
hierarchy of, 44-45
human, 44-46
physiological, 45
self-actualization, 46
Negative reinforcer, 73
Newspapers, 248
as wrap arounds, 251
Newton, Sir Isaac, 6

Non-conformists, 420
Non-conformity, 420-22
Non-literate, 10
Non-material culture, 284
Normative model of abnormality, 396
Normlessness, 426
Norms, 68, 284
cultural, 414
Nuclear family, 286
shift to, 328

Oedipus complex, 114
Oleskow, Dr. Joseph, 198
Omission training, 72-73
Operant conditioning, 17, 72-73
Opiates, 449
Opposable thumb, 279
Organic mental disorders, 404
Orphan children, Spitz study of, 66
Ostracism, 68

Parenting, 86-90
Parlby, Irene, 143, 144
Passive avoidance, 72
Patriarchal society, 139
Patterson, Dr. Francine, 222
Pay equity, 146-47, 161, 162
Ontario, 161, 162
Pavlov, Ivan Petrovich, 71
PCP (angel dust), 449
Peer, defined, 18
Peer group
approval, 18
influence of, 70
Peking fossil discoveries, 24
Perceptions, defined, 9
Personality
assessing, 103-107
birth order, 108-109
defined, 98-99
ectomorph, 102
endomorph, 102
extrovert, 101
family and, 107-110
family size and, 110
gender, influence of, 109-10
introvert, 101
inventory, 104-105
mesomorph, 102
rating, 104

sense of self, 123, 125, 126-27
traits, 99-100
types, 100-103
Personality assessment
behavioural, 106
interviews, 105
projective tests, 105-106
responses and interpretation, 106-107
Personality development
Erikson's theory of, 117-23
Freud's psychosexual theory of, 111-16
Jung's theory of, 116-17
Phobias, 403-404
Phrenology, 407
Physical anthropology 21-24
fossil discoveries, 23-24
Physical therapies, 408
Physiological needs, 45
Piaget, Jean, 74-77
Piaget's theory of learning, 74-75
concrete operations stage, 76
formal operations stage, 76-77
preoperational stage, 75-76
sensorimotor stage, 75
Pinel, Philippe, 51, 407
Play stage, role learning, 78
Polyandrous family, 323
Polygamous family, 323
Polygynous family, 323
Popenoe, Joshua, 375
Positive reinforcement, 408
Positive reinforcer, 73
Poverty and divorced women, 361-62
Powerlessness, 425
Prejudice, 134-36, 176-77, see also Discrimination
against aboriginal people of Canada, 189-90
causes of, 184-86
defined, 134
multicultural society, in 195-96
Preparatory stage, role learning, 78
Primates, 33
Printing press, invention of, 248
Projective tests, personality assessment, 105-106
Promotions equity, 148-49
Psychiatry, 10

Psychoanalysis, 18-19
Psychologists, 10
    behaviour and, 14-16
    environmental, 15
Psychology, 10
    clinical, 18-19
    educational, 17-18
    methods, 15-16
Psychotherapy, 18, 407

Race, myth of, 176-77
Racial superiority, concept of, 34
Racism, 176-77
    defined, 177
    reality of, 177-78
Radio, 244-45
Rationalization, 398
Rayner, Rosalie, 71
Reciprocal exchange, 290, 291
Reconstituted family, 323
    rise of, 329-30
Redistribution, 290, 291
Religion
    defined, 295-96
    Durkheim's views of, 12
    explanations of life and death, 4-6
    Marx's view of, 13
Religious customs, 295-96
Repression, 401-402
Reproduction, cultural differences, in, 278
Reserves, 189, 190
Reward training, 72
Rich, Joe, 362
Richards, Martin, 357
Richer, Stephen, 83
Ricker, John, 381
Rock videos, 247
Rodin, Judith, 419
Role ambiguity, 81
Role conflict, 81
Role models, 79
Roles, 37
    achieved, 80
    ascribed, 79
    front, 80
    game stage, 78-79
    importance of, 77-79
    life chances, 79
    play stage, 78

    preparatory stage, 78
    settings, 80-81
Role sets, 80
Role strain, 81
Romantic love, 345-46, 349-50

St. Lambert experiment, 373
Saving face, 398-99
Scapegoats, 136
Scapegoat reaction, 401
Schichor, Dr. Aric, 351
Schools
    in Canada, history of, 372-74
    future of, 382-84
    inequality in, 378
    Japanese, 382-83
    re-evaluating, 379-82
Science, defined, 8
Scientific method, 8-9
Seemans, Melvin, 425
Segraves, Dr. Robert, 363
Segregation, 179
Self-actualization, 46
Self-concept, 123, 125, 126-27
Self-estrangement, 425
Senior citizens, 441-43
Settings, defined, 80-81
Sexist language, 85, 236
    use of, 152-53
Sex premarital, 325, 345
Sheldon, W.H., 102
Sherif, Muzafer, 417
Sifton, Clifford, 198
Sign language, 222
Sinclair, Gordon, 444
Single-parent family, 323
    rise of, 329, 330-31
Skinner, B.F., 17
Slang, 218
Social class, language and, 219
Social-comparison theory of conformity, 417-18
Social control, 295-96
Social influence, 414
Social institutions, 10, 13
Social interaction, 288
Social isolation, 425-26
Socialization, 48-50
    conditioning, 71-73
    conform, reasons to, 69-70
    and the family, 332

    feral children, 48-49
    gender roles, 82-86
    isolates, 49-50
    learning, 71-77
    and norms, 68-70
    parenting, 86-90
    process of, 64-66
    roles, importance of, 77-81
    and stereotypes, 84-86
    support, need for, 66
    women, 154-59
Social sciences, defined, 8
Social structure
    associations, 287-88
    kinship, 285-87
Society, individual in, 2-6
Sociologist, 10
    role of 11
Sociology, 10
    current state of, 14
    development of, 12-13
Sour grapes, 399
Spanier, Graham B., 333
Spencer, Metta, 83
Spitz, René, 66
Spradley, James P., 282
Stagnation, 121-22
Stamp, Robert M., 376-77
Star, Barbara, 440
Statistical Model of abnormality, 395
Steam engine, invention of, 7
Stereotype, 83
    male, 85
    in reading materials, 85
    reinforcement of, 155-56
    sexist language, 85
Sternberg, Robert 352
Stowe, Emily, 143
Sub-culture, 420
Subordinate group, 137
Suffrage, 140
Suffragettes, 141, 143
Suicide, 426-31
    alcoholics, 429
    altruistic, 428
    anomic, 428
    depression, 429
    egotistic, 427
    elderly, 430
    fatalistic, 428

how to help, 431
native people, 430
people at risk, 428-30
preventing, 430-31
young people, 429-30
Sullivan, Annie, 233
Summerhill, 375-76
Superego, 18
Suppression, 402
Surrogate mothers (monkey studies), 67
Sweet lemon, 399-400
Symbols
   defined, 5, 215
   types of, 219-22

Tasaday tribe, 304-308
   clothing, 305
   food gathering, 304
   future of, 307-308
   religious customs and beliefs, 306-307
   social structure, 306
   technology, 305-306
Teenagers and media violence, 257-58
Television, 245
   as cool medium, 251
   illiteracy, effect on, 229
   as mass medium, 246-47
Tennov, Dorothy, 350
Theories, defined, 4
Theory, 8-9
Thorndike, Edward Lee, 72
Thought disorders, 404-405
Thriller, 247
Todas (India), 323
Traits, personality, 99-100
Tribe, 287
Troyer, Warner, 359
Tutu, Desmond, 181-82
Twins, identical, 41, 42

Ukrainian immigrants, 198-99
Unconscious self, 18

Vaughan, Diane, 342
Victor, wild boy, 51-52
Violence, domestic, 438-41
   child abuse, 440-41
   wife assault, 438-39
Violence in media, 255-59
   Gallup Poll, 258-59

Watson, John B., 71
Watson, Patrick, 249
Weitzman, Lenore, 158
Wheeler, Michael, 360
Wife assault, 438-39
Wilson, Bertha, 146
Wilson, Cairne, 145
Wirth, Louis, 138
Withdrawal symptoms, 448
Wives, changing roles of, 333-34, 349
Women
   attitudes toward, 138-39
   Cinderella complex, 158
   discrimination in the workforce, 146-50
   divorce, effect on, 360-62
   employment equity, 162-63
   equity and mental health, 151-52
   maternity leave, 161
   pay equity, 161, 162
   socialization of, 154-59
   vote, obtaining, 142
   in workforce, 328-29
   working, Canadian, 160-63
Women's Christian Temperance Union (WCTU), 141
Women's rights movement, 140-44
   early members, 144-46
World Plan of Action, 158-59
Wundt, Wilhelm, 15-16